Ralph G. Archibald
Queens College

an introduction to the THEORY OF NUMBERS

Charles E. Merrill Publishing Co.
A Bell & Howell Company
Columbus, Ohio

Merrill Mathematics Series
Erwin Kleinfeld, Editor

Copyright © 1970 by Charles E. Merrill Publishing Co., Columbus, Ohio. All rights reserved. No part of this book may be reproduced in any form, electronic or mechanical, including photocopy, recording, or any information storage and retrieval system, without permission in writing from the publisher.

ISBN-0-675-09351-1

Library of Congress Catalog Card Number: 76-118978

AMS 1969 Subject Classifications 1005, 1010, 1043

1 2 3 4 5 6 7 8 9 10—79 78 77 76 75 74 73 72 71 70

Printed in the United States of America

Preface

Although the theory of numbers has engaged the attention of leading mathematicians from earliest times, Euclid (about 300 B.C.) is generally recognized as being the first to give a systematic presentation with proofs of results. It was principally through the Arabs that Western Europe became acquainted with the work of the Greek mathematicians. The gradual development of the subject in Western Europe was heightened by the outstanding work of Fermat (1601–1665). The theory was further developed mainly by the efforts of Euler, Lagrange, and Legendre. The really classic and basic work, however, in developing the theory of numbers as a systematic body of doctrine is the *Disquisitiones Arithmeticae* (1801) of Gauss (1777–1855). Since the time of Gauss there has been an extensive development not only of the subject matter itself but also in methods of approach. Ore's *Number Theory and Its History* will be found helpful in tracing this development. The monumental three volume *History of the Theory of Numbers* by Dickson covers the extensive literature of the subject up to the time of its publication.

It has been my purpose to prepare an introductory textbook in the theory of numbers suitable for advanced undergraduate and beginning graduate students. A background in college algebra and elementary calculus should be minimal prerequisites. A knowledge of the concepts of abstract algebra is not assumed. Having constantly in mind the student beginning the study of the theory of numbers, I have emphasized basic concepts and variety of treatment, and have given a somewhat detailed exposition: this is particularly true in the first three chapters. Frequently, even capable mathematics students, I have found, experience difficulty with the subject at the beginning.

In the exposition, the development of theory is not made to depend upon results given in the problems, and, in general, proofs of theorems are not left to the reader. In the belief that theory is not fully understood until it can be effectively applied, I have placed considerable emphasis upon the working of problems. Problems at the beginning of a set are usually quite routine; those of a more demanding character will be found toward the end of the set. In some sets, later problems are based upon, or partly motivated by, earlier problems of the same set.

One feature is the series of *Notes* on each chapter. In these Notes there will often be found references to further material for a student who desires to study the subject in greater depth or more extensively; somewhat detailed proofs, particularly those by mathematical induction; and discussions of interesting sidelights which, it was felt, might detract from the main trend of the exposition if retained in the body of the text itself. For example, in the Notes there occur a Calendar Problem (in Chapter 4), a brief treatment of Decimal Fractions (in Chapter 5), and Gauss's first proof of the Quadratic Reciprocity Law (in Chapter 6).

The scope of the book is well indicated by the table of contents. For a beginning course, a selection would undoubtedly be made. After the introductory Chapter 1, Chapters 2 and 3 present the usual basic concepts of divisibility and congruence. Several proofs of the infinitude of primes are given in Section 2-15. Sections 3-7, 3-15, 3-20, and 3-23 in Chapter 3 could be omitted without destroying the continuity. In Chapter 4, more than usual attention is given to arithmetical functions and recurrence formulae. Possible omissions here would be Sections 4-3, 4-7, 4-8, 4-10, 4-11, and 4-12. In Chapter 5 different methods of approach to primitive roots are given, and possible omissions could be Sections 5-4, 5-5, and 5-6. Two proofs of the Quadratic Reciprocity Law are given in Chapter 6; possible omissions in this chapter could be Sections 6-8, 6-9, 6-10, 6-11. Chapters 7 and 8 are somewhat more advanced. Chapter 7 includes a proof of Bertrand's Postulas (Section 7-4) and an introduction to Ramanujan's Highly Composite Numbers

Preface

(Section 7-9). A large part of Chapter 8, Continued Fractions, may have to be omitted in an introductory course. Chapter 9 contains a solution of the Pythagorean equation (Section 9-2) and a proof of the four-square theorem (Section 9-11).

My indebtedness to the works of others is obvious. It is a pleasure to acknowledge receiving helpful suggestions from colleagues during the preparation of this text. I especially want to express my appreciation for help and advice from my colleagues, Professors Arthur B. Brown and Elliott Mendelson. To Professor Gordon Pall for helpful suggestions and constructive criticism I give my sincere thanks. Finally, it is a privilege to express my gratitude to Mr. Ronald M. Solomon for critically reading the greater part of the manuscript. His suggestions have materially helped to improve the exposition. It is my hope that this work may help to foster in students an interest in this fascinating subject, the theory of numbers.

Ralph G. Archibald
Queens College
Flushing, New York

Contents

1 Introduction

1-1 Nature of the Subject   1
1-2 Some Questions Considered   3
1-3 Problems   5

2 Divisibility

2-1 Introduction   7
2-2 Sundry Definitions   8
2-3 Elementary Theorems   8
2-4 Some Fundamental Principles   9
2-5 Basic Theorem   10

2-6	Mathematical Induction	11
2-7	Problems	13
2-8	Scales of Notation	14
2-9	Problems	15
2-10	Common Divisors	16
2-11	Euclid's Algorithm	18
2-12	Linear Diophantine Equations	19
2-13	Problems	21
2-14	Greatest Common Divisor and Least Common Multiple	22
2-15	Number of Primes Infinite	24
2-16	Sieve of Eratosthenes	26
2-17	Unique Factorization	26
2-18	Problems	28

3 Congruences

3-1	Residue Classes	29
3-2	Congruence Symbol	30
3-3	Properties of Congruences	30
3-4	Problems	34
3-5	Euler's ϕ-Function	35
3-6	Fermat's Theorem and Euler's Generalization	37
3-7	Pseudoprimes	41
3-8	Problems	42
3-9	Linear Congruences and Their Solution	43
3-10	Simple Continued Fractions	44
3-11	Wilson's Theorem	48
3-12	The Chinese Remainder Theorem	48
3-13	Problems	51
3-14	Identical and Conditional Congruences	52
3-15	Equivalent Congruences	53
3-16	Division of Polynomials, modulo m	54
3-17	Problems	56
3-18	Number of Solutions of a Congruence	57
3-19	Number of Solutions of Special Congruences	59
3-20	Number of Solutions of a Binomial Quadratic Congruence	61
3-21	Problems	63
3-22	Solution of the Congruence $f(x) \equiv 0 \pmod{m}$	63
3-23	Polynomials Representing Primes	68
3-24	Problems	70

4 Some Significant Functions in the Theory of Numbers

- 4-1   The Greatest Integer Function   71
- 4-2   Problems   77
- 4-3   Generalization of Euler's ϕ-Function   78
- 4-4   Functions $\tau(n)$ and $\sigma(n)$ 81
- 4-5   Problems   84
- 4-6   Perfect Numbers   85
- 4-7   Möbius μ-Function   86
- 4-8   Liouville's Function $\lambda(n)$ 91
- 4-9   Problems   93
- 4-10 Recurrence Formulae   95
- 4-11 Fibonacci's and Lucas' Sequences   99
- 4-12 Problems   99

5 Primitive Roots and Indices

- 5-1   Belonging to an Exponent   103
- 5-2   Problems   109
- 5-3   Primitive Roots   110
- 5-4   Obtaining Primitive Roots   113
- 5-5   Sum of Numbers Belonging to an Exponent   115
- 5-6   Further Consideration of Primitive Roots of p^n 118
- 5-7   Problems   120
- 5-8   Indices   122
- 5-9   Problems   126

6 Quadratic Congruences

- 6-1   A Quadratic Congruence   129
- 6-2   Quadratic Residue and Quadratic Nonresidue   130
- 6-3   Problems   132
- 6-4   Euler's Criterion   132
- 6-5   Legendre's Symbol   134
- 6-6   Quadratic Reciprocity Law   134
- 6-7   Problems   142
- 6-8   Another Proof of the Quadratic Reciprocity Law   144
- 6-9   The Jacobi Symbol   147
- 6-10 Generalized Quadratic Reciprocity Law   150
- 6-11 Problems   151

7 Elementary Considerations on the Distribution of Primes and Composites

- 7-1 Introduction 153
- 7-2 The O-notation 154
- 7-3 Problems 156
- 7-4 Bertrand's Postulate 157
- 7-5 Problems 162
- 7-6 Bounds for $\pi(x)$ 163
- 7-7 Remarks on the Prime Number Theorem 168
- 7-8 Primes in Arithmetical Progressions 169
- 7-9 Highly Composite Numbers 170
- 7-10 Relatively Highly Composite Numbers 172
- 7-11 Problems 173

8 Continued Fractions

- 8-1 Introduction 175
- 8-2 Finite Continued Fractions 177
- 8-3 Convergents and Their Limits 179
- 8-4 Problems 184
- 8-5 Representation of Irrational Numbers 184
- 8-6 Approximation by Rational Numbers 189
- 8-7 Problems 196
- 8-8 Quadratic Irrational Numbers 197
- 8-9 Periodic Continued Fractions 201
- 8-10 Problems 209
- 8-11 Pell's Equation 209
- 8-12 Problems 217
- 8-13 Farey Sequences 218
- 8-14 Problems 221

9 Certain Diophantine Equations and Sums of Squares

- 9-1 Introductory Remarks 223
- 9-2 The Pythagorean Equation 224
- 9-3 The Diophantine Equation $x^2 + 2y^2 = z^2$ 225
- 9-4 Problems 228
- 9-5 Some Fourth Degree Diophantine Equations 228

9-6	Problems	237
9-7	Solution of the Equations $X^4 - 2Y^4 = \pm Z^2$	238
9-8	Sum of Two Squares	244
9-9	Sum of Three Squares	247
9-10	Problems	249
9-11	Sum of Four Squares	250
9-12	Remarks on Waring's Problem	252
9-13	Problems	253

Notes 255

Bibliography 288

Table of Primes 291

Introduction 1

1-1 Nature of the Subject

To a greater extent than some other branches of pure mathematics, the theory of numbers is a theoretical science: many of its results are seemingly devoid of practical utility. In the course of his presidential address in 1890 before the section on Mathematics and Physics of the British Association, J. W. L. Glaisher made remarks[1] that still have relevance for us today.

> Many of the greatest masters of the mathematical sciences were first attracted to mathematical inquiry by problems relating to numbers, and no one can glance at the periodicals of the present day which contain questions for solution without noticing how singular a charm such problems still con-

---

[1]Numbered footnotes refer to the Notes on each chapter. These Notes will be found at the end of the text, indicated under the entry Footnotes in the Index.

tinue to exert. This interest in numbers seems implanted in the human mind, and it is a pity that it should not have freer scope in this country. The methods of the theory of numbers are peculiar to itself, and are not readily acquired by a student whose mind has for years been familiarized with the very different treatment which is appropriate to the theory of continuous magnitude;.... From the moment that Gauss, in his wonderful treatise[2] of 1801, laid down the true lines of the theory, it entered upon a new day, and no one is likely to be able to do useful work in any part of the subject who is unacquainted with the principles and conceptions with which he endowed it.

In addition, in his Report[3] on the Theory of Numbers, some of H. J. S. Smith's comments appear very pertinent and up-to-date.

the Theory of Numbers has acquired a great and increasing claim to the attention of mathematicians. It is equally remarkable for the number and importance of its results, for the precision and rigorousness of its demonstrations, for the variety of its methods, for the intimate relations between truths apparently isolated which it sometimes discloses, and for the numerous applications of which it is susceptible in other parts of analysis. 'The higher arithmetic,' observes Gauss†, confessedly the great master of the science, 'presents us with an inexhaustible store of interesting truths,—of truths, too, which are not isolated, but stand in a close internal connexion, and between which, as our knowledge increases, we are continually discovering new and sometimes wholly unexpected ties. A great part of its theories derives an additional charm from the peculiarity that important propositions, with the impress of simplicity upon them, are often easily discoverable by induction, and yet are of so profound a character that we cannot find their demonstration till after many vain attempts; and even then, when we do succeed, it is often by some tedious and artificial process, while the simpler methods may long remain concealed.'

As has already been observed, the theory of numbers has engaged the attention of many leading mathematicians of the past. For more than two thousand years, significant contributions have been made to the subject by both orientals and occidentals, by amateurs and professional mathematicians. On referring to a history of the development of the theory of numbers, one cannot fail to be impressed by the significant array of men of outstanding mathematical genius and insight who have left their mark in the annals of scientific and cultural progress. Gauss is reputed to have made the interesting assertion that "mathematics is the queen of the sciences, and the theory of numbers the queen of mathematics." In 1929, L. E. Dickson commented: "Recent investigations [in the theory of numbers] compare favorably with the older ones. Future discoveries will far surpass those of the past."[4] Unlike some other mathematical disciplines, many of the problems, theorems, and

---

†Preface to Eisenstein's "Mathematische Abhandlungen," Berlin, 1847.

results of the theory of numbers can be stated sufficiently clearly to arouse the interest and appreciation of those without special mathematical training as well as those scientists whose main interest lies outside this particular branch of mathematics.

In 1922, G. H. Hardy, President of Section A (Mathematics and Physics) of the British Association for the Advancement of Science, brought out this fact forcefully in the course of his presidential address.[5]

> It happens, by a fortunate accident, that the particular subject [The Theory of Numbers] which I love the most, and which presents most of the problems which occupy my own researches, is by no means overwhelmingly recondite or obscure, and indeed is sharply distinguished from almost every other branch of pure mathematics, in that it makes a direct, popular, and almost irresistible appeal to the heart of the ordinary man.

Later in his presentation he asked the question:

> *Are there infinitely many primes of the form* $2^n - 1$? I find it hard to imagine a problem more fascinating or more terribly difficult than that. It is plain, though, that this is a question which computation can never decide, and it is very unlikely that it can ever give us any data of serious value.[6]

Recently, this fact has also been pointed out by Professor Marshall Stone.

> Many of the problems of number theory can be very simply formulated in the mathematical terminology of common speech and are thus easily understood without much mathematical preparation. Among them are some of the most difficult unsolved problems in the whole of mathematics. Such problems attract the attention not only of serious mathematicians but also of amateurs and even of mere publicity seekers.... Number theory is not totally without interest for applied mathematics, but it remains largely the preserve of the pure mathematician. Some of the analytical problems, and perhaps also some of the algebraic applications, associated with number theory do indeed have a certain intrinsic interest for fields of application, but this would hardly justify teaching number theory in an applied mathematics program. On the other hand, the inclusion of elementary courses on number theory in the mathematical curriculum, whether on technical or cultural grounds, certainly needs no defense.[7]

One characteristic of the theory of numbers is that the simplicity of statement of the problem gives no indication of the difficulty or the mathematical depth of a valid solution. Let us consider a few examples.

1-2 Some Questions Considered

What is the *least* number of squares required such that we can say that *every* positive integer can be written as a sum of not more than that number of squares? By trial we see that the answer must be not less than four, since the number 7 requires four squares:

$$7 = 2^2 + 1^2 + 1^2 + 1^2.$$

Would, however, some numbers need more than four squares for their representation? The answer to this problem is now well known: four squares are sufficient for all integers.[8]

Next, how about the case of cubes? What is the least number of positive or zero cubes such that every positive integer can be represented as the sum of that number of cubes? It has been shown that nine cubes are sufficient.[9] In particular, nine cubes are required to represent 23:

$$23 = 2^3 + 2^3 + 7 \cdot 1^3.$$

Also, 239 requires nine cubes:

$$239 = 5^3 + 3 \cdot 3^3 + 4 \cdot 2^3 + 1^3.$$

These two problems regarding squares and cubes are particular cases of the famous Waring's Problem.

We might ask what odd primes can be written as the sum of two integral squares. The answer to this is also well known. An interesting theorem of Fermat states that all primes of the form $4m + 1$ (such as 5, 13, 17) can be written (in essentially one way) as the sum of two squares, while none of the primes of the form $4m + 3$ (such as 3, 7, 11) can be so expressed.

We shall now mention a few problems, simple to state but very difficult to solve. Is it true, as conjectured by Christian Goldbach, that every positive *even* integer greater than 4 can be written as the sum of two odd primes?[10] For example, we see that $6 = 3 + 3, 8 = 5 + 3, 10 = 7 + 3, 12 = 7 + 5$. Although progress has been made, this still remains an unsolved problem. Again, is there an infinite number of pairs of prime numbers which differ by two? For instance, 3 and 5, 5 and 7, 11 and 13, 17 and 19 are examples of such pairs. Moreover, is there an infinite number of triplets of primes of the form $p, p + 2, p + 6$ or triplets of the form $p, p + 4, p + 6$? Examples of the former set are: 5, 7, 11; 11, 13, 17; 41, 43, 47. Examples of the latter set are: 7, 11, 13; 13, 17, 19; 37, 41, 43. Similarly, it could be asked whether there is an infinite number of quadruplets of primes of the form $p, p + 2, p + 6, p + 8$. Examples are 5, 7, 11, 13 and 11, 13, 17, 19. The same question could be asked regarding quintuplets of primes of the form $p, p + 2, p + 6, p + 8, p + 12$. Examples are provided by 5, 7, 11, 13, 17 and 101, 103, 107, 109, 113. The answers to these questions are, as yet, not known.

Consider another as yet unsolved problem. Are there infinitely many prime numbers which are one more than a perfect square? Instances of such primes are: $2 = 1^2 + 1, 5 = 2^2 + 1, 17 = 4^2 + 1, 37 = 6^2 + 1$.

Again, is there an infinite number of primes of the form $2^p - 1$ (Mersenne numbers)? For $p = 2, 3, 5$, we obtain the primes 3, 7, 31, respectively. The answer to this question is also not known.

Even before the beginning of the Christian era, arithmeticians had been interested in "perfect" numbers; that is, positive numbers n such that the sum of all its positive divisors less than the number n itself, is equal to the number n. For example, 6 is a perfect number since its divisors, less than itself, are 1, 2, 3, and $6 = 1 + 2 + 3$. The next perfect number is 28, since its divisors, exclusive of 28 itself, are 1, 2, 4, 7, 14, and $28 = 1 + 2 + 4 + 7 + 14$. All known[11] perfect numbers are *even.* Whether there exists an *odd* perfect number still remains an open question.

Another famous unsolved problem, known as "Fermat's Last Theorem," can be stated as follows:

If the integer n is three or greater, there exist no non-zero[12] integral solutions x, y, z of the equation $x^n + y^n = z^n$. The importance of this problem is its relation to the history of mathematics. An outgrowth of the study of this problem has been the general theory of algebraic numbers.[13] The concept of ideals in rings, which is of enormous importance in modern mathematics, grew out of the study of this problem.

Enough has been said to indicate that the theory of numbers is replete with problems of intriguing interest to the enquiring mind. Moreover, the solutions of many apparently 'simple' problems still elude mathematicians. In this branch of mathematics there yet remains 'very much land to be possessed.'

1-3 Problems

1. Find nine positive numbers not greater then 20 which can be represented as the sum of three non-zero squares.

2. Prove that if Goldbach's Theorem regarding two primes were true, then every odd number greater than 7 could be written as the sum of three odd primes.

3. Prove that Goldbach's Theorem would be true if and only if every even number greater than 6 were the sum of three primes.

4. If it were true that every positive even number can be expressed in more than a finite number of ways as the difference of two primes, show that there would then be an infinite number of pairs of prime numbers differing by two.

5. Prove that if Fermat's Last Theorem were true for a specific value $n = k$, then the theorem would be true for all positive integral multiples of k.

Divisibility 2

2-1 Introduction

Although historically the theory of numbers has been primarily concerned with the positive integers or natural numbers, $1, 2, 3, \ldots n, \ldots$, we shall use the term *integer* or *whole number* to refer to a rational integer, positive, negative, or zero: that is, an *integer* will denote a member of the set $\ldots -3, -2, -1, 0, 1, 2, 3, \ldots$. Unless it is otherwise stated, italicized Roman letters will denote integers.

The basic properties of integers and the operations of addition, subtraction, and multiplication as applied to integers have been carefully and fully considered elsewhere.[1] These properties and operations we shall assume to be known. A discussion of the foundations of our number system does not fall within the scope of this text.

2-2 Sundry Definitions

Let c be an arbitrary integer and a a non-zero integer. Then if there is an integer b such that $c = ab$, we say that *a divides c, a* is a *divisor of c, a* is a *factor of c, c* is *divisible by a*, or that *c* is a *multiple of a*. Symbolically, we write $a|c$ to denote that a divides c. Obviously, if a divides c, so does $-a$ divide c. If there is no such integer b, we may write $a \nmid c$. For example, since $12 = 3 \cdot 4$ and $0 = 7 \cdot 0$, we say that 3 is a divisor of 12 and 7 is a divisor of zero; that is, $3|12$, $7|0$, but $5 \nmid 12$ since there is no integer b satisfying $12 = 5\,b$. It will be noted that 1 and -1 are *divisors of every integer* and that 0 is a *multiple of every non-zero integer*. Every integer a different from zero is divisible by 1, -1, a, and $-a$.

By the symbol $|a|$ we mean the *numerical value* of a. Thus $|2| = 2$, $|-5| = 5$, $|0| = 0$. That is, $|a| = a$ if $a \geq 0$, and $|a| = -a$ if $a < 0$. Note that the following properties[2] readily follow:

(i) $|a| > 0$ if $a \neq 0$, and $|a| = 0$ if $a = 0$;
(ii) $|a| = |-a|$;
(iii) $|ab| = |a| \cdot |b|$;
(iv) $|a + b| \leq |a| + |b|$;
(v) $|a| - |b| \leq |a + b|$.

A *prime number*, or simply a *prime*, is an integer p greater than one having only the two distinct positive integral divisors 1 and p. A *composite number*, on the other hand, is an integer n greater than one having more than the two distinct positive divisors 1 and n. For example, 2 and 5 are prime numbers, and 4 and 21 are composite numbers. We note that 5 has the positive divisors 1 and 5, and 21 has the positive divisors 1, 3, 7, 21. The integers 1 and -1, which are the only integers dividing every integer, are called *units*.

An *even integer* is one having the factor 2; an *odd integer* is one which does not have the factor 2. For example, 6 and 0 are even integers, while 15 is an odd integer. If two integers are both odd or both even, they are said to have the *same parity*; otherwise, *opposite parity*. For example, 7 and 3 or -4 and 6 are of the same parity, while 3 and 8 are of opposite parity.

An integer which divides each of two or more integers is said to be a *common divisor* of these integers. For example, 3 is a common divisor of 9 and -12, 7 is a common divisor of -21, 0, and 35.

2-3 Elementary Theorems

• **Theorem 2-1.** (i) Every non-zero integer a divides itself; that is, $a|a$ if $a \neq 0$.

(ii) Every non-zero integer a which divides c also divides every multiple of c; that is, if $a|c$, then for every k, $a|kc$.

(iii) If a divides c and if c divides d, then a divides d; that is, if $a|c$ and $c|d$, then $a|d$.

(iv) If a divides both c and d, then a divides the sum of any multiple of c and any multiple of d; that is, if $a|c$ and $a|d$, then for all integers m and n (positive, negative, or zero) $a|(mc + nd)$.

(v) If $c \neq 0$ and if $a|c$, then $|a| \leq |c|$.

(vi) If two integers divide each other, they are numerically equal; that is, if $a|c$ and $c|a$, then $|a| = |c|$.

Proofs Now $a = a \cdot 1$. From the statement $c = ab$, it follows that $kc = k(ab) = a(bk)$. Hence (i) and (ii) are obviously true. If in addition to $c = ab$ we have $d = ce$, then $d = (ab)e = a(be)$; thus (iii) follows. Also, if $c = ab$ and $d = af$, then $mc + nd = mab + naf = a(mb + nf)$. Thus (iv) is true. We see that if $c = ab \neq 0$, then $|c| = |a| \cdot |b|$ and $|c| \geq |a|$. This establishes (v). Moreover, if $a|c$ and if $c|a$, then by (v), $|a| \leq |c|$ and also $|c| \leq |a|$; consequently $|a| = |c|$.

2-4 Some Fundamental Principles

We shall assume, and use in the sequel, the following properties[3] of integers and sets of integers:

I Every set of positive integers which is not empty contains a smallest integer (The *Well-ordering Principle*).

I(a) Every non-empty set of integers positive or zero contains a smallest integer.

II If a set of positive integers contains the integer a, and also contains the integer $n + 1$ whenever it contains the integer n, then this set contains all the positive integers greater than or equal to a (The *Principle of Mathematical Induction*).

II(a) If a set of positive integers contains the integer a, and if it also contains the integer $n + 1$ whenever it contains all integers greater than or equal to a and less than or equal to n, then the set contains all integers greater than or equal to a (The *Principle of Complete Mathematical Induction*).

III If n is an arbitrary positive integer, a non-empty set of positive integers not greater than n contains a greatest integer.

IV For positive integers a and b there exists a positive integer n such that $na > b$ (*Archimedes' Principle* applied to positive integers).

2-5 Basic Theorem

The following basic theorem is ascribed to Euclid.

- **Theorem 2-2** If b is an arbitrary integer and if $a \neq 0$, there exist unique integers q and r satisfying the condition that $b = aq + r$, where $0 \leq r < |a|$.

Consider a few examples. If $a = 3$, $b = 23$, then $23 = 3 \cdot 7 + 2$ and $q = 7$ and $r = 2$, where $0 \leq r < 3$; again, if $a = -3$, $b = 7$, $7 = (-3) \cdot (-2) + 1$ and $q = -2$ and $r = 1$, where $0 \leq r < |-3| = 3$; moreover, if $a = 79$, $b = -3$, $-3 = 79(-1) + 76$ and $q = -1$ and $r = 76$, where $0 \leq r < 79$; if $a = -6$, $b = 0$, then $0 = (-6) \cdot 0 + 0$ and $q = r = 0$; if $a = -11$, $b = -93$, then $-93 = (-11)9 + 6$ and $q = 9$ and $r = 6$, where $0 \leq r < |-11| = 11$.

Proof Consider the infinite ordered set of integers:

$$(2\text{-}1) \qquad \ldots, b - 3a, b - 2a, b - a, b, b + a, b + 2a, \ldots$$

First of all, we note that some of these integers must be positive.[4] The subset S of the integers in (2-1) which are positive or zero will contain [by Principle I(a)] a least integer, r say. Let such an integer be written $r = b - qa$, whence $b = aq + r$. Obviously, $0 \leq r = b - qa$. Moreover, $r - |a|$ is a number of the set (2-1). For, if $a > 0$, $r - |a| = r - a = b - (q + 1)a$; and if $a < 0$, $r - |a| = r + a = b - (q - 1)a$. Hence, $r - |a|$ is less than r and therefore negative. Thus r satisfies the condition that $0 \leq r < |a|$. There is only one such pair of values q and r. Suppose, for example, there is a second pair q_1 and r_1 such that $b = aq_1 + r_1$, with $0 \leq r_1 < |a|$. Then, since $aq + r = aq_1 + r_1$, $a(q - q_1) = r_1 - r$. Hence, $r_1 - r$ is a multiple of a which is numerically less than $|a|$, and therefore $r_1 - r$ must be zero. Consequently, $r_1 = r$ and $q_1 = q$.

- **Corollary** If b is arbitrary and if a is non-zero, there exist unique integers q and r satisfying

$$b = aq + r, \quad \text{where} \quad -\tfrac{1}{2}|a| < r \leq \tfrac{1}{2}|a|.$$

Proof By the above theorem there exist unique integers q_0 and r_0 such that $b = aq_0 + r_0$, where $0 \leq r_0 < |a|$. If $0 \leq r_0 \leq \tfrac{1}{2}|a|$, we take $r = r_0$ and $q = q_0$; consequently $0 \leq r \leq \tfrac{1}{2}|a|$. If, however, $\tfrac{1}{2}|a| < r_0 < |a|$, we take $r = r_0 - |a|$ and $q = q_0 + 1$ if $a > 0$, and $q = q_0 - 1$ if $a < 0$; thus, in this case, $-\tfrac{1}{2}|a| < r < 0$.

Suppose b has two such representations: $b = aq + r = aq' + r'$, where both r and r' are subject to the given condition. Assume that $r' \geq r$. Then $a(q - q') = r' - r$. Since, however, $0 \leq r' - r < |a|$

Mathematical Induction **11**

and $r' - r$ is a multiple of $|a|$, $r' - r = 0$. Then, too, $q - q' = 0$, and the representation is unique.

- **Theorem 2-3.** <mark>Every number greater than one has a prime factor.</mark>

    Proof Let us assume that there is an integer greater than 1 that does not have a prime factor. Consequently, since there is then a non-empty set S of such integers, we see that by the Fundamental Principle I, the set S has a least element, n say. Thus n is not a prime and we can write $n = n_1 n_2$, where n_1 and n_2 are each less than n and have no prime factors, since their prime factors would be prime factors of n. This contradicts the fact than n is the least number greater than 1 which does not have a prime factor. Consequently, every number greater than 1 does have a prime factor.

The following simple theorem will have a number of useful applications.

- **Theorem 2-4.** <mark>The square of every odd integer is one greater than a multiple of 8.</mark>

For example, the square of the odd number 5 is $5^2 = 25 = 8 \cdot 3 + 1$; also $11^2 = 121 = 8 \cdot 15 + 1$.

Proof From Euclid's Theorem 2-2 above, we see that a number b can be written $b = 2q + r$, where $0 \leq r < 2$. But, since b is odd, r must be one: $b = 2q + 1$. Then $b^2 = (2q + 1)^2 = 4q^2 + 4q + 1 = 4q(q + 1) + 1$. If q is even, $4q$ is a multiple of 8; if q is odd, $q + 1$ is even and $4(q + 1)$ is a multiple of 8. Consequently, the theorem is established.

2-6 Mathematical Induction

The method of proof by mathematical induction (Principle II) is frequently useful in the theory of numbers. Familiarity with this type of argument is essential to subsequent work. Let us now consider an example of this method of proof.

Example 1   Show that if n is a positive integer, $1 + 3 + 5 + \ldots + (2n - 1) = n^2$.

Proof First, this statement is obviously true when $n = 1$: $1 = 1^2$. Second, let us assume the statement to be true for a specific value of n, $n = k$, say: that is, we assume that

$$1 + 3 + 5 + \ldots + (2k - 1) = k^2.$$

Add the quantity $[2(k + 1) - 1] = 2k + 1$ to each side of this equation, obtaining

$$(1 + 3 + 5 + \ldots + 2k - 1) + 2(k + 1) - 1 = k^2 + 2k + 1.$$

Hence

$$1 + 3 + 5 + \ldots + (2k - 1) + [2(k + 1) - 1] = (k + 1)^2.$$

This last equation, however, is precisely what the statement of the theorem becomes if n is replaced by $k + 1$. Hence, we have proved that *if* the statement is true for $n = k$, it *must* also be true for $n = k + 1$. Finally, let S be the set of positive integral values of n for which the statement is true. We shall see that this set contains an arbitrary positive integer N. The statement is true for $n = 1$. If we take $k = 1$ in the above argument, we see that if it is true for $n = 1$, it must be true for $n = 1 + 1 = 2$; that is, for $k + 1$. We have now established the truth of the result for $n = 1$ and $n = 2$. If we next take $k = 2$, we see that, since it is true for $n = 2$, it must be true for the next higher value of n; that is, for $n = 2 + 1 = 3$. Thus we have established the truth of the result for $n = 1, 2,$ and 3. Proceeding in the same fashion, step by step, we establish the truth of the result for $n = N$, an arbitrary positive integer. We could, of course, have applied Principle II directly to the set S to obtain our result.

Example 2 Show that for positive or zero values of n, $5^{2n} + 24n - 1$ is divisible by 48.

First Solution Let $f(n) = 5^{2n} + 24n - 1$. Since $f(0) = 0$ and $f(1) = 48$, the statement is true for $n = 0$ and for $n = 1$. Assume the statement to be true for $n = k$; that is, assume that $f(k) = 5^{2k} + 24k - 1$ is divisible by 48. Then

$$\begin{aligned} f(k + 1) - f(k) &= 5^{2k+2} + 24k + 23 - 5^{2k} - 24k + 1 \\ &= 5^{2k}(5^2 - 1) + 24 \\ &= 24\{(5^k)^2 + 1\}. \end{aligned}$$

Since the square of an odd number is of the form $8M + 1$, we have, on transposing $-f(k)$:

$$f(k + 1) = f(k) + 24\{8M + 2\}.$$

Since $f(k)$ is, by hypothesis, divisible by 48, the right member of this last equation is divisible by 48. Consequently, $f(k + 1)$ is divisible by 48. Therefore, if the statement is true for a value $n = k$, then it must be true for $n = k + 1$. But, since it is true when we take $n = 1$, it must be true also for $n =$

$1 + 1 = 2$. Consequently, the statement is now true for $n = 1$ and $n = 2$. Next, taking the value of k to be 2, we see that the statement must be true for $n = 2 + 1 = 3$. Hence, employing mathematical induction, we see that the statement is true for the set of all positive and zero integers n.

The above proof could be modified slightly as follows.

$$f(k+1) - 25f(k) = 5^2 5^{2k} + 24k + 23 - 25(5^{2k} + 24k - 1)$$
$$= 24(-24k + 2).$$

Thus,

$$f(k+1) = 25f(k) + 48(1 - 12k).$$

Since $f(k)$ is by hypothesis divisible by 48, $25f(k)$ is also, and therefore the right member of this equation is divisible by 48. The induction now proceeds as before.

Second Solution For $n = 0$, the result is obvious. For $n \geq 1$, we note that $a - b$ is a factor of $a^n - b^n$. Then

$$f(n) = 5^{2n} + 24n - 1 = (5^2)^n - 1 + 24n$$
$$= 25^n - 1^n + 24n$$
$$= (25 - 1)(25^{n-1} + 25^{n-2} + \ldots + 25 + 1) + 24n$$
$$= 24\{(25^{n-1} + 25^{n-2} + \ldots + 25 + 1 + n\}.$$

First, let n be even: $n = 2r$. Since $(25^{n-1} + 25^{n-2} + \ldots + 25 + 1)$ is the sum of an even number of odd integers, it must be even, $2K$ say. Hence, the right member becomes $24\{2K + 2r\} = 48\{K + r\}$. Thus $f(n)$ is divisible by 48. Next, let $n = 2s + 1$. Since $25^{n-1} + 25^{n-2} + \ldots + 25$ is the sum of an even number of odd integers, it must be even, $2L$ say. Hence

$$24\{(25^{n-1} + \ldots + 25) + 1\} + 24n = 24\{2L + 1\} + 24(2s + 1),$$
$$= 48(L + s) + 48.$$

Then $f(n)$ is divisible by 48. It is to be noted that this latter proof does not employ mathematical induction.

2-7 Problems

1. If m is an arbitrary integer, show that all numbers of the form $5m^3 + 7m$ are multiples of six.

2. If $n \geq 0$, show that 16 divides $83^{4n} - 2 \cdot 97^{2n} + 1$.

3. Show that 14 divides $5^{2n+1} + 3^{4n+2}$ when $n \geq 0$.

4. Prove that for a non-negative integer n, $2^{4n} - 1$ is divisible by 15.

5. Prove that for $n \geq 0$, $3^{2n} + 7$ is a multiple of 8.

6. If $3x + 2$ is a multiple of 7, show that:

   (a) $15x^2 - 11x + 14$ is a multiple of 14;
   (b) $24x^2 + 4x + 41$ is a multiple of 49.

7. Show that for non-negative values of n, $7^{2n+2} - 48n - 49$ is divisible by 2304.

8. Prove that if an integer is both a square and a cube, it must be of the form $7M$ or $7M + 1$. Give an example of each form.

9. If n is a positive integer, prove that $a^n - b^n$ is divisible by $a - b$.

10. Prove that if n is a positive odd integer, then $a^n + b^n$ is divisible by $a + b$.

11. If n is positive, show that $2^{4n+2} + 1$ is composite.

12. Prove by mathematical induction that
$$1 + 2 + 3 + \ldots + n = \frac{n(n+1)}{2}.$$

13. Prove by mathematical induction that
$$1^2 + 2^2 + 3^2 + \ldots + n^2 = \frac{n(n+1)(2n+1)}{6}.$$

14. Show that for non-negative integers n, 32 divides
$$4(2n+1)^{2n} + 5^{6n+1} + 3^{18n+3} - 4.$$

15. Show that the difference of the squares of two primes each greater than three is divisible by 24.

16. If the product of four consecutive integers is increased by 1, prove that the result is a perfect square.

17. Show that the cube of an integer can be written as the difference of two squares.

2-8 Scales of Notation

We customarily write numbers in the scale of 10, the denary system. For example, the number 3852 denotes $3852 = 3 \cdot 10^3 + 8 \cdot 10^2 + 5 \cdot 10 + 2$; the number 2016 denotes $2016 = 2 \cdot 10^3 + 0 \cdot 10^2 + 1 \cdot 10 + 6$. A different *scale* or *base* or *radix* could be used instead of 10. The modern electronic digital computers are essentially binary in character, employing the radix 2. Some, however, employ hexadecimal (that is, 16) representation for specific

Scales of Notation   **15**

purposes. The number $n = 375 = 3 \cdot 10^2 + 7 \cdot 10 + 5$ could also be written, in the binary scale, $n = 1 \cdot 2^8 + 0 \cdot 2^7 + 1 \cdot 2^6 + 1 \cdot 2^5 + 1 \cdot 2^4 + 0 \cdot 2^3 + 1 \cdot 2^2 + 1 \cdot 2 + 1$. Using a subscript to indicate the scale of notation, we could write $n = (375)_{10} = (101110111)_2$. Similarly, $(1394)_{10} = 4 \cdot 7^3 + 0 \cdot 7^2 + 3 \cdot 7 + 1 = (4031)_7$; $(37)_{10} = 1 \cdot 2^5 + 0 \cdot 2^4 + 0 \cdot 2^3 + 1 \cdot 2^2 + 0 \cdot 2 + 1 = (100101)_2$; $(13)_{10} = 1 \cdot 2^3 + 1 \cdot 2^2 + 0 \cdot 2 + 1 = (1101)_2$; $(100101)_2 + (1101)_2 = (110010)_2 = (50)_{10}$; $(100101)_2 - (1101)_2 = (11000)_2 = (24)_{10}$; $(100101)_2 \times (1101)_2 = (111100001)_2 = (481)_{10}$. Also, $(391)_{10} = (110000111)_2$; $(23)_{10} = (10111)_2$; and $(110000111)_2 \div (10111)_2 = (10001)_2 = (17)_{10}$.

The usual arithmetical operations of addition, subtraction, multiplication, and division can be effectively carried out for numbers written in any scale g greater than 1 in a manner precisely analogous to that used for numbers written in the scale of 10. For example, we could write

$$\begin{array}{cc} (125)_7 & (125)_7 \\ + \ (54)_7 & \times \ (54)_7 \\ \hline (212)_7, \quad \text{and} & 536 \\ & 664 \\ \hline & (10506)_7 \end{array}$$

Similarly, $(23)_7 + (56)_7 = (112)_7$; $(112)_7 - (54)_7 = (25)_7$; $(54)_7 \times (23)_7 = (1635)_7$; $(4031)_7 \div (46)_7 = (56)_7$.

The validity of these operations for such a scale of notation g is due to the fact that positive integers have a unique representation as stated in the following theorem.

• **Theorem 2-5.** Let g be a given integer greater than 1. Then every positive integer n can be written in one and only one way in the form[5]

$$n = a_s g^s + a_{s-1} g^{s-1} + \ldots + a_1 g + a_0,$$

where $s \geq 0$, $0 < a_s < g$, $0 \leq a_i \leq g - 1$ for $i = 0, 1, 2, \ldots, s - 1$.

2-9 Problems

1. Show that $n^{4k} + 4$ cannot be a prime when $k \geq 1$ and $n \geq 2$.

2. Prove that if, for a positive integer k, $2^k - 1$ is a prime, then k must be a prime also.

3. Evaluate the following in the scale of 3:
   (a) $(201102)_3 + (10002)_3$;
   (b) $(1000101)_3 - (22011)_3$;
   (c) $(12001)_3 \times (2101)_3$;
   (d) $(210210)_3 \div (10122)_3$.

4. Evaluate the following:
   (a) $(100101)_2 + (11001)_2$;
   (b) $(101001)_2 - (1001)_2$;
   (c) $(10101)_2 \times (1101)_2$;
   (d) $(110000001)_2 \div (1011)_2$;
   (e) $(3142)_5 \times (1123)_5$.
5. (a) In the scale of 8, add $(237012)_8$ and $(50107)_8$.
   (b) In the scale of 9, multiply $(751)_9$ by $(804)_9$.
   (c) In the scale of 7, divide $(10625)_7$ by $(64)_7$.

2-10 Common Divisors

DEFINITION *An integer d is a common divisor of a and b if d divides a and if d divides b.*

If d is a common divisor of a and b, so is $-d$. We note that, if $ab \neq 0$, every such common divisor is numerically not greater than the smaller of $|a|$ and $|b|$. Since every non-zero integer has only a finite number of positive divisors, there is only a finite number of positive common divisors of a and b, provided that not both a and b are zero. It is to be noted that if both a and b were zero, every non-zero integer would divide both. Hence, on the assumption that not both a and b are zero, among the positive common divisors of a and b (of which one such is unity) there is a greatest (by Principle III). This is called *the greatest common divisor of a and b or the highest common factor of a and b*. The abbreviations $G.C.D.$ and $H.C.F.$ are commonly used. We denote the $G.C.D.$ of a and b by the symbol (a, b). For example, $(21, 15) = 3$, $(0, 6) = 6$, $(15, 28) = 1$.

Consider, then, two integers a and b, not both zero. The set S of numbers $ax + by$ with x and y ranging independently of each other over all integers, includes the integer a (for $x = 1, y = 0$) and the integer b (for $x = 0, y = 1$). This set S also includes positive integers, since at least one of $a, -a, b, -b$ is positive. Then let n_0 denote the least positive integer in the set of all positive integers contained in the set S (using Principle I); that is, $n_0 = ax_0 + by_0$. Divide an arbitrary integer $n = ax + by$ of the set S by the integer n_0:

$$n = n_0 q + r, \quad 0 \leq r < n_0;$$

therefore,

$$ax + by = (ax_0 + by_0)q + r,$$

and
$$r = a(x - x_0 q) + b(y - y_0 q).$$

Thus r is a non-negative number of the set S less than n_0. By the definition of n_0 as the *least positive integer* of the set of positive integers in S, we see that r cannot be positive. Hence $r = 0$ and $n_0 | n$. Thus n_0 divides every integer in the set S. Hence, since both a and b are in S, $n_0 | a$ and $n_0 | b$. We can also see from the above expression for n_0 that *every* common divisor of a and b must divide n_0. Hence n_0 is the greatest common divisor of a and b. It should be noted that this is entirely an 'existence' proof. A suitable practical method of finding the greatest common divisor of a and b will be given in Section 2-11. In the foregoing, we have established the following results:

- **Theorem 2-6** If a and b are not both zero, their greatest common divisor d exists and is expressible as a linear combination, with integral coefficients, of a and b; namely,
$$d = (a, b) = ax_0 + by_0.$$
Moreover, d is the smallest positive number of the form $ax + by$.

- **Corollary** The greatest common divisor of a and b is divisible by every common divisor of a and b.

For example, 6 is the greatest common divisor of 42 and 72; and $6 = 42 \cdot (-5) + 72 \cdot 3$ or $6 = 42 \cdot 7 + 72 \cdot (-4)$.

DEFINITION *Integers a and b are said to be relatively prime if $(a, b) = 1$.*

- **Theorem 2-7** If m is an arbitrary positive integer,
$$D = (ma, mb) = m(a, b) = md.$$
Proof Since $d = (a, b)$, there exist integers x and y such that $d = ax + by$; then $md = ma \cdot x + mb \cdot y$. Since $D|ma$ and $D|mb$, we have $D|md$. There also exist integers x', y' such that $D = ma \cdot x' + mb \cdot y'$. Since $d|a$ and also $d|b$, $d|(ax' + by')$. Thus md divides $m(ax' + by')$. Therefore $md|D$. Consequently, $D = md$.

- **Corollary 1** If $a = ka_0$ and $b = kb_0$, then $(a, b) = |k| \cdot (a_0, b_0)$.

- **Corollary 2** If $d = (a, b)$ and $a = da_0$ and $b = db_0$, then a_0 and b_0 are relatively prime: $(a_0, b_0) = 1$.

- **Theorem 2-8** If a and b are each relatively prime to m, then ab is relatively prime to m.

Proof Since $(a, m) = 1$, there are integers x and y such that $1 = ax + my$; therefore $b = ab \cdot x + m \cdot by$. Thus a divisor of both ab and m must divide b. However, m and b are relatively prime. Hence $(ab, m) = 1$.

- **Corollary** If each of $a_1, a_2, \ldots, a_s$ is relatively prime to m, then $a_1 a_2 \ldots a_s$ is relatively prime to m.

- **Theorem 2-9** If $(k, b) = 1$ and if $k|ab$, then $k|a$.

  Proof There exist integers x and y such that $1 = kx + by$; consequently $a = akx + aby$. Let $ab = kc$. Then $a = k(ax + cy)$; that is, $k|a$.

- **Corollary 1** If p is a prime dividing neither a nor b, then p does not divide ab.

- **Corollary 2** If p is a prime dividing none of $a_1, a_2, \ldots, a_s$, then p does not divide the product of the a's.

  Proof If p were to divide $a_1 a_2 \ldots a_s$, then, since p does not divide a_1, it must, by Theorem 2-9, divide $a_2 a_3 \ldots a_s$. Since it does not divide a_2, it must divide $a_3 a_4 \ldots a_s$. Consequently, it must ultimately divide a_s. This, however, is a contradiction.

- **Corollary 3** An integer n divisible by two relatively prime integers, a and b, is divisible by their product ab.

  Proof Setting $n = an_0$, we see that, where $(a, b) = 1$, $b|an_0$. Hence, by Theorem 2-9, $b|n_0$; that is, $n = abn_1$.

2-11 Euclid's Algorithm

Consider two integers a and b, not both zero. If one of these integers is zero, the greatest common divisor of a and b is obviously the numerical value of the other. Since a negative integer has the same divisors as its numerical value, we may assume without loss of generality that both a and b are positive integers with, say, $a \geq b$. The method to be described derives from Euclid and is known as *Euclid's Algorithm.*

Divide a by b, obtaining the quotient q_1 and the remainder r_1. Then, if $r_1 \neq 0$, divide b by r_1 obtaining the quotient q_2 and the remainder r_2. Then, if $r_2 \neq 0$, we repeat the process and divide r_1 by r_2, obtaining the quotient q_3 and the remainder r_3. This process, or "algorithm," is repeated until we first obtain the remainder zero. Thus, applying this process we obtain the following equations:

Linear Diophantine Equations

$$a = bq_1 + r_1, \quad 0 < r_1 < b,$$
$$b = r_1 q_2 + r_2, \quad 0 < r_2 < r_1,$$
$$r_1 = r_2 q_3 + r_3, \quad 0 < r_3 < r_2,$$
$$\ldots \qquad \ldots$$
$$r_{i-3} = r_{i-2} q_{i-1} + r_{i-1}, \quad 0 < r_{i-1} < r_{i-2},$$
$$r_{i-2} = r_{i-1} q_i + r_i, \quad 0 < r_i < r_{i-1},$$
$$\ldots \qquad \ldots$$
$$r_{t-3} = r_{t-2} q_{t-1} + r_{t-1}, \quad 0 < r_{t-1} < r_{t-2},$$
$$r_{t-2} = r_{t-1} q_t + r_t, \quad 0 < r_t < r_{t-1},$$
$$r_{t-1} = r_t q_{t+1}.$$

Since $b > r_1 > r_2 > \ldots > r_{i-1} > r_i \ldots \geq 0$, it is evident that after a certain number of steps we must reach a zero remainder. Let us suppose that after $t + 1$ applications of this process, we first obtain the remainder $r_{t+1} = 0$. Now, r_t can be shown to be the greatest common divisor of a and b. From the first equation, it is seen that every common divisor, k, say, of a and b must divide r_1; from the second equation, every common divisor of b and r_1 must divide r_2. Hence k divides a, b, r_1, and r_2. Continuing this process, we see that k divides r_{t-2}, r_{t-1}, and r_t. Considering the above equations in reverse order, we see that r_t divides $r_{t-1}, r_{t-2}, r_{t-3}, \ldots, r_3, r_2, r_1, b, a$. Thus $r_t | a$ and $r_t | b$. We have also seen that every common divisor k of a and b divides r_t. Therefore, r_t is the greatest common divisor of a and b.

We observe, too, that from these equations we obtain:

$$r_1 = a - q_1 b,$$
$$r_2 = b - r_1 q_2 = -q_2 a + (1 + q_1 q_2) b,$$
$$r_3 = (1 + q_2 q_3) a - (q_1 + q_3 + q_1 q_2 q_3) b.$$

Likewise, we see that r_i ($i = 1, 2, \ldots, t$) is of the form

$$r_i = A_i a + B_i b,$$

for suitable integers A_i and B_i. Consequently, as we saw previously, the greatest common divisor of a and b, namely r_t, can be written as a linear combination of a and b with integral coefficients.

2-12 Linear Diophantine Equations

A *Diophantine equation* (named after the Greek mathematician, Diophantus of Alexandria, fl. c. 250–275 A. D.) is an equation in one or more unknowns which is usually to be solved for integral values of the unknowns. Diophantus, however, often admitted rational solutions.

Consider the linear Diophantine equation in two unknowns:

$$ax + by = n.$$

If it has an integral solution for x and y, then obviously $d = (a, b)$ must divide n. If d does not divide n, there can be no integral solution. Hence, let $a = da_0$, $b = db_0$, $n = dn_0$, where now $(a_0, b_0) = 1$. Thus solving this equation is equivalent to the problem of solving an equation of similar form:

$$a_0 x + b_0 y = n_0,$$

where $(a_0, b_0) = 1$. A solution of one of these equations is a solution of the other.

· **Theorem 2-10** If $(a, b) = 1$ and if x_0, y_0 is a particular solution of the Diophantine equation

$$ax + by = n,$$

then, when t ranges over all integers, a complete solution is given by

$$x = x_0 + tb, \qquad y = y_0 - ta.$$

Proof If the values x and y satisfy $ax + by = n$ and if also $ax_0 + by_0 = n$, then

$$ax + by = ax_0 + by_0 \quad \text{and} \quad a(x - x_0) = -b(y - y_0).$$

Let us first assume that $ab \neq 0$. Since $(a, b) = 1$, $b|(x - x_0)$. Let $x - x_0 = tb$; that is, $x = x_0 + tb$. Then $a \cdot tb = -b(y - y_0)$ and $y - y_0 = -ta$; consequently, $y = y_0 - ta$. Since for an arbitrary integer t, these expressions for x and y satisfy the given equation $ax + by = n$, we see that all solutions are given by these expressions for x and y as t ranges over all integers, positive, negative, and zero. Later we shall consider methods of obtaining a particular solution x_0, y_0 of the given Diophantine equation.

If we had $a = \pm 1$ and $b = 0$, a particular solution is $x_0 = \pm n$ and $y_0 = 0$; consequently, a complete solution is $x = \pm n + t \cdot 0$, $y = 0 \mp t$. Likewise, if $a = 0$ and $b = \pm 1$, we could take $x_0 = 0$ and $y_0 = \pm n$, and have a complete solution $x = 0 \pm t$, $y = \pm n - t \cdot 0$.

Example 1 Since

$$6x + 35y = 1$$

has a solution $x_0 = 6$, $y_0 = -1$, a complete solution is given by $x = 6 + 35t$, $y = -1 - 6t$, where t ranges over all integers.

Linear Diophantine Equations

Example 2 Since
$$14x - 33y = 7$$
has a solution $x_0 = 17$, $y_0 = 7$, a complete solution is given by $x = 17 - 33t$, $y = 7 - 14t$ for arbitrary t.

Example 3 Since $(6, 51) = 3$ and 3 does not divide 22, there is no integral solution of
$$6x + 51y = 22.$$

Example 4 Now,
$$10x + 15y = 190$$
has the same solutions as
$$2x + 3y = 38.$$
Since $x_0 = 1$ and $y_0 = 12$ gives us a solution, a complete solution is given by $x = 1 + 3t$, $y = 12 - 2t$. All pairs of positive values of x and y are determined by the conditions $1 + 3t > 0$ and $12 - 2t > 0$. That is, $0 \leq t \leq 5$. This yields the following pairs of values of x, y: 1, 12; 4, 10; 7, 8; 10, 6; 13, 4; 16, 2.

Example 5 If a and b are relatively prime integers, we wish to find the greatest common divisor of $a^2 + b^2$ and $a + b$.

Let $D = (a^2 + b^2, a + b)$. Now $a^2 + b^2 = (a + b)(a - b) + 2b^2$. Every possible prime factor of D must divide $2b^2$. Because a and b are relatively prime, this prime factor cannot divide b^2 (since then it would divide b as well as $a + b$, and hence divide their difference, a). Consequently, the only possible prime factor of D is 2. Since a and b are relatively prime, D could not have the factor 2^2. Thus $D = 1$ or 2. If a and b are both odd, $D = 2$; otherwise, $D = 1$.

2-13 Problems

1. (a) Find the G. C. D. of 819 and 1430.
   (b) Find the G. C. D. of 227 and 659.
   (c) Find the G. C. D. of 584 and 1606.
2. Solve completely the following Diophantine equations:
   (a) $2x + 11y = 5$;
   (b) $8x + 12y = 6$;
   (c) $18x - 21y = 15$.

3. If $(a, b) = 1$, find possible values of the G.C.D. of $a + 3b$ and $a^2 - b^2$. Illustrate each possibility by giving specific values for a and b.

4. If $(a, b) = 1$, find the G.C.D. of $a^2 + b^2$ and $a^2 - b^2$.

5. Find the numbers a and b and also their G.C.D. if $a + b^2 = 3505678$ and $a - b = 6908$.

6. If $(m, n) = 1$, find the G.C.D. of $m^3 + n^3$ and $m^3 - n^3$.

7. If $(m, n) = 1$, find the G.C.D. of $m^2 - n^2$ and $m^3 - n^3$.

8. If $(m, n) = 1$, find the G.C.D. of $m + kn$ and $m^3 + n^3$.

9. Given that $(m, n) = 1$, find the G.C.D. of $m^2 + n^2$ and $m^3 + n^3$ if: (a) m and n are both odd; (b) m and n are of opposite parity.

10. Show that for integral values of m, $m^2 - m + 1$ and $3m^3 + m^2 + m + 2$ are relatively prime.

11. Show that $n^2 + 3n + 1$ and $7n^3 + 18n^2 - n - 2$ are relatively prime.

12. If $(m, n) = 1$, show that the G.C.D. of $7m + 3n$ and $2m - n$ is either 1 or 13.

13. Show that the G.C.D. of $m^2 + m + 2$ and $5m^3 - m^2 + 4m + 2$ is even, and that it is 14 if and only if $m = 7k + 3$.

14. Show that the G.C.D. of $m^3 + 3m^2 - 4m$ and $5m^4 - m^3 + m^2 + 7m + 6$ is a multiple of 6.

15. Prove that $\sqrt[3]{2}$ is irrational.

16. Find all positive integral solutions x, y, and z of $x! + y! = z!$.

17. If $|st| \neq 1$, show that $s^4 + 4t^4$ cannot be a prime number.

2-14 Greatest Common Divisor and Least Common Multiple

If not all the integers $a_1, a_2, \ldots, a_n$ are zero, let a_k be non-zero. Consider the set of all positive integers which simultaneously divide $a_1, a_2, \ldots, a_n$. Each member of this non-empty set of positive integers $\leq |a_k|$. Hence there is a greatest integer of this set of positive divisors (Principle III).

DEFINITION *If not all of $a_1, a_2, \ldots, a_n$ (where $n \geq 2$) are zero, the greatest positive integer which simultaneously divides $a_1, a_2, \ldots, a_n$ is called the greatest common divisor of $a_1, a_2, \ldots, a_n$. In symbols, we denote it by $(a_1, a_2, \ldots, a_n)$.*

Greatest Common Divisor and Least Common Multiple

If the greatest common divisor is the integer one, the integers $a_1, a_2, \ldots, a_n$, where $n \geq 2$, are said to be *relatively prime*: $(a_1, a_2, \ldots, a_n) = 1$.

Note that we have our former definition of the greatest common divisor in case $n = 2$.

If *every* pair of the integers $a_1, a_2, \ldots, a_n$, where $n \geq 2$, has the greatest common divisor one, we say that $a_1, a_2, \ldots, a_n$ are *relatively prime in pairs*. If n integers are relatively prime in pairs, they are relatively prime, but not conversely. As examples, note that the three numbers 6, 14, and 21 are relatively prime, but they are not relatively prime in pairs; on the other hand, the numbers 6, 25, 77 are relatively prime in pairs and consequently are relatively prime.

Consider the set of positive integers which are multiples of each of the non-zero integers $a_1, a_2, \ldots, a_n$, where $n \geq 2$. One such positive integer is $|a_1 a_2 \cdots a_n|$. Then (by Principle I) there is a smallest positive integer of this set. We call it the *least common multiple* of $a_1, a_2, \ldots, a_n$, abbreviated to L.C.M. of $a_1, a_2, \ldots, a_n$. For example, the L.C.M. of 6, 14, and 21 is 42. This L.C.M. divides every multiple of each of $a_1, a_2, \ldots, a_n$. For, if the L.C.M. is denoted by L_n and if M is an arbitrary multiple of each of these a's, then dividing M by L_n we get a quotient q and a remainder r satisfying $M = qL_n + r$, $0 \leq r < L_n$. Hence the non-negative number $r = M - qL_n$ is less than L_n and is divisible by each of $a_1, a_2, \ldots, a_n$. This number r cannot be positive since that would contradict the definition of L_n. Hence $r = 0$, and the L.C.M. of $a_1, a_2, \ldots, a_n$ divides every common multiple of $a_1, a_2, \ldots, a_n$.

Let a and b be positive integers such that $d = (a, b)$, $a = da_0$, $b = db_0$, and $(a_0, b_0) = 1$. Every common multiple M (assumed to be positive) of a and b is a multiple of a, and hence must be of the form ma. But, since M is also a multiple of b, $b|ma$; that is, $db_0|mda_0$. Hence $b_0|m$ and $m = b_0 m_0$. Thus, where $m_0 \geq 1$, $M = ma = b_0 m_0 da_0 = m_0(da_0 b_0)$. Now $da_0 b_0$ is a positive integer which is a multiple of both $a = da_0$ and $b = db_0$. It is also the least positive number which is a multiple of both a and b. Hence, $da_0 b_0 = ab/d$ is the L.C.M. of a and b. We have just established the following results:

- **Theorem 2-11** If a and b are positive integers and if $d = (a, b)$, the L.C.M. of a and b is ab/d.

- **Corollary 1** If a and b are relatively prime, their L.C.M. is ab.

- **Corollary 2** The product of two positive integers a and b is equal to the product of their G.C.D. and their L.C.M.

Note that for more than two integers, Corollary 2 is not valid. For example, $(6, 14, 21) = 1$ and the L.C.M. of 6, 14, and 21 is 42. Thus, the product $6 \cdot 14 \cdot 21$ is not equal to $1 \cdot 42$. We can, however, establish the following result.

- **Theorem 2-12** If L_{n-1} denotes the L.C.M. of the set of integers $a_1, a_2, \ldots, a_{n-1}$, where $n \geq 3$, then the L.C.M. of the set of integers $a_1, a_2, \ldots, a_{n-1}, a_n$ is the L.C.M. of L_{n-1} and a_n.

Proof Let L_n denote the L.C.M. of the set $a_1, a_2, \ldots, a_{n-1}, a_n$. Also let L denote the L.C.M. of L_{n-1} and a_n. Thus L is a common multiple of $a_1, a_2, \ldots, a_{n-1}, a_n$. Hence L is a multiple of L_n. Every common multiple of $a_1, a_2, \ldots, a_{n-1}$ is a multiple of their L.C.M., L_{n-1}. If an integer is a multiple of L_{n-1} and also a multiple of a_n, it is a multiple of the L.C.M. of L_{n-1} and a_n, namely of L. Now L_n, being a common multiple of $a_1, a_2, \ldots, a_{n-1}$, is a multiple of L_{n-1} and is also a multiple of a_n. Thus, L_n is a multiple of L, the L.C.M. of L_{n-1} and a_n. Consequently $L = L_n$.

2-15 Number of Primes Infinite

We shall give several proofs of the following theorem.

- **Theorem 2-13** The number of primes is infinite.

First Proof[6] **(from Euclid)** Suppose there is a *last prime*, p_n. In increasing sequence, let $p_1, p_2, \ldots, p_n$ be all the existing primes. Consider the integer

$$N = p_1 p_2 \cdots p_n + 1.$$

Now N, being greater than one, must be divisible by a prime. It cannot be a prime since it is larger than the greatest prime, p_n. Moreover, it is not divisible by any existing prime. For, if prime p_i divides N, then p_i must divide

$$N - p_1 p_2 \cdots p_i \cdots p_n = 1.$$

This is a contradiction, since a prime cannot divide one. Consequently, there is no greatest prime, and the number of primes is infinite.

Second Proof[7] (from T. J. Stieltjes) Assuming p_n to be the largest prime, express the product of the (distinct) existing primes $p_1, p_2, \ldots, p_n$ as a product $A \cdot B$ of two factors in any way whatever. Since each of the distinct primes $p_1, p_2, \ldots, p_n$ divides either A or B (but *not both A and B*), $A + B$ is not divisible by any of the primes $p_1, p_2, \ldots, p_n$. However, since $A + B$ is greater than one, it must have a prime factor. Consequently, there must exist a prime greater than p_n, the largest prime. Thus, the number of primes is infinite. Note that if $A = 1$, we have the above proof of Euclid.

Third Proof[8] (from J. Braun) As before, let p_n be the greatest prime, and assume $p_1 < p_2 < \ldots < p_n$. Assume also that $p_n \geq 5$. Now,

$$\frac{1}{p_1} + \frac{1}{p_2} + \ldots + \frac{1}{p_n} = \frac{N}{D},$$

where $N = p_2 p_3 \ldots p_n + p_1 p_3 \ldots p_n + \ldots + p_1 \ldots p_{n-1}$ and $D = p_1 p_2 \ldots p_n$. Since $\frac{1}{2} + \frac{1}{3} + \frac{1}{5} > 1$, this rational number N/D is greater than one. Consequently, the numerator N, being greater than the denominator, is itself greater than one and therefore has a prime factor. However, no prime p_i divides N, for p_i divides every term in the sum for N except the term $p_1 p_2 \ldots p_{i-1} p_{i+1} \ldots p_n$. Consequently, N must have a prime factor greater than p_n, the greatest prime. Thus, again, there must be an infinite number of primes.

Fourth Proof[9] (from E. Kummer) Assume that all the primes, in increasing order of magnitude, are $p_1, p_2, \ldots, p_n$, the greatest prime being p_n, where $n \geq 2$. Since every integer greater than one has a prime factor p_i, the only positive integer less than $D = p_1 p_2 \ldots p_n$ which is relatively prime to D is the number one. However, $D - 1$, being greater than one and less than D, has no prime factor in common with D. For, if p_i divides both D and $D - 1$, it divides their difference one—an impossibility. Hence, there are now at least two positive numbers less than D and relatively prime to D: namely, 1 and $D - 1$. This contradicts the above result that 1 is the only positive number of this character. Hence the supposition that there is a greatest prime is false.

Fifth Proof[10] A direct proof of the existence of an infinite number of primes may be given as follows. First, we know that every number greater than one has a prime factor. Consider the following infinite sequence of numbers $n_i (i = 1, 2, \ldots)$: $n_1 = 2, n_2 = n_1 + 1, n_3 = n_1 n_2 + 1, n_4 = n_1 n_2 n_3 + 1, \ldots, n_{i+1} = n_1 n_2 \cdots n_i + 1, \ldots$. Since each n_i is divisible by a prime and since no two n's have the same prime factor (since they are relatively prime to each other), the number of primes must be at least as great as the number of n's; that is, the number of primes must be infinite.

2-16 Sieve of Eratosthenes

Eratosthenes (fl. c. 230 B.C.) employed a simple but useful device for obtaining all primes not exceeding a positive integer n. Consider the sequence of consecutive integers $1, 2, \ldots, n$. Passing over 1, we note that 2 is the first prime of this sequence. Let us now strike out every second number after 2; that is, the numbers $2 \cdot 2 = 4, 3 \cdot 2 = 6, \ldots$. Since the first number after 2 which is not struck out is 3, this 3 is the next prime. Then, retaining 3, we strike out every third number after 3; that is, the numbers $2 \cdot 3 = 6, 3 \cdot 3 = 9, \ldots$. The next number after 3 not struck out is the prime 5. We next strike out every fifth number after 5. We proceed in this fashion until the next remaining number after that whose multiples were struck out is greater than $\sqrt{n}$. The numbers exceeding one which remain are primes. The existent tables of primes have been constructed by modification of this process.

For example, let $n = 30$. Consider the sequence: 1 2 3 4 5 6 7 8 9 10 11 12 13 14 15 16 17 18 19 20 21 22 23 24 25 26 27 28 29 30. First, we delete the multiples of 2: $4, 6, 8, \ldots, 30$. Next, we delete the multiples of 3: $6, 9, 12, \ldots, 30$. Then we delete the multiples of 5: $10, 15, 20, 25, 30$. The next remaining number after 5 is the prime 7. However, since 7 exceeds $\sqrt{30}$, the multiples of 7 (greater than 7 itself) have already been deleted. Consequently, the numbers (greater than unity) retained in this sequence are all the primes ≤ 30. These primes are: 2, 3, 5, 7, 11, 13, 17, 19, 23, 29. An adaptation of this sieve method of Eratosthenes has been applied to the problem of estimating the number of integers in a specified interval having certain divisibility properties with regard to a designated set of primes.

2-17 Unique Factorization

- **Theorem 2-14** Every integer greater than unity is either a prime or a product of primes.

    Proof Consider the positive integer $n > 1$. To prove the existence of such a representation of n, we shall proceed by complete mathematical induction on n. For $n = 2$ the theorem is true since n is then a prime. Assuming the existence of such a representation for the integers 2, 3, 4, $\ldots, k$, we shall show that there is such a representation for the integer $n = k + 1$. If $k + 1$ is itself a prime, the theorem is established for this value of n. If $k + 1$ is composite, $k + 1 = n_1 n_2$, where $1 < n_1 < k + 1$ and $1 < n_2 < k + 1$. By the hypothesis of the induction, both n_1 and n_2 can be represented as a product of primes; hence $k + 1$ can be so represented. Thus the existence of a representation of n as a product of primes is established.

Unique Factorization

In Theorem 2-14, we have considered merely the *existence* of a representation of n as a product of primes. In some number systems the representation of an "integer" as a product of "undecomposable integers" may not be unique; that is, it may be possible to write an "integer" in more than one way as a product of "undecomposable integers." For rational integers, however, we have the following theorem, known as the *Fundamental Theorem of Arithmetic*.

- **Theorem 2-15** Every integer greater than unity is either a prime or is, apart from the order of its prime factors, uniquely representable as a product of primes.

    Proof From the preceding theorem we know that such a representation exists for integers n greater than 1. In order to prove the uniqueness of this representation, we shall assume that an integer greater than one has two representations as the product of primes: $p_1 p_2 \ldots p_r$ and $q_1 q_2 \cdots q_s$, where $p_1 \leq p_2 \leq \ldots \leq p_r$ and $q_1 \leq q_2 \leq \ldots \leq q_s$. We wish to show that $s = r$ and $q_i = p_i$ for $i = 1, 2, \ldots, r$. We shall proceed by complete mathematical induction on n. We note that when n is a prime, the representation is obviously unique. As hypothesis of the induction, assume uniqueness of representation for each of the integers $2, 3, \ldots, k$. We desire to show that the representation is unique for $n = k + 1$. If $k + 1$ is a prime, uniqueness of representation is established. If $k + 1$ is composite, and if $k + 1 = p_1 p_2 \ldots p_r = q_1 q_2 \ldots q_s$, where $p_1 \leq p_2 \leq \ldots \leq p_r$ and $q_1 \leq q_2 \leq \ldots \leq q_s$ with $r \geq 2$ and $s \geq 2$, then $p_1 | q_1 q_2 \ldots q_s$ and $q_1 | p_1 p_2 \ldots p_r$. Then, since the p's and q's are primes, $q_1 = p_i$ and $p_1 = q_j$ for some p_i and some q_j, where $1 \leq i \leq r$ and $1 \leq j \leq s$. Since $q_1 = p_i \geq p_1 = q_j \geq q_1$, we have $q_1 = p_1$. Moreover, $1 < (k + 1)/p_1 = p_2 p_3 \ldots p_r = q_2 q_3 \ldots q_s < k$. By the hypothesis of the induction, $(k + 1)/p_1$ has a unique representation as a product of primes. Hence $r - 1 = s - 1$ and $p_i = q_i$ ($i = 2, 3, \ldots, r$). Thus $k + 1$ has a unique representation as a product of primes: $k + 1 = p_1 p_2 \ldots p_r$. Consequently, by complete mathematical induction, the theorem is established.

If we bring equal primes together, we see that every integer n greater than unity can be written in the *canonical form* (or the *canonical decomposition*),

$$n = p_1^{a_1} p_2^{a_2} \cdots p_r^{a_r},$$

where the p_i are in increasing order of magnitude and the a_i are positive integers. For example, we would write the number $17640 = 3 \cdot 5 \cdot 7 \cdot 2 \cdot 3 \cdot 7 \cdot 2 \cdot 2$ as $2^3 \cdot 3^2 \cdot 5 \cdot 7^2$.

2-18 Problems

1. Employing the sieve of Eratosthenes, obtain all primes less than one hundred.

2. Given that the product of two positive numbers is 22122100 and that their least common multiple is 850850, find the possible values of these two numbers if each number exceeds 3000.

3. Show that an integer can be represented as the difference of two squares if and only if it is an odd integer or a multiple of four.

4. If $2^n + 1$ is an odd prime number, prove that the integer n is a power of 2.

5. If P denotes the product of the first n primes of the form $4M - 1$, by considering the number $4P - 1$, show that there exists still a further prime of the form $4M - 1$. Hence, prove that there exist more than a finite number of primes of this form.

6. Prove that there is an infinitude of primes of the form $6M - 1$.

7. If p_n denotes the nth prime, prove that, when $n > 1$, $p_{n+1} < p_n^n + 1$.

8. Where p_n denotes the nth prime, prove that $p_n < 2^{2^n}$.

9. Find all positive integers x such that:
   (a) $x(x + 30)$ is a perfect square;
   (b) $x(x + 42)$ is a perfect square;
   (c) $x(x + 12)$ is both a square and a cube.

10. (a) For Fermat's numbers F_n, defined as $F_n = 2^{2^n} + 1$ where $n \geq 1$, prove that $(F_m, F_n) = 1$ if $m \neq n$.

    (b) Hence, using the result from part (a), give another proof that the number of primes is infinite.

11. Show that a composite number m exceeding four divides $(m - 1)!$.

Congruences 3

3-1 Residue Classes

Let m be an integer greater than one. By an earlier theorem, we know that every integer n may be written uniquely in the form

$$n = mq + r, \quad \text{where} \quad 0 \leq r < m.$$

Since there are exactly m possibilities for r (namely, $0, 1, 2, \ldots, m - 1$), every integer with respect to the number m can be assigned to exactly one class. All integers, and only those integers, having the same value of r will be in the same class. Consequently, if two integers differ by a multiple of m, they belong to the same class. Moreover, two integers of the same class must necessarily differ by a multiple of m. These m classes are called *Residue Classes*. For example, if $m = 6$, then the possible values of r would be 0, 1, 2, 3, 4, and 5. Consequently, for $m = 6$, there are six residue classes. The numbers $\ldots, -12, -6, 0, 6, 12, 18, \ldots$ would be in the same class, corresponding to $r = 0$. Also, ..., $-11, -5, 1, 7, 13, 19, ...$ would be in a second class, corresponding to $r = 1$. Finally, the numbers ..., $-7, -1, 5, 11, 17, ...$ would be in a sixth class, corresponding to $r = 5$. With respect to the number 6, every integer would now be classified in one, and only one, of these six classes. The number -22 would be in the same class as the number 2; 101, in the same class as 5.

DEFINITION *A set of m numbers, one from each of the m residue classes, is called a complete set of residues, modulo m.*

For example, the set 0, 1, 2, 3, 4, as well as the set 0, 6, -3, 3, 4, forms a complete set of residues, modulo 5. The numbers $0, 1, 2, ..., m-1$ form *a complete set of least residues*, modulo m.

3-2 Congruence Symbol

In his *Disquisitiones Arithmeticae*, C. F. Gauss (1777–1855) introduced into number theory a useful symbol, the congruence symbol. According to Gauss, two integers a and b are *congruent*, modulo m when their difference $a - b$ is divisible by the integer m. This he expressed symbolically as follows: $a \equiv b \pmod{m}$. The m classes in the preceding section will now be referred to as *residue classes, modulo m*. If a and b are in the same residue class, modulo m, they correspond to the same value of r, and consequently their difference is a multiple of m; that is, a is congruent to b, modulo m. If a and b are in different residue classes, modulo m, their difference cannot be a multiple of m; consequently, in this case, a is *incongruent* to b, modulo m: $a \not\equiv b \pmod{m}$. It can also be seen that every integer from the same residue class, modulo m, has with the integer m the same greatest common divisor.

3-3 Properties of Congruences

This congruence relation is an *equivalence relation* possessing the usual reflexive, symmetric, and transitive properties, as stated in Theorem 3-1.

• **Theorem 3-1** (i) $a \equiv a \pmod{m}$; (ii) If $a \equiv b \pmod{m}$, then $b \equiv a \pmod{m}$; (iii) If $a \equiv b \pmod{m}$ and if $b \equiv c \pmod{m}$, then $a \equiv c \pmod{m}$.

Proofs Now (i) is obvious from the fact that $a - a = 0 \cdot m$. Let $a - b$ be a multiple of m: $a - b = Mm$. Then $b - a = -Mm$. Consequently (ii) is immediately established. Next, if $a - b = Mm$ and if $b - c = Nm$,

Properties of Congruences

then by the addition of these equations we have $a - c = (M + N)m$. Consequently, by the definition of congruent numbers, (iii) is established. We are now able to prove the following useful results.

- **Theorem 3-2**

  (i) If $a \equiv b \pmod{m}$, then for every integer k
  $$ka \equiv kb \pmod{m}.$$

  (ii) If $a \equiv b \pmod{m}$ and if $c \equiv d \pmod{m}$, then
  $$a + c \equiv b + d \pmod{m},$$
  $$a - c \equiv b - d \pmod{m}.$$

  (iii) If $a \equiv b \pmod{m}$ and $c \equiv d \pmod{m}$,
  $$ac \equiv bd \pmod{m}.$$

  (iv) If $a \equiv b \pmod{m}$, then
  $$a^n \equiv b^n \pmod{m} \quad \text{for} \quad n \geq 0.$$

  (v) If $e_i \equiv f_i \pmod{m}$, $i = 1, 2, \ldots, n$, then
  $$\sum_{i=1}^{n} e_i \equiv \sum_{i=1}^{n} f_i \pmod{m}.$$

  (vi) If $a \equiv b \pmod{m}$ and if $f(x) = a_0 x^n + \ldots + a_{n-1} x + a_n$, where the a's are integers and n is a non-negative integer, then
  $$f(a) \equiv f(b) \pmod{m}.$$

Examples Since $2 \equiv -4 \pmod 6$, then, in (i) $2k \equiv -4k \pmod 6$ for the arbitrary integer k. Since $-3 \equiv 9 \pmod 6$, then, in (ii), $2 \pm (-3) \equiv -4 \pm 9 \pmod 6$. Also, in (iii), since $2 \equiv -4 \pmod 6$ and $-3 \equiv 3 \pmod 6$, $2(-3) \equiv (-4)3 \pmod 6$, or $-6 \equiv -12 \pmod 6$. Since $2 \equiv -4 \pmod 6$, we have, by (iv), $2^n \equiv (-4)^n \pmod 6$. Further, in (vi), for $f(x) = x^2 - x + 5$, $f(2) \equiv f(-4) \pmod 6$, or $7 \equiv 25 \pmod 6$.

Proofs As before, let $a - b = Mm$. Then $ka - kb = k(a - b) = kMm$. Thus part (i) is proved. If we let $c - d = M_1 m$, then $(a - b) \pm (c - d) = Mm \pm M_1 m$; or $a \pm c - (b \pm d) = (M \pm M_1)m$. Therefore part (ii) follows. By part (i) of this theorem, since $a \equiv b \pmod m$, $ac \equiv bc \pmod m$; and, since $c \equiv d \pmod m$, $bc \equiv bd \pmod m$. Hence, by the transitive property of the congruence symbol, $ac \equiv bd \pmod m$, and part (iii) follows. We shall prove part (iv) by mathematical induction. Now (iv) is obviously true if $n = 0$ or 1. Assume that (iv) is true for $n = k$; that is, assume $a^k \equiv b^k \pmod m$. By part (iii) just proved, $aa^k \equiv bb^k \pmod m$, or $a^{k+1} \equiv b^{k+1} \pmod m$. But part (iv) is true for

$n = 0$ and 1. Taking k to be 1, we see by what has just been shown that part (iv) must be true for $n = 1 + 1 = 2$. Next we see, by using $k = 2$, that (iv) must be true for $n = 2 + 1 = 3$. Thus, step by step, we establish the truth of part (iv) for an arbitrary value of n, $n = N$ say. Part (v) follows directly from part (ii) by induction on n. Now, since $a \equiv b$ (mod m), we have, for $n \geq i \geq 0$, $a^{n-i} \equiv b^{n-i}$ (mod m). Also, $a_i a^{n-i} \equiv a_i b^{n-i}$ (mod m). By part (v) of the theorem,

$$\sum_{i=0}^{n} a_i a^{n-i} \equiv \sum_{i=0}^{n} a_i b^{n-i} \pmod{m};$$

that is,

$$f(a) \equiv f(b) \pmod{m};$$

therefore, part (vi) is established.

- **Theorem 3-3** If $ka \equiv kb$ (mod m) and if $d = (k, m)$, $k = dk_0$, and $m = dm_0$, then $a \equiv b$ (mod m_0).

  Proof Since $ka - kb = Mm$, $dk_0 a - dk_0 b = Mdm_0$; or $dk_0(a - b) = dMm_0$. Hence m_0 divides $k_0(a - b)$, where $(k_0, m_0) = 1$. Therefore $m_0 | (a - b)$, or $a \equiv b$ (mod m_0).

- **Corollary** If $(k, m) = 1$, and if $ka \equiv kb$ (mod m), then $a \equiv b$ (mod m). That is to say, if each member of a congruence is divisible by a number relatively prime to the modulus, this factor may be cancelled. The proof is obvious since the d of the theorem is unity.

If, however, (k, m) is greater than 1, we may not always cancel the factor k: for example, although $2 \cdot 5 \equiv 2 \cdot 3$ (mod 4), $5 \not\equiv 3$ (mod 4).

Consider the following examples. Given that $6 \cdot 7 \equiv 6 \cdot 2$ (mod 15), we see that here $k = 6$, $d = (6, 15) = 3$, and $m_0 = 5$. It follows from the theorem that $7 \equiv 2$ (mod 5). Next, if we are given $77 \equiv -28$ (mod 15), we have $7 \cdot 11 \equiv 7 \cdot (-4)$ (mod 15). Since 7 and 15 are relatively prime, the corollary states that $11 \equiv -4$ (mod 15).

- **Theorem 3-4** If $a \equiv b$ (mod p^r), where $r \geq 1$, then $a^{p^s} \equiv b^{p^s}$ (mod p^{r+s}), where $s \geq 0$.

  Proof The theorem is obviously true if $s = 0$. Assume the theorem true for $s = k \geq 0$: $a^{p^k} \equiv b^{p^k}$ (mod p^{r+k}); that is, $a^{p^k} = b^{p^k} + Mp^{r+k}$. Raise both members of the equation to the power p:

$$(a^{p^k})^p = (b^{p^k} + Mp^{r+k})^p$$
$$= b^{p^{k+1}} + \frac{p}{1}(b^{p^k})^{p-1} Mp^{r+k} + \frac{p(p-1)}{1 \cdot 2}(b^{p^k})^{p-2} M^2 p^{2r+2k}$$
$$+ \cdots + M^p(p^{r+k})^p.$$

Properties of Congruences 33

Since the exponents of the powers of p, namely $r + k + 1, 2r + 2k, 3r + 3k, \ldots, pr + pk$ are each $\geq r + k + 1$, we have
$$a^{p^{k+1}} \equiv b^{p^{k+1}} \pmod{p^{r+k+1}}.$$
Consequently, if the theorem is true for $s = k$, it must be true also for $s = k + 1$. The theorem then follows by mathematical induction on s. In the applications of this theorem, p is frequently a prime.

Example 1 Show that for numbers written in the denary scale, the number itself and the sum of its digits differ by a multiple of 9. (This leads to the process known as "casting out nines.")

Solution Let the number N in the scale of 10 be $N = (a_n a_{n-1} \ldots a_1 a_0)_{10} = a_n 10^n + a_{n-1} 10^{n-1} + \ldots + a_2 10^2 + a_1 10 + a_0$. Since $10 \equiv 1 \pmod 9$, $10^k \equiv 1^k \equiv 1 \pmod 9$. By part (vi) of Theorem 3-2,
$$\sum_{k=0}^{n} a_k 10^k \equiv \sum_{k=0}^{n} a_k \cdot 1 \pmod 9.$$
Thus, since the left member of this congruence is N,
$$N \equiv a_n + a_{n-1} + \ldots + a_1 + a_0 \pmod 9.$$
Hence, the difference between N and the sum of its digits (in the scale of ten) is a multiple of 9. For example $3741 = 3 \cdot 10^3 + 7 \cdot 10^2 + 4 \cdot 10 + 1$; and $3741 - (3 + 7 + 4 + 1) = 3726$, which is divisible by 9.

Example 2 Show that every number of the form $n^3 - 3n^2 + 2n$ is divisible by 6 if n is odd, and by 24 if n is even.

First Solution Let
$$f(n) = n^3 - 3n^2 + 2n = n(n^2 - 3n + 2) = (n - 2)(n - 1)n.$$
If n is odd, $(n - 1)$ is even, and $(n - 2)(n - 1)n$ is then divisible by 2. If n is even, n is double an odd number or double an even number: thus, $n = 2(2k + 1) = 4k + 2$ or $n = 2(2k) = 4k$. If $n = 4k + 2$, $n - 2 = 4k$; if $n = 4k$, $n - 2 = 4k - 2 = 2(2k - 1)$. In either case, $(n - 2)n$ is divisible by 8. Since n must be of one of the forms $n = 3t + 0, 3t + 1, 3t + 2$, one of the integers $n - 2, n - 1, n$ must always be divisible by 3. Hence, $f(n) = (n - 2)(n - 1)n$ is always divisible by 3. Thus, if n is odd, $f(n)$ is divisible by $2 \cdot 3 = 6$; if n is even, $f(n)$ is divisible by $8 \cdot 3 = 24$.

Second Solution For $n = -1, 0, 1, 2$ the statement is obviously true. Assume the statement to be true for $n = k$; that is, assume that $f(k)$ is divisible by 6 when k is odd, and by 24 when k is even. Since $f(k + 2) - f(k) = 6k^2$, we have $f(k + 2) = f(k) + 6k^2$. We note that $k - 2, k$, and

$k + 2$ all have the same parity. If k is positive and odd, we see (by induction on n) that the statement is true for $n = 1, 3, \ldots, k, k + 2, \ldots$. If k is non-negative and even, we see likewise that the statement is true for $n = 0, 2, 4, \ldots, k, k + 2, \ldots$. Since $f(k - 2) - f(k) = -6k^2 + 24(k - 1)$, $f(k - 2) = f(k) - 6k^2 + 24(k - 1)$. Thus (by induction) we see that if k is negative and odd, the statement is true for $n = -1, -3, -5, \ldots, k, k - 2, \ldots$. Moreover, if k is non-positive and even, the statement is true for $n = 0, -2, -4, \ldots, k, k - 2, \ldots$. By mathematical induction, therefore, the statement is true for all integral values of n.

3-4 Problems

1. Show that all numbers of the form $2m^3 + 7m$ are multiples of 3.

2. Show that $a^2 - b^2$ is divisible by 4 if a and b are both even, and is divisible by 8 if both a and b are odd.

3. Show that every number of the form $3n(n + 3)$ is divisible by 6.

4. Show that if a number n greater than one is both a uth power and a vth power, then $n = (p_1^{k_1} p_2^{k_2} \cdots p_r^{k_r})^t$, where the p_i are the distinct prime factors of n, the k_i positive integers, and t the least common multiple of u and v.

5. (a) Show that every integer which is both a square and cube must be of one of the four forms: $36m$, $36m + 1$, $36m + 9$, $36m + 28$.

   (b) Obtain at least one such integer (which is both a square and a cube) of each of these four forms.

6. Show that no cube has the form $9m \pm 2$, $9m \pm 3$, or $9m \pm 4$.

7. (a) Show that if a number is both a square and a cube, it cannot be of either of the forms $5m + 2$ or $5m + 3$.

   (b) Show by examples that there are numbers of the forms $5m$, $5m + 1$, $5m + 4$ which are both a square and a cube.

8. (a) If an integer is simultaneously a fourth power and a fifth power, show that it must be of one of the forms $40m, 40m + 1, 40m + 16$, or $40m + 25$.

   (b) Determine whether there exist such numbers (both a fourth power and a fifth power) of *each* of these forms.

9. (a) If a number is simultaneously a cube and a fourth power, show that it must be of the form $5m$ or $5m + 1$.

   (b) Show by examples that there are numbers of each of the forms $5m$ and $5m + 1$ which are both a cube and a fourth power.

Euler's ϕ-Function

10. Prove that $11^{26^5} \equiv 15^{26^5} \pmod{676^3}$.

11. If $a^n - 1$ is a prime number p where $n > 1$ and $a > 1$, prove that a is equal to 2 and n is a prime.

12. Show that

$$\frac{1}{2} + \frac{1}{5} + \frac{1}{8} + \cdots + \frac{1}{3n-1}$$

cannot be an integer.

13. Show that if p_i denotes the ith prime, $\sum_{i=1}^{n} \frac{1}{p_i}$ can never be an integer.

3-5 Euler's ϕ-Function

DEFINITION *For a positive integer m, $\phi(m)$ is defined as the number of positive integers not exceeding m which are relatively prime to m.*

For this function $\phi(m)$, E. Prouhet proposed (in 1845) the term *indicator*; J. J. Sylvester (in 1879) introduced the term *totient*. For example, for $m = 1$, there is only one positive integer (namely, unity itself) not exceeding one and relatively prime to one; hence $\phi(1) = 1$. For $m = 2$, there is likewise only one positive integer (unity itself) not exceeding 2 and relatively prime to 2; hence $\phi(2) = 1$. Also, $\phi(3) = \phi(4) = \phi(6) = 2$ and $\phi(5) = 4$.

If from a complete set of residues, modulo m, we delete those which are not relatively prime to m, there will remain a set of $\phi(m)$ numbers each of which is relatively prime to m. We call such a set of $\phi(m)$ numbers a *reduced set of residues, modulo m*. If, for example, from the complete set of residues $1, 2, 3, \ldots, m$, modulo m, we delete those which have a common factor with m greater than unity, we obtain $\phi(m)$ numbers which form a reduced set of residues, modulo m. Obviously, if m is a prime p, then $1, 2, \ldots, p-1$ form a reduced set of residues, modulo p. We note then that $\phi(p) = p - 1$.

- **Theorem 3-5** If p is a prime and $e > 0$, $\phi(p^e) = p^{e-1}(p-1)$.

For example, $\phi(2^3) = 2^2(2-1) = 4$. The numbers in the set $1, 2, \ldots, 8$ which are relatively prime to 8 are the numbers 1, 3, 5, 7.

Proof Those numbers of the set of p^e numbers $1, 2, \ldots, p, p+1, \ldots, 2p, \ldots, 3p, \ldots p^{e-1} \cdot p$ which have a factor in common with p^e are the p^{e-1} numbers $p, 2p, 3p, \ldots p^{e-1} \cdot p$. Hence by the definition of $\phi(p^e)$,

$$\phi(p^e) = p^e - p^{e-1} = p^{e-1}(p-1) = p^e\left(1 - \frac{1}{p}\right).$$

- **Theorem 3-6** If a and b are relatively prime positive integers,

$$\phi(ab) = \phi(a)\phi(b).$$

For example, $\phi(12) = \phi(3)\phi(4) = 2 \cdot 2 = 4$.

Proof If either a or b is equal to one, the result is obviously true. Assume $a > 1$ and $b > 1$. Arrange the ab numbers from 1 to ab in b rows of a numbers each, as in the following array:

$$\begin{array}{ccccccc}
1 & 2 & 3 & \cdots & r & \cdots & a \\
a+1 & a+2 & a+3 & \cdots & a+r & \cdots & 2a \\
2a+1 & 2a+2 & 2a+3 & \cdots & 2a+r & \cdots & 3a \\
\vdots & \vdots & \vdots & \vdots & \vdots & \vdots & \vdots \\
(b-1)a+1 & (b-1)a+2 & (b-1)a+3 & \cdots & (b-1)a+r & \cdots & ba.
\end{array}$$

We see that each row in this array exhibits a complete set of residues, modulo a, and each column a complete set of residues, modulo b. For, in considering the row of a integers:

$$qa + 1, qa + 2, \ldots, qa + a,$$

we see that, if two of these integers were congruent, modulo a, then $qa + s \equiv qa + t \pmod{a}$; thus, $s \equiv t \pmod{a}$. This, however, is impossible if s and t are distinct. Similarly, a column of b integers,

$$r, a + r, 2a + r, \ldots, (b - 1)a + r,$$

evidently form a complete set of residues, modulo b. For, if $ua + r \equiv va + r \pmod{b}$, $ua \equiv va \pmod{b}$; then, since $(a, b) = 1$, $u - v \equiv 0 \pmod{b}$. Again, this is impossible if u and v are distinct.

Now $qa + r$, where $0 \leq q < b$ and $1 \leq r \leq a$, is relatively prime to a if and only if r is relatively prime to a. Let r denote a definite one of these $\phi(a)$ integers r. Then the column $r, a + r, 2a + r, \cdots, (b - 1)a + r$, being a complete set of residues, modulo b, will contain exactly $\phi(b)$ numbers which are relatively prime to b. Hence, since in each of the $\phi(a)$ columns of integers relatively prime to a, there are $\phi(b)$ integers which are relatively prime also to b, the number of positive integers not exceeding ab which are relatively prime to both a and b (and therefore to the product, ab) is $\phi(a)\phi(b)$.

- **Theorem 3-7** Where the a_i are positive integers and the p_i distinct primes ($i = 1, 2, \ldots, r$), if $n = p_1^{a_1} p_2^{a_2} \ldots p_r^{a_r}$, then

$$\phi(n) = n \left(1 - \frac{1}{p_1}\right)\left(1 - \frac{1}{p_2}\right) \cdots \left(1 - \frac{1}{p_r}\right).$$

For example, $\phi(360) = \phi(2^3 \cdot 3^2 \cdot 5) = 360(1 - \frac{1}{2})(1 - \frac{1}{3})(1 - \frac{1}{5}) = 96$.

Fermat's Theorem and Euler's Generalization

Proof If $r = 1$ or 2, the theorem follows immediately from the preceding Theorems 3-5 and 3-6. Assume the present theorem to be true for $r = k$. Then, since $p_{k+1}^{a_{k+1}}$ is relatively prime to $p_1^{a_1} p_2^{a_2} \cdots p_k^{a_k}$, we have

$$\phi(p_1^{a_1} p_2^{a_2} \cdots p_k^{a_k} p_{k+1}^{a_{k+1}}) = \phi(p_1^{a_1} \cdots p_k^{a_k})\phi(p_{k+1}^{a_{k+1}}),$$

which by hypothesis

$$= p_1^{a_1} \cdots p_k^{a_k}\left(1 - \frac{1}{p_1}\right)\cdots\left(1 - \frac{1}{p_k}\right) \cdot \phi(p_{k+1}^{a_{k+1}})$$

$$= p_1^{a_1} \cdots p_k^{a_k} p_{k+1}^{a_{k+1}}\left(1 - \frac{1}{p_1}\right)\cdots\left(1 - \frac{1}{p_k}\right)\left(1 - \frac{1}{p_{k+1}}\right).$$

Hence, by mathematical induction on r, the theorem is established.

3-6 Fermat's Theorem and Euler's Generalization

- **Theorem 3-8** (Fermat) If the integer a is not divisible by the prime p, then

$$a^{p-1} \equiv 1 \pmod{p}.$$

For example, $4^{5-1} \equiv 1 \pmod 5$.

Proof Consider the set of $p - 1$ integers:

$$1 \cdot a, 2 \cdot a, \ldots, (p-1)a.$$

Each of these $p - 1$ numbers is relatively prime to p, and no two of them are congruent, modulo p. For, if $sa \equiv ta \pmod{p}$, then $(s - t)a \equiv 0 \pmod{p}$. This is impossible unless $s = t$. Consequently, $1 \cdot a, 2 \cdot a, \ldots, (p-1)a$ form a reduced set of residues, modulo p. Hence, each of these numbers is congruent to one and only one number in the reduced set of residues $1, 2, \ldots, (p-1)$. Consequently, the product of the numbers in one set is congruent to the product of the numbers in the other reduced set of residues, modulo p. Thus,

$$1 \cdot a \quad 2 \cdot a \ldots (p-1)a \equiv 1 \cdot 2 \ldots \cdot (p-1) \pmod{p},$$

and

$$1 \cdot 2 \ldots \cdot (p-1)a^{p-1} \equiv 1 \cdot 2 \ldots \cdot (p-1) \pmod{p}.$$

Cancelling $1 \cdot 2 \ldots \cdot (p-1)$ from each side, we have the result:

$$a^{p-1} \equiv 1 \pmod{p}.$$

We can now readily establish, in similar fashion, Euler's Generalization of Fermat's Theorem.

- **Theorem 3-9** If $(a, m) = 1$, then
$$a^{\phi(m)} \equiv 1 \pmod{m}.$$

For example, $2^{\phi(15)} \equiv 2^8 \equiv 1 \pmod{15}$.

Proof Let $a_1, a_2, \ldots, a_{\phi(m)}$ be a reduced set of residues modulo m. Then $a_1 a, a_2 a, \ldots, a_{\phi(m)} a$ also form a reduced set of residues modulo m. Obviously, each of the $a_i a$ ($i = 1, 2, \ldots, \phi(m)$) is relatively prime to m. Moreover, no two are congruent modulo m. For, if $a_i a \equiv a_j a \pmod{m}$, then $a_i \equiv a_j \pmod{m}$. This is a contradiction if $i \neq j$. Hence the $\phi(m)$ integers $a_1 a, \ldots, a_{\phi(m)} a$ are, in some order, congruent to the $\phi(m)$ integers $a_1, \ldots, a_{\phi(m)}$, modulo m. That is,

$$a_1 a \equiv a_{i_1} \pmod{m}$$
$$a_2 a \equiv a_{i_2} \pmod{m}$$
$$\cdots$$
$$a_{\phi(m)} a \equiv a_{i_{\phi(m)}} \pmod{m}.$$

Therefore,
$$a^{\phi(m)} a_1 a_2 \cdots a_{\phi(m)} \equiv a_1 a_2 \cdots a_{\phi(m)} \pmod{m};$$

this yields
$$a^{\phi(m)} \equiv 1 \pmod{m}.$$

- **Theorem 3-10** If $a^{m-1} - 1$ is divisible by m and if $a^k - 1$ is not divisible by m when k is a divisor of $m - 1$ such that $1 \leq k \leq m - 2$, then m is a prime. (This converse of Fermat's Theorem 3-8 was given by E. Lucas in 1876.)

Proof When $m = 2$, the theorem is obviously true for $a \equiv 1 \pmod{m}$; when $m = 3$, the theorem follows for $a \equiv 2 \pmod{m}$. Let us assume m to be composite; that is, $m = m_1 m_2$ where $1 < m_1 < m$ and $1 < m_2 < m$. Since m_1 is not relatively prime to m, $\phi(m) < m - 1$. Let $d = (m - 1, \phi(m))$. Then $1 \leq d \leq \phi(m)$. By Theorem 2-6, there exist integers r and s such that $d = r(m - 1) + s\phi(m)$. Only one of the integers r and s must and can be positive. If $r > 0$ and $s \leq 0$, $d + (-s)\phi(m) = r(m - 1)$ and $a^d \cdot a^{(-s)\phi(m)} \equiv a^{r(m-1)} \pmod{m}$. Since, by Theorem 3-9, $a^{\phi(m)} \equiv 1 \pmod{m}$, $a^d \equiv 1 \pmod{m}$. Since $1 \leq d \leq m - 2$, this is a contradiction. Similarly, if $s > 0$ and $r \leq 0$, $d + (-r) \cdot (m - 1) = s\phi(m)$; and, since $a^d \cdot a^{(-r)(m-1)} \equiv a^{s\phi(m)} \pmod{m}$, $a^d \equiv 1 \pmod{m}$. Again, this is a contradiction. Hence m must be a prime.

- **Theorem 3-11** If $d_1, \ldots, d_r$ are all the distinct positive divisors of the positive integer n, then $\phi(d_1) + \ldots + \phi(d_r) = n$.

First Proof For $n = 1$, the theorem is obviously true. First, let n be a power of a single prime, $n = p^a$. Then the possible divisors are $1, p, p^2, \ldots, p^a$.

$$\phi(1) = 1,$$
$$\phi(p) = p - 1,$$
$$\phi(p^2) = p^2 - p,$$
$$\phi(p^3) = p^3 - p^2,$$
$$\vdots \quad \vdots \quad \vdots$$
$$\phi(p^{a-1}) = p^{a-1} - p^{a-2},$$
$$\phi(p^a) = p^a - p^{a-1}.$$

Adding these equations and noting cancellations on the right, we get $\phi(1) + \phi(p) + \phi(p^2) + \ldots + \phi(p^a) = p^a$, the desired result. Next, let n be written in the canonical form, $n = p_1^{a_1} \ldots p_r^{a_r}$. In view of the case already treated, we know that

$$p_i^{a_i} = \phi(1) + \phi(p_i) + \ldots + \phi(p_i^{a_i}),$$

where $i = 1, 2, \ldots, r$. Then

$$\{\phi(1) + \phi(p_1) + \phi(p_1^2) + \ldots + \phi(p_1^{a_1})\} \times$$
$$\times \{\phi(1) + \phi(p_2) + \ldots + \phi(p_2^{a_2})\} \times \ldots$$
$$\times \{\phi(1) + \phi(p_r) + \phi(p_r^2) + \ldots + \phi(p_r^{a_r})\}$$
$$= p_1^{a_1} p_2^{a_2} \ldots p_r^{a_r} = n.$$

However, since the p_i are distinct primes, the expansion of the left member gives

$$\sum_{d|n} \phi(d) = n,$$

where the summation is extended over all positive divisors d of n.

Second Proof Since $1 = \phi(1)$, the theorem is obviously true for $n = 1$. Assume that $n > 1$. Suppose m to be any integer of the set $1, 2, 3, \ldots, n$. Then $(m, n) = d$, say. We may then write $m = m_0 d$, $n = n_0 d$, where $(m_0, n_0) = 1$ and, since $m \leq n$, $1 \leq m_0 \leq n_0$. Thus, for a given value m, a unique value of d is determined. However, for a given divisor d of n, there may be several values of m corresponding. In fact, since $(m_0, n_0) = 1$ and $1 \leq m_0 \leq n_0$, there are $\phi(n_0) = \phi(n/d)$ values of m_0 (and hence of m) corresponding to a given value d. As already noted, each integer m of the set $1, 2, \ldots, n$ has with n a *unique* value d as greatest common divisor. Moreover, as d ranges over all divisors of n, n/d likewise ranges over all divisors of n. Thus, since there are altogether n possible values of m,

$$n = \sum_{d|n} \phi\left(\frac{n}{d}\right) = \sum_{d|n} \phi(d),$$

where the summation is extended to all divisors d of n.

Example 1 Show that for a non-negative integer n
$$f(n) = 3^{5n+2} + 5^{5n+3} - 2$$
is divisible by 22.

First Solution Since each of the first two terms is odd, their sum is even. Consequently, $f(n)$ is divisible by 2. Next, since $3^2 \equiv 9 \equiv -2 \pmod{11}$, $3^4 \equiv 4 \pmod{11}$ and $3^5 \equiv 12 \equiv 1 \pmod{11}$. Also, $5^2 \equiv 3 \pmod{11}$, $5^4 \equiv 9 \equiv -2 \pmod{11}$, and $5^5 \equiv -10 \equiv 1 \pmod{11}$. Hence, since $f(n) = (3^5)^n \cdot 3^2 + (5^5)^n \cdot 5^3 - 2$, $f(n) \equiv 1^n(-2) + 1^n \cdot 4 - 2 \equiv -2 + 4 - 2 \equiv 0 \pmod{11}$. Thus, since $f(n)$ is divisible by both 2 and 11, it is divisible by 22.

Second Solution For $n = 0$, $f(n) = 9 + 125 - 2 = 132 = 22 \cdot 6$. Assume the result to be true for $n = k$: $f(k) = 3^{5k+2} + 5^{5k+3} - 2 = 22 \cdot s$, say. Then, since $f(k+1) = 3^{5k+7} + 5^{5k+8} - 2$ and $f(k) = 3^{5k+2} + 5^{5k+3} - 2$,
$$f(k+1) - f(k) = 3^{5k+2}(3^5 - 1) + 5^{5k+3}(5^5 - 1)$$
and
$$f(k+1) = 3^{5k+2}(3^5 - 1) + 5^{5k+3}(5^5 - 1) + 22 \cdot s.$$
Since 22 divides each of the even numbers $3^5 - 1$ and $5^5 - 1$, we see that $f(k+1)$ is divisible by 22. Consequently the result follows by mathematical induction.

Example 2 If the integer n is positive or zero, prove that $7^{2n+2} - 48n - 49$ is divisible by 2304.

First Solution Let $f(n) = 7^{2n+2} - 48n - 49$. Since $f(0) = 0$, the result is true for $n = 0$. Noting that $2304 = 48^2$, we assume the result to be true for $n = k$: that is, $f(k) = 7^{2k+2} - 48k - 49 = 2304t$. Now, $f(k+1) - f(k) = 7^{2k+2}(7^2 - 1) - 48 = 48(7^{2k+2} - 1) = 48((7^2)^{k+1} - 1) = 48(49^{k+1} - 1)$. Since $49 - 1 = 48$ divides the bracket, $f(k+1) - f(k) = 48^2 u$ or $f(k+1) = 48^2 u + 2304t$. The result follows by mathematical induction.

Second Solution
$$\begin{aligned}
f(n) &= (7^2)^{n+1} - 48n - 49 \\
&= 49^{n+1} - 49 - 48n \\
&= 49\{49^n - 1\} - 48n \\
&= 48\{49^n - 1\} + 49^n - 1 - 48n \\
&= 48(49 - 1)\{49^{n-1} + 49^{n-2} + \ldots + 49 + 1\} \\
&\quad + (49 - 1)\{49^{n-1} + 49^{n-2} + \ldots + 49 + 1\} - 48n \\
&= 48^2\{49^{n-1} + 49^{n-2} + \ldots + 49 + 1\} \\
&\quad + 48\{49^{n-1} + 49^{n-2} + \ldots + 49 + 1 - n\}.
\end{aligned}$$

Pseudoprimes

Since $49^k \equiv 1 \pmod{48}$,
$$49^{n-1} + 49^{n-2} + \ldots + 49 + 1 - n$$
$$\equiv \underbrace{1 + 1 + \ldots + 1}_{n \text{ times}} - n \equiv 0 \pmod{48}.$$

Hence,
$$f(n) \equiv 0 \pmod{48^2}.$$

Third Solution The method employed in the first solution could be modified slightly. Knowing that the result is true for $n = 0$ and assuming the result to be true for $n = k$, we could consider $f(k+1) - 7^2 f(k) = 48^2 k + 2304 \equiv 0 \pmod{2304}$. Thus, as before, the result follows by mathematical induction.

Example 3 Prove that if ab is relatively prime to 42,
$$a^6 - b^6 \equiv 0 \pmod{168}.$$

Solution Now $168 = 2^3 \cdot 3 \cdot 7$. Since a and b are both odd, $a^2 \equiv 1 \pmod{8}$ and $b^2 \equiv 1 \pmod 8$; thus $a^6 \equiv 1 \equiv b^6 \pmod 8$. Therefore $a^6 - b^6 \equiv 0 \pmod 8$. Also by Fermat's Theorem, $a^2 \equiv 1 \equiv b^2 \pmod 3$, which yields $a^6 \equiv b^6 \pmod 3$. Similarly, $a^6 \equiv 1 \equiv b^6 \pmod 7$. Thus $a^6 - b^6$ is divisible by $8 \cdot 3 \cdot 7$.

3-7 Pseudoprimes

We know, from Fermat's Theorem, that if p is a prime, $2^p \equiv 2 \pmod p$. It is not true, however, that when $2^n \equiv 2 \pmod n$, n must be a prime.[1] For example, $2^{341} \equiv 2 \pmod{341}$, where $341 = 11 \cdot 31$.

A *Pseudoprime* is a composite number n such that $2^n \equiv 2 \pmod n$.

The following are known to be pseudoprimes: $n = 341, 561, 645, 1105, 1387, 1729, 1905, 2047$. These can be verified from the fact that: for $341 = 11 \cdot 31$, $2^{10} \equiv 1 \pmod{341}$; for $561 = 3 \cdot 11 \cdot 17$, $2^{40} \equiv 1 \pmod{561}$; for $645 = 3 \cdot 5 \cdot 43$, $2^{28} \equiv 1 \pmod{645}$; for $1105 = 5 \cdot 13 \cdot 17$, $2^{24} \equiv 1 \pmod{1105}$; for $1387 = 19 \cdot 73$, $2^{18} \equiv 1 \pmod{1387}$; for $1729 = 7 \cdot 13 \cdot 19$, $2^{36} \equiv 1 \pmod{1729}$; for $1905 = 3 \cdot 5 \cdot 127$, $2^{28} \equiv 1 \pmod{1905}$; for $2047 = 23 \cdot 89$, $2^{11} \equiv 1 \pmod{2047}$.

Some interesting results have been obtained in the study of pseudoprimes. It has been shown that there is an unlimited number of such numbers. Only seven pseudoprimes have been found that are less than 2000, all of which are odd. In 1951, it was proved[2] that there exist infinitely many even pseudoprimes. Also, the infinite arithmetical progression $b, b + a, b + 2a, b + 3a, \ldots$

contains an infinitude of pseudoprimes[3] when a is positive and relatively prime to b.

One extension of the concept of pseudoprime is given in the following definition. An *absolutely pseudoprime number* is a composite number n such that for every integer a the number $a^n - a$ is divisible by n. One such number is the pseudoprime 561. It is readily seen that not every pseudoprime is an absolutely pseudoprime number. Whether there is an unlimited number of absolutely pseudoprime numbers is an open question.

3-8 Problems

1. If n is an integer not divisible by 3, show that $5n^6 + 3n^4 - 3n^2 - 5$ is divisible by 72.

2. Show that if a has none of the primes 2, 3, 5, 7 as a factor, then $a^{12} - 1$ is divisible by 840.

3. If n is relatively prime to 35, show that $n^{12} \equiv 1 \pmod{35}$.

4. (a) If k is an odd integer relatively prime to 5, show that $2k^4 + 158$ is divisible by 160.

   (b) If k is an odd number relatively prime to 3, show that $k^2 \equiv 1 \pmod{24}$.

5. If a and b are each relatively prime to 66, show that $a^{10} - b^{10} \equiv 0 \pmod{264}$.

6. Show that if ab is relatively prime to 91, then $a^{12} - b^{12}$ is divisible by 91.

7. If $(a, 30) = 1$, show that 240 divides $a^4 - 1$.

8. If x is a positive integer, prove that in the scale of ten, the last digit on the right (that is, the unit's digit) is the same for x^5 as it is for x.

9. If a and b are integers not divisible by the prime p, show that if:
   (a) $a^p \equiv b^p \pmod{p}$, then $a \equiv b \pmod{p}$;
   (b) $a^p \equiv b^p \pmod{p}$, then $a^p \equiv b^p \pmod{p^2}$.

10. Show that for non-negative integers n, $6 \cdot 17^{4n} - 5 \cdot 13^{4n^2} - 1$ is divisible by 240.

11. Show that for all integral values of n, $7^{12n^2+1} + (210n + 1)^4 - 8$ is divisible by 1680.

12. If n is a non-negative integer, show that $3^{2n+5} - 352n^2 - 56n + 269$ is divisible by 512.

13. Show that if n is non-negative, 104 divides $3^{6n} + 5^{4n+2} - 11^{12n+6} - 1$.

Linear Congruences and Their Solution

14. If a is an odd integer and relatively prime to 3, show that 192 divides $a^4 + 14a^2 - 96a + 81$.

15. If n is a non-negative integer, show that $7 \cdot 3^{22n} + 9 \cdot 5^{26n} - 16^{n+1}$ is divisible by 752.

16. If n is a positive integer, show that $5 \cdot 7^{12n} - 7 \cdot 5^{4n+1} - 2 \cdot 3^{12n} + 3 \cdot 2^{4n}$ is divisible by 464.

17. If n is a positive odd integer, show that $5^{12n+9} + 3(3^{3n-1} + 8^{3n}) - 2^4$ is divisible by $152 = 8 \cdot 19$.

18. Show that if $(n, 6) = 1$, $3n^6 - n^4 - 7n^2 + 5$ is divisible by 4608.

19. If n is a non-negative integer, show that 3969 divides $8^{2n+2} - 63n - 64$.

20. Let m and n be integers not divisible by the prime p. Moreover, where $r \geq 1$, let p^r be the highest power of p such that $m \equiv n \pmod{p^r}$.

   (a) If p is an odd prime, prove that where $r \geq 1$, p^{r+1} is the highest power of p dividing $m^p - n^p$.

   (b) If $p = 2$ and if $r > 1$, prove that 2^{r+1} is the highest power of the prime 2 dividing $m^2 - n^2$.

   (c) If $p = 2$ and if $r = 1$, show that there exists an infinitude of pairs of integers m and n such that an arbitrary positive power of 2 divides $m^2 - n^2$.

21. Prove that $161038 = 2 \cdot 73 \cdot 1103$ is a pseudoprime.

22. (a) If p and q are distinct primes and if $a^p \equiv a \pmod{q}$ and $a^q \equiv a \pmod{p}$, prove that $a^{pq} \equiv a \pmod{pq}$.

   (b) Show that $a = 2$, $p = 11$, $q = 31$ satisfy the conditions stated in this problem (See Section 3-7).

3-9 Linear Congruences and Their Solution

By a *linear congruence* proposed for solution, we mean a congruence of the form $ax - b \equiv 0 \pmod{m}$, where m does not divide a. Not every linear congruence has a solution. For example, $6x \equiv 2 \pmod{9}$ has no solution. For, if there were a solution x, then $6x - 2 = 9M$; but 3 divides every term but -2. We shall now state and prove an important theorem on linear congruences.

• **Theorem 3-12** If $d = (a, m)$, then the congruence $ax \equiv b \pmod{m}$ is solvable if and only if d divides b. Moreover, if d does divide b, there are exactly d incongruent solutions, modulo m.

Proof If there is a solution, there exist integers x and M such that $ax - b = Mm$. Since d divides both a and m, d must divide b; if d does not divide b, there is no solution. Hence, let $d|b$ and set $a = da_0$, $b = db_0$, and $m = dm_0$. Then the given congruence has a solution if and only if

$$a_0 x \equiv b_0 \pmod{m_0} \tag{3-1}$$

has a solution. Since $(a_0, m_0) = 1$, multiply both members of the congruence (3-1) by $a_0^{\phi(m_0)-1}$. Then $a_0^{\phi(m_0)} x \equiv a_0^{\phi(m_0)-1} b_0 \pmod{m_0}$. Since $a_0^{\phi(m_0)} \equiv 1 \pmod{m_0}$, it follows that $x \equiv a_0^{\phi(m_0)-1} b_0 \pmod{m_0}$ is the solution of congruence (3-1) and, consequently, is a solution of the given congruence. Let this solution be designated x_0. Every solution of the given congruence is obviously a solution of congruence (3-1), and every x satisfying congruence (3-1) satisfies the given congruence. It is seen that every solution X of the given congruence differs from x_0 by a multiple of m_0. For integers k and k', we note that the two numbers $x_0 + km_0$ and $x_0 + k'm_0$ satisfy not only the congruence (3-1) but also the given congruence. These two numbers would, however, be congruent, modulo m (and therefore be considered the same solution, modulo m) if and only if $k - k'$ is a multiple of d. Thus, we note that the d numbers $x_0, x_0 + m_0, x_0 + 2m_0, \ldots, x_0 + (d-1)m_0$ are incongruent, modulo m, and yield all the incongruent solutions, modulo m, of the given congruence.

- **Corollary** Every linear congruence $ax \equiv b \pmod{p}$ with a prime modulus has exactly one solution.

Proof Since by the definition of a linear congruence $(a, p) = 1$, the statement obviously follows.

As an example, let us solve the congruence

$$6x \equiv 15 \pmod{21}.$$

Solving this congruence is equivalent to solving $2x \equiv 5 \pmod{7}$. Multiplying both members by 2^5, we get $2^6 \cdot x \equiv 2^5 \cdot 5 \pmod{7}$; that is, $x \equiv 6 \pmod{7}$. The given congruence has three incongruent solutions: $6, 6 + 1 \cdot 7, 6 + 2 \cdot 7$; that is, $6, 13, 20$. The congruence $3x \equiv 2 \pmod{5}$ may be solved by multiplying both members by 3^3; thus we obtain the (unique) solution $x \equiv 3^3 \cdot 2 \equiv 4 \pmod{5}$. Another method of solution will be explained in Section 3-10.

3-10 Simple Continued Fractions

We shall here develop the subject of (finite) simple continued fractions merely to the point of obtaining an application to the solving of linear congruences and linear Diophantine equations in two unknowns.

Simple Continued Fractions **45**

By a *finite simple continued fraction* is meant a fraction of the form

(3-2)
$$a_1 + \cfrac{1}{a_2 + \cfrac{1}{a_3 + \cfrac{1}{a_4 + \cfrac{1}{a_5 + \cfrac{\ddots}{\quad + \cfrac{1}{a_n}}}}}},$$

where the integer $a_1 \geqq 0$, and the integers $a_2, a_3, \ldots, a_n$ are positive.

For example, the fraction

$$1 + \cfrac{1}{1 + \cfrac{1}{2 + \cfrac{1}{6}}}$$

has the value $\frac{32}{19}$; the fraction

$$-3 + \cfrac{1}{4 + \cfrac{1}{1 + \cfrac{1}{4}}}$$

has the value $-\frac{67}{24}$.

Now, a_k will be called the kth *partial quotient*. The kth *complete quotient* is

$$a_k + \cfrac{1}{a_{k+1} + \cfrac{1}{a_{k+2} + \cfrac{\ddots}{\quad + \cfrac{1}{a_n}}}}$$

If we stop with the partial quotient a_k, we get the kth *convergent*, C_k, to the value of the continued fraction. In the first example above,

$$C_1 = \frac{1}{1}, C_2 = \frac{2}{1}, C_3 = \frac{5}{3}, C_4 = \frac{32}{19}.$$

The a_1 will be zero, of course, when the value of the fraction is positive but less than one.

For the simple continued fraction (3-2), we make the following definitions:

$$p_1 = a_1, p_2 = a_2 a_1 + 1, q_1 = 1, q_2 = a_2;$$

and

$$p_i = a_i p_{i-1} + p_{i-2} \quad \text{and} \quad q_i = a_i q_{i-1} + q_{i-2}$$

when $i \geqq 3$. By direct computation we see that the first three convergents are

$$C_1 = \frac{a_1}{1} = \frac{p_1}{q_1}, \quad C_2 = \frac{a_2 a_1 + 1}{a_2} = \frac{p_2}{q_2}, \quad C_3 = \frac{a_3(a_2 a_1 + 1) + a_1}{a_3 a_2 + 1} = \frac{p_3}{q_3}.$$

We shall now prove the following result.

- **Theorem 3-13** The nth convergent, C_n, of a simple continued fraction has the value p_n/q_n when $n \geq 1$.

    Proof From the above remarks we note that the theorem is true for $n = 1, 2, 3$. Assume the truth of the theorem for $n = k \geq 3$. Then

    $$C_k = \frac{p_k}{q_k} = \frac{a_k p_{k-1} + p_{k-2}}{a_k q_{k-1} + q_{k-2}}.$$

    The $(k+1)$st convergent, C_{k+1}, differs from C_k only in the fact that instead of a_k, we employ $a_k + 1/a_{k+1}$. Hence

    $$\begin{aligned} C_{k+1} &= \frac{(a_k + 1/a_{k+1})p_{k-1} + p_{k-2}}{(a_k + 1/a_{k+1})q_{k-1} + q_{k-2}} \\ &= \frac{\{a_{k+1}(a_k p_{k-1} + p_{k-2}) + p_{k-1}\}/a_{k+1}}{\{a_{k+1}(a_k q_{k-1} + q_{k-2}) + q_{k-1}\}/a_{k+1}} \\ &= \frac{a_{k+1} p_k + p_{k-1}}{a_{k+1} q_k + q_{k-1}} \\ &= \frac{p_{k+1}}{q_{k+1}}. \end{aligned}$$

    Hence, if the theorem is true for $n = k$, it must be true for $n = k + 1$. Since it is true for $n = k = 3$, then it is true, by mathematical induction, for $n = k + 1 = 4$, and thus for all positive integers n. We shall state our next result in the following theorem.

- **Theorem 3-14** For $n \geq 2$, $p_n q_{n-1} - p_{n-1} q_n = (-1)^n$.

    Proof Now, since $p_2 = a_2 a_1 + 1$, $p_1 = a_1$ and $q_2 = a_2$, $q_1 = 1$, we see that for $n = 2$, the theorem is true. Assume the theorem to be true for $n = k \geq 2$; namely,

    $$p_k q_{k-1} - p_{k-1} q_k = (-1)^k.$$

    Now

    $$\begin{aligned} p_{k+1} q_k - p_k q_{k+1} &= (a_{k+1} p_k + p_{k-1}) q_k - p_k(a_{k+1} q_k + q_{k-1}) \\ &= -(p_k q_{k-1} - p_{k-1} q_k) = -(-1)^k = (-1)^{k+1}. \end{aligned}$$

    Hence, if the result be true for $n = k$, it must be true also for $n = k + 1$. Thus the theorem is established by mathematical induction.

- **Corollary** Every convergent is a fraction in its lowest terms.

Simple Continued Fractions

Proof The first convergent is $a_1/1$. The nth convergent, being p_n/q_n, must be in its lowest terms since any prime factor of both p_n and q_n, where $n \geq 2$, must divide $(-1)^n$.

Example 1 Using a simple continued fraction, let us solve the linear congruence

(3-3) $$79x \equiv 3 \pmod{103}.$$

The problem of solving this congruence is equivalent to the problem of solving the linear Diophantine equation

(3-4) $$79x + 103y = 3.$$

We shall obtain a solution of (3-4) by first securing a solution of

(3-5) $$79x + 103y = 1.$$

Write $79/103$ as a simple continued fraction. The partial quotients are $a_1 = 0, a_2 = 1, a_3 = 3, a_4 = 3, a_5 = 2, a_6 = 3$. The convergents are

$$C_1 = \frac{0}{1}, \quad C_2 = \frac{1}{1}, \quad C_3 = \frac{3}{4}, \quad C_4 = \frac{10}{13}, \quad C_5 = \frac{23}{30}, \quad C_6 = \frac{79}{103}.$$

Hence, since $p_6 q_5 - p_5 q_6 = (-1)^6$, $79 \cdot 30 - 23 \cdot 103 = 1$. Thus, a solution of (3-5) is $x = 30, y = -23$. A solution of (3-4) is $x_0 = 90, y_0 = -69$. Consequently, the complete solution of (3-4) is $x = 90 + 103t, y = -69 - 79t$. The solution of (3-3), therefore, is $x \equiv 90 \pmod{103}$.

Example 2 Solve the linear congruence

(3-6) $$71x \equiv 4 \pmod{55}.$$

First, obtain a solution of

(3-7) $$71x + 55y = 4$$

by solving

(3-8) $$71x + 55y = 1.$$

Consider the fraction $\frac{71}{55}$ written as a simple continued fraction:

$$\frac{71}{55} = 1 + \cfrac{1}{3 + \cfrac{1}{2 + \cfrac{1}{3 + \cfrac{1}{2}}}}.$$

Here $a_1 = 1, a_2 = 3, a_3 = 2, a_4 = 3, a_5 = 2$, Moreover, $C_1 = \frac{1}{1}, C_2 = \frac{4}{3}, C_3 = \frac{9}{7}, C_4 = \frac{31}{24}, C_5 = \frac{71}{55}$. Thus $p_5 = 71, p_4 = 31, q_5 = 55, q_4 = 24$. Since

$p_5 q_4 - p_4 q_5 = (-1)^5$, $71 \cdot 24 + (-31) \cdot 55 = -1$. Thus, $71 \cdot (-24) + 55 \cdot (31) = 1$, and $71 \cdot (-96) + 55 \cdot (124) = 4$. Hence, $x_0 = -96$, $y_0 = 124$ are solutions of (3-7). Thus (3-6) has the solution $x \equiv -96 \equiv 14$ (mod 55).

3-11 Wilson's Theorem[4]

- **Theorem 3-15** If p is a prime, then $(p-1)! + 1$ is divisible by p.

    Proof If $p = 2$, $1! + 1 = 1 + 1 = 2$ is divisible by 2. If $p = 3$, $2! + 1 = 2 + 1 = 3$ is divisible by 3. The theorem is therefore true for the primes 2 and 3. Let the prime p be greater than 3. Let a denote any integer whatever of the set

    (3-9) $\qquad\qquad\qquad 2, 3, \ldots, (p-2)$.

    Corresponding to each such a, there is exactly one integer b of the complete set of residues $0, 1, 2, \ldots, (p-1)$, satisfying $ab \equiv 1$ (mod p). Obviously, $b \not\equiv 0$ (mod p). Moreover, if $b \equiv \pm 1$ (mod p), then $a \equiv \pm 1$ (mod p). That is, $a \equiv 1$ (mod p) or $a \equiv p - 1$ (mod p). Both of these cases are impossible for a. Hence b must be an integer of the same set (3-9). Also, $b \neq a$. For, if $b = a$, $a^2 \equiv 1$ (mod p). Then the prime p must divide $a - 1$ or $a + 1$. Since a is in the set (3-9), both of these are impossible for a. Therefore, to each integer a in the set (3-9) there corresponds exactly one value of b, different from a, in the same set (3-9) such that $ab \equiv 1$ (mod p). Conversely, since $ab = ba$, we see that a is uniquely determined by b. The $p - 3$ integers in (3-9) can be classified into $(p-3)/2$ pairs so that the product of each pair is congruent to 1, modulo p. Consequently,

    $$2 \cdot 3 \ldots \cdot (p-2) \equiv 1 \ (\text{mod } p),$$

    which implies that $(p-1)! \equiv p - 1 \equiv -1$ (mod p). Thus $(p-1)! + 1 \equiv 0$ (mod p).

3-12 The Chinese Remainder Theorem

A rule for solving the problem of finding a number with specified remainders when divided by given integers, was known to the Chinese as early as the first century A. D.

- **Theorem 3-16** If the moduli are relatively prime in pairs, then there exists a value x simultaneously satisfying the set of congruences:

The Chinese Remainder Theorem

$$x \equiv a_1 \pmod{m_1},$$
$$x \equiv a_2 \pmod{m_2},$$
$$\cdots\cdots\cdots\cdots$$
$$x \equiv a_r \pmod{m_r}.$$

Moreover, any two simultaneous solutions are congruent modulo $m_1 m_2 \ldots m_r$.

Proof Let $m = m_1 m_2 \ldots m_r$ and $M_i = m/m_i$ for $i = 1, 2, \ldots, r$. Then, since $(M_i, m_i) = 1$, there exists a solution m'_i of the linear congruence $M_i m'_i \equiv 1 \pmod{m_i}$ for $i = 1, 2, \ldots, r$. Consider now the integer $x = m'_1 M_1 a_1 + m'_2 M_2 a_2 + \ldots + m'_r M_r a_r$. We see that this integer x satisfies $x \equiv a_i \pmod{m_i}$, where $i = 1, 2, \ldots, r$. For, $M_j \equiv 0 \pmod{m_i}$ if $j \neq i$. Thus $x \equiv m'_i M_i a_i \pmod{m_i}$. But since $m'_i M_i \equiv 1 \pmod{m_i}$, $x \equiv a_i \pmod{m_i}$ for $i = 1, 2, \ldots, r$. If x' is also a simultaneous solution of the given set of congruences, we have

$$x' \equiv a_i \equiv x \pmod{m_i} \qquad \text{for } i = 1, 2, \ldots, r.$$

Consequently,

$$x' \equiv x \pmod{m}.$$

Thus, any two simultaneous solutions are congruent, modulo m.

The statement of the Chinese Remainder Theorem has been generalized as follows.[5]

- **Theorem 3-17** The set of simultaneous congruences

$$x \equiv a_1 \pmod{m_1},$$
$$x \equiv a_2 \pmod{m_2},$$
$$\cdots\cdots\cdots\cdots$$
$$x \equiv a_r \pmod{m_r},$$

has a solution if and only if $a_i \equiv a_j \pmod{d_{ij}}$, where $d_{ij} = (m_i, m_j)$, $i = 1, 2, \ldots, r$, and $j = 1, 2, \ldots, r$. Since $a_j \equiv a_i \pmod{d_{ji}}$ follows from $a_i \equiv a_j \pmod{d_{ij}}$, we may assume that $i < j$.

Example 1 Simultaneously solve the congruences:

$$x \equiv 2 \pmod{6},$$
$$x \equiv 3 \pmod{5},$$
$$x \equiv 7 \pmod{11}.$$

Solution Here $m = 6 \cdot 5 \cdot 11 = 330$, $M_1 = 5 \cdot 11 = 55$, $M_2 = 6 \cdot 11 = 66$, $M_3 = 6 \cdot 5 = 30$. We must solve for m'_1, m'_2, m'_3 in the following congruences: $55m_1' \equiv 1 \pmod{6}$, $66m_2' \equiv 1 \pmod 5$, $30m_3' \equiv 1 \pmod{11}$. Here $m_1' \equiv 1 \pmod 6$, $m_2' \equiv 1 \pmod 5$, $m_3' \equiv -4 \pmod{11}$. Hence a simultaneous solution is

$$x \equiv 1 \cdot 55 \cdot 2 + 1 \cdot 66 \cdot 3 + (-4) \cdot 30 \cdot 7 \pmod{330};$$

that is,

$$x \equiv 128 \pmod{330}.$$

Example 2 If possible, solve simultaneously the following set of congruences:

$$5x \equiv 2 \pmod{24},$$
$$3x \equiv -26 \pmod{88},$$
$$x \equiv 28 \pmod{99}.$$

Multiplying each member of the first two congruences by 5 and 59, respectively, we obtain the three equivalent congruences:

$$x \equiv 10 \pmod{24},$$
$$x \equiv 50 \pmod{88},$$
$$x \equiv 28 \pmod{99}.$$

This last set of congruences is equivalent to the set:

$$x \equiv 10 \pmod 8,$$
$$x \equiv 10 \pmod 3,$$
$$x \equiv 50 \pmod 8,$$
$$x \equiv 50 \pmod{11},$$
$$x \equiv 28 \pmod 9,$$
$$x \equiv 28 \pmod{11}.$$

Now, $(24, 88) = 8$ and $10 \equiv 50 \pmod 8$; $(24, 99) = 3$ and $10 \equiv 28 \pmod 3$; also $(88, 99) = 11$ and $50 \equiv 28 \pmod{11}$. Hence, this last set of six congruences may be replaced by the set:

$$x \equiv 2 \pmod 8,$$
$$x \equiv 6 \pmod{11},$$
$$x \equiv 1 \pmod 9.$$

Here $m = 8 \cdot 11 \cdot 9 = 792$, $M_1 = 99$, $M_2 = 72$, $M_3 = 88$; and $m_1' 99 \equiv 1 \pmod 8$, $m_2' 72 \equiv 1 \pmod{11}$, $m_3' 88 \equiv 1 \pmod 9$ have solutions $m_1' \equiv 3 \pmod 8$, $m_2' \equiv 2 \pmod{11}$, $m_3' \equiv 4 \pmod 9$. Hence, a simultaneous solution of these last three congruences is

$$x \equiv 3 \cdot 99 \cdot 2 + 2 \cdot 72 \cdot 6 + 4 \cdot 88 \cdot 1 \equiv 1810 \equiv 226 \pmod{792}.$$

Therefore, the three given congruences have the simultaneous solution

$$x \equiv 226 \pmod{792}.$$

3-13 Problems

1. If $(n, 30) = 1$, show that 240 divides $n^4 + 239$.

2. If $(n, 42) = 1$, then show that 504 divides $n^6 - 1$.

3. If n is an odd integer not divisible by 3, show that $n^7 + 3n^6 - 3n^5 - 9n^4 + 3n^3 + 9n^2 - n - 3$ is divisible by 27648.

4. Show that when 17 is divided into 15!, the remainder is 1.

5. If $(n, 6) = 1$, show that 4608 divides $(n^4 - 1)^2(3n + 1)$.

6. Show that when 255! is divided by the prime 257, the remainder is 1.

7. Show that $7(125!) + 5!$ is divisible by 127.

8. (a) Expand in a simple continued fraction $\frac{119}{32}$ and $\frac{118}{303}$.
   (b) Find the rational number represented by the partial quotients $a_1 = 3, a_2 = 1, a_3 = 1, a_4 = 4, a_5 = 1, a_6 = 3$.

9. (a) Expand in a simple continued fraction $-\frac{503}{187}$ and $-\frac{125}{198}$.
   (b) Solve the congruence $187x \equiv 2 \pmod{503}$.

10. Using simple continued fractions, find a complete solution of the following congruences:
    (a) $79x \equiv 2 \pmod{153}$;
    (b) $182x \equiv 7 \pmod{203}$.

11. If $(n, 6) = 1$, show that 3456 divides $5n^6 - 9n^4 + 3n^2 + 1$.

12. Find the least positive number which, when divided by 13 leaves a remainder 5, when divided by 12 leaves a remainder 3, and when divided by 35 leaves a remainder 2.

13. From a certain number of marbles a boy can make 6 equal piles and have 5 left over, or 21 equal piles and have 8 left over, or 35 equal piles and have 1 left over. Find the smallest number of marbles the boy must have.

14. The twenty-third Mersenne prime is $2^{11213} - 1$. Find the last two digits of this prime number.

15. Find a simultaneous solution of $x \equiv 3 \pmod{6}$, $x \equiv -1 \pmod{25}$, $x \equiv 1 \pmod{7}$.

16. Find the least positive integer x which simultaneously satisfies $x \equiv -2 \pmod{11}$, $x \equiv 13 \pmod{28}$, and $x \equiv 7 \pmod{45}$.

17. Find the least positive integer x which simultaneously satisfies the congruences: $5x \equiv 2 \pmod{13}$, $x \equiv 2 \pmod{35}$, $3x \equiv 13 \pmod{77}$, and $x \equiv 7 \pmod{20}$.

18. Show that if $(n-1)! + 1 \equiv 0 \pmod{n}$, then n is a prime (Converse of Wilson's Theorem).

19. Show that an integer $p > 3$ is a prime when and only when $2\{(p-3)!\} + 1 \equiv 0 \pmod{p}$.

20. If r and s are arbitrary positive integers, prove that there exist r consecutive integers such that each is divisible by an sth power of an integer greater than one.

21. If p is an odd prime and if

$$\frac{1}{1} + \frac{1}{2} + \frac{1}{3} + \cdots + \frac{1}{p-1} = \frac{N}{D},$$

then prove that N is divisible by p.

3-14 Identical and Conditional Congruences

Consider a polynomial $f(x_1, x_2, \ldots, x_r)$ with integral coefficients. We say that $f(x_1, x_2, \ldots, x_r)$ is *identically congruent to zero*, modulo m, if the coefficient of each term of the polynomial is divisible by the modulus m. To indicate an identical congruence we may write $f(x_1, x_2, \ldots, x_r) \equiv 0 \pmod{m}$. Two polynomials are *identically congruent* to each other, modulo m, if the coefficients of like terms in the two polynomials are congruent to each other, modulo m. Thus, if $f(x) = 5x^3 - 2x^2 + x + 5$ and $g(x) = 6x^4 - x^3 + 10x^2 - 5x - 1$, $f(x)$ is identically congruent to $g(x)$, modulo 6: in symbols, $f(x) \equiv g(x) \pmod{6}$. If, however, $f(x_1, x_2, \ldots, x_r)$ is not identically congruent to zero, modulo m, we may refer to the congruence $f(x_1, x_2, \ldots, r_r) \equiv 0 \pmod{m}$ as a *conditional congruence*.

Obviously, an identical congruence is satisfied by all integral values of the variables. However, a congruence which is satisfied by all integral values of the variables is not necessarily an identical congruence. For example, $x^7 - x \equiv 0 \pmod{7}$ is satisfied for all values of x, but it is not an identical congruence: it is a conditional congruence. From Fermat's Theorem, we know that if p is a prime, $x^p - x \equiv 0 \pmod{p}$ is satisfied by all values of x, but this congruence, by definition, is a conditional congruence.

The integral polynomial $f(x)$ may be said to be of *degree n*, modulo m, if n is the degree of that term in $f(x)$ of highest degree whose coefficient is not divisible by m. We shall also refer to the congruence $f(x) \equiv 0 \pmod{m}$ as of degree n. If $f(x)$ is a constant not divisible by m, we say the degree is zero. A zero polynomial (that is, a polynomial with all coefficients divisible by m) will be considered to have no degree, modulo m.

Equivalent Congruences

For example,
$$20x^4 + 5x^2 - 2 \equiv 0 \pmod{10},$$
$$2x^2 + 2x \equiv 0 \pmod{5},$$
and
$$42x^6 + 0 \cdot x^5 - 12x^3 + x^2 + x - 3 \equiv 0 \pmod{6}$$
are each of degree 2, with respect to the indicated modulus.

3-15 Equivalent Congruences[6]

We have already considered a number of the properties of the congruence relation. If any operation listed among the properties of the congruence relation is applied to a congruence to obtain a second congruence, and if there is also a corresponding operation among these properties which when applied to the second congruence restores it to the original congruence, we say that the first operation is reversible.

More specifically, we shall consider the following reversible operations on a congruence:

(1) Multiplying or, when possible, dividing the coefficients of each member of a congruence by a number relatively prime to the modulus.

(2) Multiplying or, when possible, dividing the coefficients of each member *and* the modulus by the same non-zero number.

(3) Adding to each member of a congruence expressions that are identically congruent for the given modulus.

(4) Replacing an integral polynomial $h(x_1, x_2, \ldots, x_r)$ occurring either as a factor of either member or as a factor of any term in either member, by an integral polynomial $h_1(x_1, x_2, \ldots, x_r)$ identically congruent to $h(x_1, x_2, \ldots, x_r)$ for the given modulus.

If any finite sequence of the above reversible operations be performed on a congruence $f(x_1, x_2, \ldots, x_r) \equiv 0 \pmod{m}$ to yield the final congruence $g(x_1, x_2, \ldots, x_r) \equiv 0 \pmod{m_1}$, we may say that the first congruence is *equivalent* to the second congruence.[7] This "equivalence" is readily seen to be reflexive, symmetric, and transitive.

Consider a few examples: $2x + 9 \equiv 0 \pmod{5}$ is equivalent to $2x + 4 \equiv 0 \pmod{5}$; $3x \equiv 15 \pmod{33}$ is equivalent to $x \equiv 5 \pmod{11}$; $2x_1^2 + 4x_1x_2 - 8x_1 \equiv 10 \pmod{11}$ is equivalent to $x_1^2 + 2x_1x_2 + 7x_1 \equiv 5 \pmod{11}$; $(2x^2 + 3)(x^2 + 1) \equiv 0 \pmod{7}$ is equivalent to $(9x^2 + 7x - 4)(x^2 + 1) \equiv 0 \pmod{7}$; $x^2 - 2x + 3 \equiv 0 \pmod{5}$ is equivalent to $3x^2 + 9x + 9 \equiv 0 \pmod{15}$.

It is obvious that if two conditional congruences are equivalent, a set of integral values satisfying the one congruence will necessarily satisfy the other.

For example, the equivalent congruences $x^2 - 2 \equiv 0$ (mod 7) and $7x^3 - 6x^2 + 5 \equiv 0$ (mod 7) both have solutions $x \equiv 3$ (mod 7) and $x \equiv 4$ (mod 7). However, two congruences may have the same solutions for a given modulus and yet not be equivalent congruences; for example, $2x^3 + 1 \equiv 0$ (mod 3) (which is equivalent to $2(x-1)^3 \equiv 0$ (mod 3)) and $2x + 1 \equiv 0$ (mod 3) are not equivalent although they are satisfied by the same number $x \equiv 1$ (mod 3). Also, $x^2 - x + 4 \equiv 0$ (mod 10) (which is equivalent to $(x-3)(x-8) \equiv 0$ (mod 10)) and $2x - 6 \equiv 0$ (mod 10) are not equivalent, yet both have the solutions $x \equiv 3$ (mod 10) and $x \equiv 8$ (mod 10).

Occasionally, a linear congruence can be easily solved by using equivalent congruences. For example, the problem of solving the congruence $13x \equiv 5$ (mod 38) may be replaced by that of solving the equivalent congruences:

$$3 \cdot 13x \equiv 3 \cdot 5 \text{ (mod 38)},$$
$$39x \equiv 15 \text{ (mod 38)},$$
$$x \equiv 15 \text{ (mod 38)}.$$

Also the problem of solving $15x \equiv 2$ (mod 23) may be replaced by the problem of solving the following sequence of equivalent congruences:

$$3 \cdot 15x \equiv 6 \text{ (mod 23)},$$
$$45x \equiv 6 \text{ (mod 23)},$$
$$-x \equiv 6 \text{ (mod 23)},$$
$$x \equiv -6 \equiv 17 \text{ (mod 23)}.$$

3-16 Division of Polynomials, modulo m

Let m be an integer greater than 1, and let $f(x), g(x), q(x)$, and $r(x)$ be polynomials with integral coefficients (that is, integral polynomials) such that $g(x)$ is different from a zero polynomial, modulo m, and $r(x)$ is either a polynomial of lower degree than that of $g(x)$ or a zero polynomial, modulo m. Then, if

$$f(x) \equiv g(x)q(x) + r(x) \text{ (mod } m)$$

is an identical congruence, we may speak of $q(x)$ and $r(x)$ as a *quotient* and a *remainder*, respectively, in the division of $f(x)$ by $g(x)$, modulo m. When $r(x) \equiv 0$ (mod m) identically, we may say that the division of $f(x)$ by $g(x)$ is *exact*, modulo m, and that $g(x)$ is a *factor* or *divisor*, modulo m, of $f(x)$. We may also speak of $f(x)$ as being a *multiple*, modulo m, of $g(x)$.

When the modulus is a prime p and the polynomials $f(x)$ and $g(x)$ are not merely integral constants, it is readily seen (Example 5) that the division, modulo p, of $f(x)$ by $g(x)$ can always be performed. Moreover, the $q(x)$ and $r(x)$ are, indeed, uniquely determined, modulo p.

Division of Polynomials, modulo m

Let us consider a few examples where the modulus is not always a prime.

Example 1 Is $2x + 1$ a factor, modulo 10, of $4x^2 - x + 1$?

Here we seek an integral polynomial $ax + b$ such that we have, identically,
$$4x^2 - x + 1 \equiv (2x + 1)(ax + b) \pmod{10}$$
$$\equiv 2ax^2 + (a + 2b)x + b \pmod{10}.$$

Hence,
$$2a \equiv 4 \pmod{10},$$
$$a + 2b \equiv -1 \pmod{10},$$
$$b \equiv 1 \pmod{10}.$$

That is, $a \equiv 2$ or $7 \pmod{10}$. If we take $a \equiv 2 \pmod{10}$, the condition $a + 2b \equiv -1 \pmod{10}$ cannot be satisfied; but this condition can be satisfied if $a \equiv 7 \pmod{10}$. Hence, if we attempt to divide by $2x + 1$ and take $2x$ as the first term of the quotient, we get a remainder $-3x + 1$; but there exists no integer k such that $2k \equiv -3 \pmod{10}$. Hence, we cannot divide $-3x + 1$ by $2x + 1$, modulo 10. Consequently,
$$4x^2 - x + 1 \equiv (2x + 1)2x - 3x + 1 \pmod{10}.$$

Moreover,
$$4x^2 - x + 1 \equiv 14x^2 + 9x + 1 \pmod{10}$$
$$\equiv (2x + 1)(7x + 1) \pmod{10}.$$

Example 2 Divide $5x^3 + x + 1$ by $5x^2 + 1$, modulo 10.

We can, of course, get many different quotients and remainders:
$$5x^3 + x + 1 \equiv (5x^2 + 1)(x) + 1 \pmod{10},$$
$$5x^3 + x + 1 \equiv (5x^2 + 1)(3x) + 8x + 1 \pmod{10},$$
$$5x^3 + x + 1 \equiv (5x^2 + 1)(5x) + 6x + 1 \pmod{10},$$
$$5x^3 + x + 1 \equiv (5x^2 + 1)(-3x) + 14x - 9 \pmod{10},$$
$$5x^3 + x + 1 \equiv (5x^2 + 1)(9x) + 2x + 1 \pmod{10}.$$

Example 3 For the prime modulus 7, divide $5x^3 + x + 1$ by $3x + 1$.

Here the quotient $4x^2 + x$ and remainder 1 are unique, modulo 7:
$$5x^3 + x + 1 \equiv (3x + 1)(4x^2 + x) + 1 \pmod{7}.$$

Example 4 If the attempt is made to divide $5x^2 - x + 1$ by $2x + 1$, modulo 10, it will be found to be impossible.

For, if $5x^2 - x + 1 \equiv (2x + 1)(ax + b) + c \pmod{10}$, then, for the coefficient of x^2, $5 \equiv 2a \pmod{10}$. Obviously, no such integer a exists since $(2, 10)$ does not divide 5.

Example 5 If the modulus is a prime p, the division of $f(x) = a_0 x^n + a_1 x^{n-1} + \ldots + a_n$, $n \geq 1$, by $g(x) = b_0 x^m + b_1 x^{m-1} + \ldots + b_m$, $m \geq 1$, leads to a unique quotient $q(x)$ and a unique remainder $r(x)$, modulo p, when $(a_0 b_0, p) = 1$.

Proof If $m > n$, we take $q(x) \equiv 0 \pmod{p}$ and $r(x) \equiv f(x) \pmod{p}$, identically. Assume, therefore, that $n \geq m$. There exists, uniquely modulo p, an integer c_0 relatively prime to p such that $b_0 c_0 \equiv a_0 \pmod{p}$. Continuing in this manner, we obtain, therefore, $f(x) \equiv g(x)h(x) + r(x) \pmod{p}$, where $h(x) \equiv c_0 x^{n-m} + c_1 x^{n-m-1} + \ldots + c_{n-m} \pmod{p}$ and $r(x) \equiv d_0 x^{m-1} + \ldots + d_{m-1} \pmod{p}$. If there were another pair of such integral polynomials $h_1(x)$ and $r_1(x)$ so that

$$f(x) \equiv g(x)h_1(x) + r_1(x) \pmod{p},$$

then, subtracting these two congruences, we obtain

$$g(x)\{h(x) - h_1(x)\} \equiv r_1(x) - r(x) \pmod{p},$$

identically. If $h(x) - h_1(x) \not\equiv 0 \pmod{p}$ identically, let the leading coefficient be $u \not\equiv 0 \pmod{p}$. Then the leading coefficient of $g(x) \cdot \{h(x) - h_1(x)\}$ will be $b_0 u \not\equiv 0 \pmod{p}$, and the degree of $g(x)\{h(x) - h_1(x)\}$ is at least as great as that of $g(x)$; while, on the other hand, the degree of $r_1(x) - r(x)$ must be less than that of $g(x)$. Since this is not possible, it follows that $h(x) - h_1(x) \equiv 0 \pmod{p}$ identically, and consequently $r_1(x) - r(x) \equiv 0 \pmod{p}$ identically.

3-17 Problems

1. Divide $2x^3 + 3x^2 - x + 1$ by $5x - 1$, modulo 12.
2. Divide $2x^3 - 5x^2 + 3x + 2$ by $5x - 2$, modulo 7.
3. Explain why it is possible to divide $8x^3 + x + 1$ by $4x - 3$, modulo 12.
4. Explain why $x^2 + 5$ cannot be divided by $2x - 1$, modulo 6.
5. Show that the congruence $2x^2 + 5x + 1 \equiv 0 \pmod{8}$ is equivalent to the congruence $2x^2 + x - 3 \equiv 0 \pmod{8}$.
6. Show that the congruence $x^3 - 2x^2 - x + 2 \equiv 0 \pmod{12}$ and the congruence $x^3 - 19x - 30 \equiv 0 \pmod{12}$ are not equivalent.

7. Determine whether the congruences $3x^3 - x^2 + x + 2 \equiv 0 \pmod{15}$ and $6x^3 - 2x^2 - 13x + 3 \equiv 0 \pmod{15}$ are equivalent.

3-18 Number of Solutions of a Congruence

If the modulus is a prime, a theorem of Lagrange states the maximum number of incongruent solutions a congruence can have; namely, for a prime modulus, a congruence of degree n cannot have more than n incongruent solutions. First, however, we shall prove the following result.

- **Theorem 3-18** Let $f(x) = a_0 x^n + a_1 x^{n-1} + \ldots + a_{n-1} x + a_n$ be an integral polynomial. If $x \equiv r \pmod{m}$ is a solution of the congruence $f(x) \equiv 0 \pmod{m}$, then and only then is $(x - r)$ a factor of $f(x)$, modulo m.

Proof From the remainder theorem in algebra, we know that the integer $f(r)$ is the remainder when $f(x)$ is divided by $(x - r)$. Consequently, we have the identity

$$f(x) = (x - r)q(x) + f(r).$$

Since the coefficients in the left member of the equation

$$f(x) - f(r) = (x - r)q(x)$$

are integers, it follows from the process of long division that $q(x)$ is an integral polynomial. Hence, if $f(r) \equiv 0 \pmod{m}$,

$$f(x) \equiv (x - r)q(x) \pmod{m},$$

identically. Thus $(x - r)$ is a factor, modulo m, of $f(x)$. Conversely, it is evident that if $x - r$ is a factor of $f(x)$, modulo m, then $f(x) \equiv 0 \pmod{m}$ has $x \equiv r \pmod{m}$ as a solution. For, putting $x = r$ in $f(x) \equiv (x - r)q(x) \pmod{m}$, we have $f(r) \equiv (r - r)q(r) \equiv 0 \pmod{m}$.

It is to be noted that if $f(x)$ is divisible, modulo m, by $(x - r_1)$ and also by $(x - r_2)$, $r_2 \not\equiv r_1 \pmod{m}$, then it does not necessarily follow that $f(x)$ is divisible, modulo m, by the product $(x - r_1)(x - r_2)$. Consider, for example, the congruence $2x - 6 \equiv 0 \pmod{10}$ which has the two solutions $x \equiv 3 \pmod{10}$ and $x \equiv 8 \pmod{10}$. Now $(x - 8)$, as well as $(x - 3)$, is a factor, modulo m; but obviously $(x - 3)(x - 8)$ is not a factor of $2x - 6$. If, however, the modulus were a prime, we would be able to draw the conclusion that the product of two such factors is also a factor. We shall now consider a relevant theorem of J. L. Lagrange.

- **Theorem 3-19** Let $f(x) = a_0 x^n + a_1 x^{n-1} + \ldots + a_{n-1} x + a_n$, $n \geq 1$, be an integral polynomial in which $(a_0, p) = 1$. Then, where p is prime, the congruence $f(x) \equiv 0 \pmod{p}$ has at most n incongruent solutions, modulo p.

It is to be noted that the theorem does *not* state that the congruence *will* have a solution. For example, the congruence $x^2 + x + 2 \equiv 0 \pmod 5$ has no solution.

Proof We shall proceed by mathematical induction on n. For the case where $n = 1$, we see that the congruence $a_0 x + a_1 \equiv 0 \pmod p$, where $(a_0, p) = 1$, has precisely one solution. Assume the truth of the theorem for $n = k \geq 1$; namely, that for every polynomial $g(x)$ of degree k, modulo p, the congruence $g(x) \equiv 0 \pmod p$ has at most k incongruent solutions. We note that the theorem is obviously true for a congruence (of positive degree) which has *no* solution. If $f(x)$ of degree $n = k + 1$ modulo p has a solution $x \equiv r \pmod p$, then

$$f(x) \equiv (x - r)q(x) \pmod p,$$

where $q(x)$ is of degree k, modulo p. Next, let $f(x) \equiv 0 \pmod p$ have a solution $s \not\equiv r \pmod p$. Then $0 \equiv f(s) \equiv (s - r)q(s) \pmod p$. Consequently, s must be a solution of $q(x) \equiv 0 \pmod p$. Since $q(x) \equiv 0 \pmod p$ has, by hypothesis, at most k incongruent solutions, $f(x)$ has at most $k + 1$ incongruent solutions. Thus, by induction on n, the theorem is established.

Note that when $x \equiv r \pmod p$ is a solution of $f(x) \equiv 0 \pmod p$, so that then $f(x) \equiv (x - r)q(x) \pmod p$, it might happen that this same value $x \equiv r \pmod p$ is a solution of $q(x) \equiv 0 \pmod p$. In such a case, let us suppose that $f(x) \equiv (x - r)^h q_1(x) \pmod p$, where $q_1(r) \not\equiv 0 \pmod p$. We will then say that $x \equiv r \pmod p$ is a solution of *multiplicity h* of the congruence $f(x) \equiv 0 \pmod p$. If $h = 1$, we say that r is a *simple* solution. If $r_1, r_2, \ldots, r_k$ are *all the incongruent solutions* of $f(x) \equiv 0 \pmod p$ and if $q_k(x) \equiv 0 \pmod p$ has no solution, we may write the identical congruence $f(x) \equiv (x - r_1)^{h_1} \cdot (x - r_2)^{h_2} \ldots (x - r_k)^{h_k} q_k(x) \pmod p$, where $h_i \geq 1$ for $i = 1, 2, \ldots, k$ and where $q_k(x)$ has the leading term $a_0 x^{n_k}$. In the case where $q_k(x)$ is the constant a_0, we have $n = h_1 + h_2 + \ldots + h_k$; in the case where $q_k(x)$ is of positive degree n_k, modulo p, we have $n - n_k = h_1 + h_2 + \ldots + h_k$. Hence, we can say that when $f(x)$ is of degree n, modulo a prime p, $f(x) \equiv 0 \pmod p$ has not more than n solutions, multiplicities of solutions being counted.

Example 1 Let $f(x) = x^5 - 5x^4 + x^3 - x^2 - 2x - 1$. We find the congruence $f(x) \equiv 0 \pmod 7$ has simple solutions $x \equiv 1 \pmod 7$, $x \equiv 2 \pmod 7$, $x \equiv 4 \pmod 7$, but a solution of multiplicity 2 for $x \equiv 6 \pmod 7$.

Number of Solutions of Special Congruences **59**

Consequently,
$$f(x) \equiv (x-1)(x-2)(x-4)(x-6)^2 \pmod{7},$$
identically.

Example 2 Let $f(x) = x^8 + 4x^7 + 4x^6 + 5x^5 + 2x^4 + 2x^3 - 2x^2 + 3x - 5$. We find that the congruence $f(x) \equiv 0 \pmod{7}$ has $x \equiv 1 \pmod{7}$ as a solution of multiplicity 4, $x \equiv 2 \pmod{7}$ as a solution of multiplicity 2, as well as the simple solutions $x \equiv 3 \pmod{7}$ and $x \equiv 6 \pmod{7}$. Counting multiplicities, we see that the given congruence has 8 solutions, but only 4 incongruent solutions, modulo 7. Therefore,
$$f(x) \equiv (x-1)^4(x-2)^2(x-3)(x-6) \pmod{7},$$
identically.

If we had used the fact that for every x, $x^7 \equiv x \pmod{7}$, we would have obtained the congruence (but not an equivalent congruence)
$$4x^6 + 5x^5 + 2x^4 + 2x^3 - x^2 - 5 \equiv 0 \pmod{7},$$
which we find has only the simple solutions $x \equiv 1 \pmod{7}$, $x \equiv 2 \pmod{7}$, $x \equiv 3 \pmod{7}$, and $x \equiv 6 \pmod{7}$. Consequently,
$$4x^6 + 5x^5 + 2x^4 + 2x^3 - x^2 - 5$$
$$\equiv (x-1)(x-2)(x-3)(x-6)(4x^2 + 4x + 2) \pmod{7},$$
identically.

3-19 Number of Solutions of Special Congruences

We can utilize Theorem 3-8 in conjunction with Lagrange's Theorem 3-19 to determine precisely the number of incongruent solutions of certain special congruences.

• **Theorem 3-20** If p is an odd prime and if $f(x)$ and $g(x)$ are integral polynomials of respective degrees m and n such that
$$x^{p-1} - 1 \equiv f(x)g(x) \pmod{p}$$
identically, then the congruences
$$f(x) \equiv 0 \pmod{p}$$
and
$$g(x) \equiv 0 \pmod{p}$$
have m and n incongruent solutions, respectively, modulo p.

Proof We know from Fermat's Theorem 3-8 that
$$x^{p-1} - 1 \equiv 0 \pmod{p}$$
has precisely $p - 1$ incongruent solutions, modulo p; namely, $x \equiv 1, 2, 3, \ldots, p - 1 \pmod{p}$. If $x^{p-1} - 1$ can be factored, modulo p, into integral polynomials $f(x)$ and $g(x)$, we see that $f(x) \equiv 0 \pmod{p}$ cannot (by Theorem 3-19) have more than m incongruent solutions, nor can $g(x) \equiv 0 \pmod{p}$ have more than n incongruent solutions. However, since $f(x)g(x) \equiv 0 \pmod{p}$ has exactly $p - 1 = m + n$ solutions, neither $f(x) \equiv 0 \pmod{p}$ nor $g(x) \equiv 0 \pmod{p}$ can have fewer than its maximum possible number of incongruent solutions, modulo p. Moreover, $f(x) \equiv 0 \pmod{p}$ and $g(x) \equiv 0 \pmod{p}$ cannot have a solution in common. Thus, the theorem is proved.

- **Corollary 1** For an odd prime p, each of the congruences $x^{\frac{p-1}{2}} - 1 \equiv 0 \pmod{p}$ and $x^{\frac{p-1}{2}} + 1 \equiv 0 \pmod{p}$ has precisely $(p - 1)/2$ incongruent solutions, modulo p.

  Proof Since
  $$x^{p-1} - 1 \equiv (x^{\frac{p-1}{2}} - 1)(x^{\frac{p-1}{2}} + 1) \pmod{p}$$
  identically, each of the congruences $x^{\frac{p-1}{2}} - 1 \equiv 0 \pmod{p}$ and $x^{\frac{p-1}{2}} + 1 \equiv 0 \pmod{p}$ must have $(p - 1)/2$ incongruent solutions.

- **Corollary 2** If p is a prime and if $p - 1$ has the factor e, then $x^e - 1 \equiv 0 \pmod{p}$ has precisely e incongruent solutions, modulo p.

  Proof If $p = 2$, the result is obvious. Let p be odd. If $p - 1 = ek$,
  $$x^{p-1} - 1 = (x^e)^k - 1 = (x^e - 1)(x^{e(k-1)} + x^{e(k-2)} + \ldots + x^e + 1);$$
  which shows that
  $$x^e - 1 \equiv 0 \pmod{p}$$
  must have e incongruent solutions, and
  $$x^{e(k-1)} + x^{e(k-2)} + \ldots + x^e + 1 \equiv 0 \pmod{p}$$
  must have $p - 1 - e$ incongruent solutions, modulo p.

Example 1 For $p = 7$, $x^6 - 1 \equiv 0 \pmod{p}$ has the six incongruent solutions $x \equiv 1, 2, 3, 4, 5, 6 \pmod{7}$. Now, $x^3 - 1 \equiv 0 \pmod{7}$ has the three solutions $x \equiv 1, 2, 4 \pmod{7}$, and $x^3 + 1 \equiv 0 \pmod{7}$ has the solutions $x \equiv 3, 5, 6 \pmod{7}$.

Example 2 Since
$$x^{p-1} - 1 = (x - 1)(x^{p-2} + x^{p-3} + \ldots + x + 1),$$

for the odd prime p,
$$x^{p-2} + x^{p-3} + \ldots + x + 1 \equiv 0 \pmod{p}$$
has $p - 2$ incongruent solutions, modulo p.

3-20 Number of Solutions of a Binomial Quadratic Congruence

- **Theorem 3-21**
  (i) If p is an odd prime not dividing d, then for a positive integer n

(3-10) $$x^2 \equiv d \pmod{p^n}$$

has no solution or precisely two (incongruent) solutions.

(ii) The number of (incongruent) solutions of (3-10) is the same for every positive integer n.

Proof (i) If (3-10) has a solution x_0, *any* solution x of (3-10) satisfies $x^2 \equiv x_0^2 \pmod{p^n}$; that is, $(x - x_0)(x + x_0) \equiv 0 \pmod{p^n}$. If both factors on the left were divisible by p, then p would divide their difference $2x_0$. That would imply that p divides x_0 and, as a result, p would divide d. This is a contradiction. Hence, p does not divide both $x - x_0$ and $x + x_0$. Therefore, $x \equiv \pm x_0 \pmod{p^n}$, and, since x_0 and $-x_0$ are incongruent, modulo p, (3-10) has two incongruent solutions, modulo p^n, if it has any solution. Thus (i) is proved.

(ii) Suppose

(3-11) $$x^2 \equiv d \pmod{p^a}, \quad \text{where } a \geq 1,$$

has exactly two (incongruent) solutions, x_1 and x_2. Then $x_1^2 = d + M_1 p^a$ and $x_2^2 = d + M_2 p^a$. Every number x satisfying (3-11) is of the form $x_1 + k_1 p^a$ or $x_2 + k_2 p^a$, where k_1 and k_2 are integers. However, every solution of

(3-12) $$x^2 \equiv d \pmod{p^{a+1}}$$

satisfies (3-11). Hence, all possible (incongruent) solutions of (3-12) will be found among the numbers $x_1 + k_1 p^a$, $k_1 = 0, 1, \ldots, p - 1$, and $x_2 + k_2 p^a$, $k_2 = 0, 1, \ldots, p - 1$. Now, $x_1 + k_1 p^a$ is a solution of (3-12) if and only if $(x_1 + k_1 p^a)^2 - d \equiv 0 \pmod{p^{a+1}}$; that is, $x_1^2 - d + 2x_1 \cdot p^a k_1 + k_1^2 p^{2a} \equiv 0 \pmod{p^{a+1}}$, and $p^a(M_1 + 2x_1 k_1 + k_1^2 p^a) \equiv 0 \pmod{p^{a+1}}$. That is, it is a solution of (3-12) if and only if k_1 satisfies $2x_1 k_1 \equiv -M_1 \pmod{p}$. There is a unique value k_1, where $0 \leq k_1 \leq p - 1$, satisfying this linear congruence. Hence, for a solution of (3-11) there is precisely one solution of (3-12). Similarly, for the solution x_2 of

(3-11) there is a unique solution $x_2 + k_2 p^a$ of (3-12), and $x_1 + k_1 p^a$ and $x_2 + k_2 p^a$ are incongruent, modulo p^{a+1}.

If (3-11) has two (incongruent) solutions, so does (3-12) have two (incongruent) solutions; and, if (3-12) has a solution, so does (3-11). By part (i), each of (3-11) and (3-12) has either no solution or two solutions; that is (3-11) and (3-12) have the same number of incongruent solutions. By induction on n, we see that for all positive integers n, (3-10) has the same number of solutions as does $x^2 \equiv d \pmod{p}$.

• **Theorem 3-22** If $(d, 2) = 1$ and $n \geq 1$, consider the number of solutions of the congruence

(3-13) $$x^2 \equiv d \pmod{2^n}.$$

(i) If $n = 1$, there is one solution, $x \equiv 1 \pmod 2$.

(ii) If $n = 2$, there are two solutions if $d \equiv 1 \pmod 4$, and no solution if $d \equiv 3 \pmod 4$.

(iii) If $n \geq 3$, there is no solution if $d \not\equiv 1 \pmod 8$, and exactly four (incongruent) solutions if $d \equiv 1 \pmod 8$.

Proof Since the square of an odd number (Theorem 2-4) is of the form $8K + 1$, the cases (i) and (ii) are evident. Moreover, when $n \geq 3$ and $d \not\equiv 1 \pmod 8$, (3-13) obviously has no solution.

Consider case (iii), where $n \geq 3$ and $d \equiv 1 \pmod 8$. As an example, we see that for $n = 3$, there are the four solutions $x \equiv 1, 3, 5, 7 \pmod 8$.

If (3-13) has one solution, we shall see that it must have four solutions. Assuming (3-13) has a solution x_0, we see that for *any* solution x of (3-13), $x^2 \equiv x_0^2 \pmod{2^n}$; that is, $(x - x_0)(x + x_0) \equiv 0 \pmod{2^n}$. Since d is odd, both x and x_0 must be odd. Consequently, $(x - x_0)/2$ and $(x + x_0)/2$ are two integers whose product is divisible by the even number 2^{n-2} and whose sum is the odd number x. Thus, one of the numbers $(x - x_0)/2$ and $(x + x_0)/2$ is odd and the other even (being divisible by 2^{n-2}). Consequently, when x is a solution of (3-13), either $x \equiv x_0 \pmod{2^{n-1}}$ or $x \equiv -x_0 \pmod{2^{n-1}}$. That is, the solutions of (3-13) are to be found among $x_0, x_0 + 2^{n-1}, -x_0, -x_0 + 2^{n-1}$. However, since x_0 was assumed to be a solution of (3-13), these four numbers, incongruent modulo 2^n, are all solutions of (3-13). That is, if (3-13) has one solution, it must have four.

Let q be a positive odd integer $< 2^n$. The four numbers $q, q + 2^{n-1}, -q, -q + 2^{n-1}$, being incongruent, modulo 2^n, are each congruent, modulo 2^n, to a positive odd number $< 2^n$. Also, the square of each of these four numbers is in the same residue class, modulo 2^n; moreover, the square of no odd number incongruent, modulo 2^n, to these four is

Solution of the Congruence $f(x) \equiv 0 \pmod{m}$

in this same residue class, modulo 2^n. Consequently, each of the 2^{n-1} positive odd numbers $< 2^n$ can be assigned to a set consisting of four numbers each, such that the square of each of the four numbers has the same least positive residue, modulo 2^n. There will thus be $\frac{1}{4} 2^{n-1} = 2^{n-3}$ such sets of four numbers each. There will be, then, exactly 2^{n-3} numbers d for which (3-13) is solvable; moreover, if it is solvable, $d \equiv 1 \pmod{8}$. There are, however, precisely $\frac{1}{8} 2^n = 2^{n-3}$ positive odd integers d less than 2^n of the form $8M + 1$. Hence, the distributing process described above yields every such positive odd number $d < 2^n$ of the form $8M + 1$. Part (iii) is thus established.

3-21 Problems

1. Determine all the solutions of the congruence $x^5 - 3x^4 + 2x^3 - x^2 + 3x - 2 \equiv 0 \pmod{11}$.

2. Determine all the solutions of the congruence $x^5 - 6x^4 - 3x^3 - 2x^2 + 3x \equiv 0 \pmod{9}$.

3. Find all the solutions of the congruence $x^4 - x^2 - 6 \equiv 0 \pmod{7}$.

4. Find all the solutions of the congruence $x^5 + x^3 + x^2 + 2 \equiv 0 \pmod{15}$.

5. Determine all the solutions of the congruence $x^8 - 3x^6 + 4x^5 - x^4 - x^2 + 4x + 4 \equiv 0 \pmod{8}$.

6. Determine all the solutions of the congruence $x^3 - 3x^2 + 2x - 6 \equiv 0 \pmod{6}$.

3-22 Solution of the Congruence $f(x) \equiv 0 \pmod{m}$

Consider an integral polynomial $f(x)$ of degree $n \geq 1$ modulo m: $f(x) = b_0 x^n + b_1 x^{n-1} + \ldots + b_{n-1} x + b_n$. We shall see that the problem of solving the congruence

(3-14) $\qquad f(x) \equiv 0 \pmod{m}$

can be reduced to the problem of solving the congruence with respect to prime moduli together with the solving of certain linear congruences. Let m be written in the canonical form $m = p_1^{a_1} p_2^{a_2} \ldots p_r^{a_r}$, where the p_i's are distinct primes and the a_i's positive integers.

Every solution of (3-14) is obviously a solution of the congruences

(3-15)
$$f(x) \equiv 0 \pmod{p_1^{a_1}},$$
$$f(x) \equiv 0 \pmod{p_2^{a_2}},$$
$$\ldots$$
$$f(x) \equiv 0 \pmod{p_r^{a_r}}.$$

Conversely, if x_i is a solution of $f(x) \equiv 0 \pmod{p_i^{a_i}}$, $i = 1, 2, \ldots, r$, then by the Chinese Remainder Theorem we can obtain an integer X satisfying the congruences:

$$X \equiv x_1 \pmod{p_1^{a_1}},$$
$$X \equiv x_2 \pmod{p_2^{a_2}},$$
$$\ldots$$
$$X \equiv x_r \pmod{p_r^{a_r}}.$$

This X is a solution of each of the r congruences (3-15), and consequently is a solution of (3-14). For, $f(X) \equiv f(x_i) \equiv 0 \pmod{p_i^{a_i}}$, where $i = 1, 2, \ldots, r$.

Consider then the problem of solving a congruence of the type

(3-16) $$f(x) \equiv 0 \pmod{p^a},$$

where a is greater than 1 and p is a prime. Every x satisfying (3-16) must also satisfy

(3-17) $$f(x) \equiv 0 \pmod{p^{a-1}}.$$

Now, let $x \equiv x_0 \pmod{p^{a-1}}$ be a solution of (3-17), so that

$$f(x_0) = M \cdot p^{a-1}$$

for a suitable integer M. Since $f(x)$ is a polynomial, it has derivatives of all orders; and the Taylor's expansion, in powers of h, of $f(x_0 + h)$ is terminating. Then

$$f(x_0 + kp^{a-1}) = f(x_0) + kp^{a-1} f'(x_0)$$
$$+ k^2 p^{2a-2} \frac{f''(x_0)}{2!} + \ldots + k^n p^{na-n} \frac{f^{(n)}(x_0)}{n!}.$$

Note that $f^{(r)}(x_0)/r!$ is an integer because the coefficient of *each* term in $f^{(r)}(x_0)$ has a factor which is the product of r consecutive positive integers. Moreover, when s is a non-negative integer,

$$\frac{(s+1)(s+2)\ldots(s+r)}{1 \cdot 2 \ldots \cdot (r-1) \cdot r}$$

is the number of combinations of $s + r$ things taken r at a time; that is, the product of r consecutive positive integers[8] is divisible by $r!$. Since $a \geq 2$, the exponent $ta - t$ of the power of p, p^{ta-t}, is at least as great as a if $t \geq 2$. Consequently, we have

$$f(x_0 + kp^{a-1}) \equiv f(x_0) + kp^{a-1} f'(x_0) \pmod{p^a}.$$

Therefore $x_0 + kp^{a-1}$ is a solution of (3-16) if we can obtain a value of k satisfying the linear congruence

$$M \cdot p^{a-1} + k \cdot p^{a-1} \cdot f'(x_0) \equiv 0 \pmod{p^a};$$

that is, if k satisfies

(3-18) $$k \cdot f'(x_0) \equiv -M \pmod{p}.$$

This last congruence (3-18) has a solution if and only if the greatest common divisor of $f'(x_0)$ and p divides $-M$.

Thus, if we consider *each* solution $x \equiv x_0 \pmod{p^{a-1}}$ of (3-17) and determine the values of k for each such x_0, we then obtain *all* solutions of (3-16): $x \equiv x_0 + k \cdot p^{a-1} \pmod{p^a}$. Similarly, we could secure all solutions of (3-17) if we knew all solutions of

(3-19) $$f(x) \equiv 0 \pmod{p^{a-2}}.$$

Consequently, we are eventually led to solve the congruence

(3-20) $$f(x) \equiv 0 \pmod{p}.$$

Example 1 Where $f(x) = 2x^3 - 3x^2 + 2x + 122$, completely solve the congruence

$$f(x) \equiv 0 \pmod{675}.$$

Solution Since $675 = 3^3 \cdot 5^2$, we may replace this problem by that of solving simultaneously the two congruences

$$f(x) \equiv 0 \pmod{3^3}$$

and

$$f(x) \equiv 0 \pmod{5^2}.$$

Considering the first of these last two congruences, we first solve $f(x) \equiv 0 \pmod{3}$; that is, we will solve the congruence $2x^3 - 3x^2 + 2x + 122 \equiv 0 \pmod{3}$. Here the only solution is $x \equiv 1 \pmod{3}$. Now $f(1) = 123 = 41 \times 3$, $f'(1) = 2$. In order to get solutions of $f(x) \equiv 0 \pmod{3^2}$, we determine k so that $k \cdot f'(x_0) \equiv -M \pmod{3}$; that is, $k \cdot 2 \equiv -41 \pmod{3}$; therefore $k \equiv 2 \pmod{3}$. Thus, $1 + k \cdot 3 = 7$ is the (only) solution of $f(x) \equiv 0 \pmod{3^2}$. Now, $f(7) = 675 = 75 \cdot 9$, $f'(7) = 254$. We now determine k so that $k \cdot f'(x_0) \equiv -M \pmod{3}$; that is, $k \cdot 254 \equiv -75 \pmod{3}$. Hence $k \equiv 0 \pmod{3}$. Therefore, $x \equiv 7 + k \cdot 3^2 \equiv 7 \pmod{3^3}$ is the only solution of $f(x) \equiv 0 \pmod{3^3}$.

Consider next the congruence $f(x) \equiv 0 \pmod{5^2}$. First, we observe that the only solutions of $f(x) \equiv 2x^3 - 3x^2 + 2x + 122 \equiv 0 \pmod{5}$ are $x \equiv 2 \pmod{5}$, $x \equiv 3 \pmod{5}$, and $x \equiv 4 \pmod{5}$. Employing first $x \equiv 2 \pmod{5}$, we see that $f(2) = 130 = 26 \cdot 5$, $f'(2) = 14$. We find that $k \equiv 1 \pmod{5}$ satisfies the condition $k \cdot 14 \equiv -26 \pmod{5}$. Therefore, $2 + 1 \cdot 5 = 7$ is a solution of $f(x) \equiv 0 \pmod{5^2}$. Next, using $x \equiv 3 \pmod 5$, we find $f(3) = 155 = 31 \cdot 5$, $f'(3) = 38$; therefore $k \cdot 38 \equiv -31 \pmod 5$ and $k \equiv 3 \pmod 5$. Then $x = 3 + 3 \cdot 5 = 18$ is a second solution of $f(x) \equiv 0 \pmod{5^2}$. Finally, employ $x \equiv 4 \pmod 5$, $f(4) = 210 = 42 \cdot 5$, $f'(4) = 74$. Then determine k so that $k \cdot 74 \equiv -42 \pmod 5$; thus $k \equiv 2 \pmod 5$. This determines a third solution of $f(x) \equiv 0 \pmod{5^2}$; namely, $x \equiv 4 + 2 \cdot 5 \equiv 14 \pmod{5^2}$. By the Chinese Remainder Theorem, we determine simultaneous solutions of each of the three pairs of congruences:

$$x \equiv 7 \pmod{3^3}, \qquad x \equiv 7 \pmod{3^3}, \qquad x \equiv 7 \pmod{3^3},$$
$$x \equiv 7 \pmod{5^2}; \qquad x \equiv 18 \pmod{5^2}; \qquad x \equiv 14 \pmod{5^2}.$$

The solutions of these pairs are, respectively, $x \equiv 7, 493, 439 \pmod{675}$. These are, therefore, the solutions of the given congruence $f(x) \equiv 0 \pmod{675}$.

Example 2 Where $f(x) = x^4 - 115x^3 + 33x^2 + 27x - 162$, solve completely the congruence

(3-21) $\qquad\qquad f(x) \equiv 0 \pmod{216}$.

Solution We note that $216 = 8 \cdot 27 = 2^3 \cdot 3^3$.

Every integer x satisfying (3-21) automatically satisfies both

(3-22) $\qquad\qquad f(x) \equiv 0 \pmod{2^3}$

and

(3-23) $\qquad\qquad f(x) \equiv 0 \pmod{3^3}$.

Also, if $x \equiv x_1 \pmod 8$ is a solution of (3-22) and if $x \equiv x_2 \pmod{27}$ is a solution of (3-23), we can readily obtain, by the Chinese Remainder Theorem, a common solution X of the congruences $X \equiv x_1 \pmod 8$ and $X \equiv x_2 \pmod{27}$. Since $f(X) \equiv 0 \pmod 8$ and also $f(X) \equiv 0 \pmod{27}$, this X is a solution of congruence (3-21). If either of the congruences (3-22) or (3-23) fails to have a solution, then obviously there is no solution of congruence (3-21). If congruence (3-22) has r solutions modulo 8 and if congruence (3-23) has s incongruent solutions modulo 27, then congruence (3-21) has rs incongruent solutions modulo $216 = 8 \cdot 27$. We shall find that congruence (3-21) has $5 \cdot 10 = 50$ incongruent solutions modulo 216.

We desire to solve congruence (3-22). First, we find that the solutions of $f(x) \equiv 0 \pmod 2$ are $x \equiv 0 \pmod 2$ and $x \equiv 1 \pmod 2$. Next, we desire the solutions of $f(x) \equiv 0 \pmod{2^2}$.

Using, *in the first place*, $x \equiv 0 \pmod 2$, we see that $f(0) = -162 = (-81)2$ and $f'(0) = 27$. The congruence $k \cdot f'(x_0) \equiv -M \pmod 2$ becomes $k \cdot 27 \equiv 81 \pmod 2$; hence $k \equiv 1 \pmod 2$. Thus $x = 0 + 1 \cdot 2 = 2$ is

Solution of the Congruence f(x) = 0 (mod m)

a solution of $f(x) \equiv 0 \pmod{2^2}$. To get solutions of $f(x) \equiv 0 \pmod{2^3}$ we use $x \equiv 2 \pmod{2^2}$, where $f(2) = -880 = (-220)\,4$ and $f'(2) = -1189$. We then choose k so that $k(-1189) \equiv 220 \pmod 2$; namely, $k \equiv 0 \pmod 2$. Hence a solution of $f(x) \equiv 0 \pmod{2^3}$ is $x = 2 + 0 \cdot 2^2 = 2$.

Using, *in the second place*, $x \equiv 1 \pmod 2$, we see that $f(1) = -216 = (-108)\,2$ and $f'(1) = -248$. The congruence $k \cdot f'(x_0) \equiv -M \pmod 2$ becomes $k\,(-248) \equiv 108 \pmod 2$. Hence the values of k are $k \equiv 0 \pmod 2$ and $k \equiv 1 \pmod 2$. This yields the values $x = 1 + 0 \cdot 2 = 1$ and $x = 1 + 1 \cdot 2 = 3$ as solutions of $f(x) \equiv 0 \pmod{2^2}$. To get solutions of $f(x) \equiv 0 \pmod{2^3}$ we first use $x \equiv 1 \pmod{2^2}$. Since $f(1) = -216 = (-54)\,2^2$ and $f'(1) = -248$, we obtain k so that $k\,(-248) \equiv 54 \pmod 2$. Hence $k = 0$ and $k = 1$. Solutions of $f(x) \equiv 0 \pmod{2^3}$ are then $x = 1 + 0 \cdot 2^2 = 1$ and $x = 1 + 1 \cdot 2^2 = 5$. We next use $x \equiv 3 \pmod{2^2}$. Since $f(3) = -2808 = (-702)2^2$, $f'(3) = -2772$, we obtain k so that $k(-2772) \equiv 702 \pmod 2$; that is, $k = 0$ and $k = 1$. Solutions of $f(x) \equiv 0 \pmod{2^3}$ are then $x = 3 + 0 \cdot 2^2 = 3$ and $x = 3 + 1 \cdot 2^2 = 7$. Combining these results, we see that the only solutions of $f(x) \equiv 0 \pmod{2^3}$ are $x \equiv 1, 2, 3, 5, 7 \pmod 8$. These solutions are simple solutions modulo 8 except 1 and 5, which both have multiplicity two.

We desire next to solve congruence (3-23). First, we find that the solutions of $f(x) \equiv 0 \pmod 3$ are $x \equiv 0 \pmod 3$ and $x \equiv 1 \pmod 3$.

Using, *in the first place*, $x \equiv 0 \pmod 3$ and noting that $f(0) = -162$, $f'(0) = 27$, we obtain k so that $k \cdot 27 \equiv 54 \pmod 3$. Hence $k \equiv 0, 1, 2 \pmod 3$. Therefore, solutions of $f(x) \equiv 0 \pmod{3^2}$ are the three values: $x = 0 + 0 \cdot 3 = 0$, $x = 0 + 1 \cdot 3 = 3$, $x = 0 + 2 \cdot 3 = 6$. Using first $x \equiv 0 \pmod{3^2}$, we note that $f(0) = -162 = (-18)\,3^2$, $f'(0) = 27$. We determine k so that $k \cdot 27 \equiv 18 \pmod 3$. In this case $k = 0, 1, 2$. Consequently, solutions of $f(x) \equiv 0 \pmod{3^3}$ are:

$$0 + 0 \cdot 3^2 = 0,$$
$$0 + 1 \cdot 3^2 = 9,$$
$$0 + 2 \cdot 3^2 = 18.$$

Using next $x \equiv 3 \pmod{3^2}$ and recalling that $f(3) = -2808 = (-312)\,3^2$ and $f'(3) = -2772$, we find the values of k satisfying $k(-2772) \equiv 312 \pmod 3$ are $k \equiv 0, 1, 2 \pmod 3$. Hence, solutions of $f(x) \equiv 0 \pmod{3^3}$ are:

$$3 + 0 \cdot 3^2 = 3,$$
$$3 + 1 \cdot 3^2 = 12,$$
$$3 + 2 \cdot 3^2 = 21.$$

Using $x \equiv 6 \pmod{3^2}$ and noting that $f(6) = -22356 = (-2484)\,3^2$, $f'(6) = -11133$, we find the values of k satisfying $k(-11133) \equiv 2484 \pmod 3$ are $k \equiv 0, 1, 2 \pmod 3$; thus solutions of $f(x) \equiv 0 \pmod{3^3}$ are:

$$6 + 0 \cdot 3^2 = 6,$$
$$6 + 1 \cdot 3^2 = 15,$$
$$6 + 2 \cdot 3^2 = 24.$$

Using, *in the second place*, $x \equiv 1 \pmod{3}$ and recalling that $f(1) = -216 = (-72)3 = (-24)3^2$ and $f'(1) = -248$, we see that the only value of k satisfying $kf'(1) \equiv 0 \pmod 3$ is $k \equiv 0 \pmod 3$. Hence, $x = 1$ is a solution not merely of $f(x) \equiv 0 \pmod{3^2}$ but also of $f(x) \equiv 0 \pmod{3^3}$.

If e_i is a solution of $f(x) \equiv 0 \pmod{2^3}$ and if f_j is a solution of $f(x) \equiv 0 \pmod{3^3}$, the solutions of $f(x) \equiv 0 \pmod{2^3 3^3}$ are now found by employing the Chinese Remainder Theorem to solve simultaneously the following pair of congruences for each combination of e_i and f_j:

$$x \equiv e_i \pmod{2^3}, \quad \text{for } i = 1, 2, \cdots, 5;$$
$$x \equiv f_j \pmod{3^3}, \quad \text{for } j = 1, 2, \cdots, 10.$$

The e_i and f_j are given as follows:

$$e_1 = 1, \; e_2 = 2, \; e_3 = 3, \; e_4 = 5, \; e_5 = 7;$$
$$f_1 = 0, \; f_2 = 9, \; f_3 = 18, \; f_4 = 3, \; f_5 = 12,$$
$$f_6 = 21, \; f_7 = 6, \; f_8 = 15, \; f_9 = 24, \; f_{10} = 1.$$

The following 50 solutions, incongruent modulo 216 (Table 3-1), are obtained for the indicated pair of values of the subscripts i and j for e_i and f_j. Let the smallest positive simultaneous solution of $x \equiv e_i \pmod{2^3}$ and $x \equiv f_j \pmod{3^3}$ be denoted by x_{ij}. The $5 \cdot 10 = 50$ values of x_{ij} have been calculated and entered in the accompanying table. For example, the simultaneous solution of $x \equiv 2 \pmod{2^3}$ and $x \equiv 12 \pmod{3^3}$ is $x_{25} = 66$, and is indicated in the body of Table 3-1 in the position $i = 2$ and $j = 5$.

Table 3-1

i \ j	1	2	3	4	5	6	7	8	9	10
5	135	63	207	111	39	183	87	15	159	55
4	189	117	45	165	93	21	141	69	213	109
3	27	171	99	3	147	75	195	123	51	163
2	162	90	18	138	66	210	114	42	186	82
1	81	9	153	57	201	129	33	177	105	1

3-23 Polynomials Representing Primes

Some polynomials represent primes for many consecutive values of the variable. For example,

Polynomials Representing Primes

$$x^2 + x + 41 = x(x + 1) + 41$$

yields a prime for $x = 0, 1, 2, \ldots, 39$, but is composite for $x = 40$ and 41. Also

$$x^2 - x + 41 = (x - 1)x + 41$$

yields a prime for $x = 0, 1, 2, \ldots, 40$. Hence, since

$$f(x) = x^2 - 79x + 1601 = (x - 40)^2 + (x - 40) + 41,$$
$$= (40 - x)^2 - (40 - x) + 41,$$

$f(x)$ is a prime for $0 \leq x - 40 \leq 39$ and for $0 \leq 40 - x \leq 40$. Thus $f(x)$ is prime for $x = 0, 1, 2, \ldots, 79$.

It has been proved that functions exist that represent primes exclusively[9], but it has been shown that no non-constant rational function of x represents primes only. In the following theorem we shall establish this fact for integral polynomials.

- **Theorem 3-23** There exists no integral polynomial $f(x)$ of positive degree which yields only primes for every x or for every x greater than a specified positive value.

  Proof Now, for a polynomial of positive degree s with real coefficients,

  $$g(x) = b_0 x^s + b_1 x^{s-1} + \cdots + b_{s-1} x + b_s$$
  $$= x^s \left(b_0 + \frac{b_1}{x} + \cdots + \frac{b_{s-1}}{x^{s-1}} + \frac{b_s}{x^s} \right),$$

  we can take x sufficiently large numerically so that $g(x)$ has the same sign as its leading term, $b_0 x^s$. Moreover, $|g(x)|$ increases without limit as x increases without limit.

  Consider the integral polynomial of positive degree n:

  $$f(x) = a_0 x^n + a_1 x^{n-1} + \cdots + a_{n-1} x + a_n.$$

  Let $a_0 > 0$, and for the integer x_0 suppose that $f(x_0) = q > 1$. Choose $M > 0$ so that $f(x) - q > 0$ when the integer $x > M$. Employing Taylor's expansion, we note that

  $$f(x_0 + kq) = f(x_0) + kq\,f'(x_0) + k^2 q^2 \frac{1}{2!} f''(x_0) +$$
  $$\cdots + k^n q^n \frac{1}{n!} f^{(n)}(x_0).$$

  Since $(1/t!)f^{(t)}(x)$ has only integral coefficients,[10] we see that for integral values of k,

  $$f(x) - q = q \left\{ kf'(x_0) + k^2 q \frac{1}{2!} f''(x_0) + \right.$$
  $$\left. \cdots + k^n q^{n-1} \frac{1}{n!} f^{(n)}(x_0) \right\} = qN$$

is a multiple of q and positive when $x = x_0 + kq > M$. Since $f(x_0 + kq) = q(1 + N)$, we have shown that $f(x_0 + kq)$ has a proper factor $q > 1$, and is therefore composite when $kq > M - x_0$.

3-24 Problems

1. Find all solutions of the congruence $x^2 + 3x + 2 \equiv 0 \pmod{12}$.

2. Find all solutions of the congruence $x^5 + 6x^3 - 6x^2 - 7x - 6 \equiv 0 \pmod{21}$.

3. Find all solutions of the congruence $7x^3 + 2x^2 + 7x + 20 \equiv 0 \pmod{63}$.

4. Find all solutions of the congruence $2x^4 + x^2 + 1 \equiv 0 \pmod{24}$.

5. (a) Find all integral values of k, modulo 20, for which the congruence $5x^3 + 3x^2 + 5x + k \equiv 0 \pmod{20}$ is solvable.
   (b) For each such value of k, obtain all solutions of the congruence.

6. Completely solve each of the following congruences:
   (a) $2x^3 + x^2 - x + 2 \equiv 0 \pmod{35}$;
   (b) $2x^3 + x^2 - x + 1 \equiv 0 \pmod{35}$.

7. Completely solve the congruence $2x^3 + x^2 - x + 3 \equiv 0 \pmod{35}$.

8. Completely solve the congruence $x^2 + 6x + 8 \equiv 0 \pmod{225}$.

9. Completely solve the congruence $3x^2 + x + 4 \equiv 0 \pmod{441}$.

10. Completely solve the congruence $x^4 + 6x^2 + 5 \equiv 0 \pmod{16}$.

11. Completely solve the congruence $x^4 + 6x^2 + 5 \equiv 0 \pmod{60}$.

12. Completely solve the congruence $x^3 + 4x^2 - x - 2 \equiv 0 \pmod{3025}$.

13. Completely solve the congruence $x^3 + 4x^2 - 2x + 15 \equiv 0 \pmod{3025}$.

Some Significant Functions in the Theory of Numbers

4

4-1 The Greatest Integer Function

A function found to be of special importance in connection with questions of divisibility is the function $[\alpha]$ defined, for an arbitrary real number α, as *the greatest integer not exceeding* α. In other words, for the real quantity α, the integer $[\alpha]$ satisfies the condition

$$[\alpha] \leq \alpha < [\alpha] + 1.$$

This is equivalent to saying that

$$\alpha = [\alpha] + \theta, \quad \text{where } 0 \leq \theta < 1.$$

Here θ is called the *fractional part* of α. Thus, $[\alpha]$ is defined for all real values of α and takes on integral values only.

For a real quantity α and an integer a, it follows from $[a + \alpha] \leq a + \alpha < [a + \alpha] + 1$ that $[a + \alpha] - a \leq \alpha < [a + \alpha] - a + 1$; consequently $[\alpha] = [a + \alpha] - a$. That is, $[a + \alpha] = a + [\alpha]$.

71

Consider the following examples: $[5] = 5$, $[27/4] = 6$, $[0] = 0$, $[\sqrt{10}] = 3$, $[-12/5] = -3$, $[-0.3] = -1$, $[-\sqrt{10}] = -4$.

We shall derive some properties of this function, and, applying these properties, obtain some results of particular usefulness in the theory of numbers.

- **Theorem 4-1** For a real quantity α and an arbitrary positive integer n, we have

$$\left[\frac{[\alpha]}{n}\right] = \left[\frac{\alpha}{n}\right].$$

For example, $\left[\dfrac{[\sqrt{17}]}{3}\right] = \left[\dfrac{\sqrt{17}}{3}\right] = 1.$

Proof Let $\alpha/n = [\alpha/n] + \theta$, where $0 \leq \theta < 1$. Then $\alpha = n[\alpha/n] + n\theta$, where $0 \leq n\theta < n$.

Since $n[\alpha/n]$ is an integer,

$$[\alpha] = n\left[\frac{\alpha}{n}\right] + [n\theta].$$

Therefore, $\dfrac{[\alpha]}{n} = \left[\dfrac{\alpha}{n}\right] + \dfrac{[n\theta]}{n}.$

Now, since $0 \leq [n\theta] \leq n\theta < n,$ $\quad 0 \leq \dfrac{[n\theta]}{n} < 1.$

Thus, $\left[\dfrac{[\alpha]}{n}\right] = \left[\dfrac{\alpha}{n}\right].$

- **Corollary** For positive integers a, m, and n,

$$\left[\frac{[a/m]}{n}\right] = \left[\frac{a}{mn}\right].$$

- **Theorem 4-2** If n denotes an integer greater than one and α a real quantity not less than one, then

$$[\alpha] > \left[\frac{\alpha}{n}\right].$$

Proof Let $\alpha/n = [\alpha/n] + \theta_1$, where $0 \leq \theta_1 < 1$.

Then $\alpha = n\left[\dfrac{\alpha}{n}\right] + n\theta_1,$

$[\alpha] = n\left[\dfrac{\alpha}{n}\right] + [n\theta_1],$

and $[\alpha] \geq 2\left[\dfrac{\alpha}{n}\right].$

The Greatest Integer Function

Consequently,
$$[\alpha] > \left[\frac{\alpha}{n}\right].$$

- **Theorem 4-3** For positive integers a, b, and n,
$$\left[\frac{ab}{n}\right] \geq a\left[\frac{b}{n}\right].$$

Proof Let $b/n = [b/n] + \theta$, where $0 \leq \theta < 1$.

Therefore,
$$\frac{ab}{n} = a\left[\frac{b}{n}\right] + a\theta.$$

Since $a[b/n]$ is an integer, we have
$$\left[\frac{ab}{n}\right] \geq a\left[\frac{b}{n}\right].$$

- **Theorem 4-4** Let $\alpha_1, \alpha_2, \ldots, \alpha_r$ be real quantities such that $\alpha = \alpha_1 + \alpha_2 \ldots + \alpha_r$. Then $[\alpha] \geq [\alpha_1] + [\alpha_2] + \ldots + [\alpha_r]$.

Proof Let
$$\begin{aligned}
\alpha_1 &= [\alpha_1] + \theta_1, & \text{where } 0 \leq \theta_1 < 1, \\
\alpha_2 &= [\alpha_2] + \theta_2, & \text{where } 0 \leq \theta_2 < 1, \\
&\vdots \\
\alpha_r &= [\alpha_r] + \theta_r, & \text{where } 0 \leq \theta_r < 1.
\end{aligned}$$

Then $\alpha = \alpha_1 + \alpha_2 + \ldots + \alpha_r = [\alpha_1] + [\alpha_2] + \ldots + [\alpha_r] + \theta_1 + \theta_2 + \ldots + \theta_r$.

Therefore, $[\alpha] = [\alpha_1] + [\alpha_2] + \ldots + [\alpha_r] + [\theta_1 + \theta_2 + \ldots + \theta_r]$.

Hence, $[\alpha] \geq [\alpha_1] + [\alpha_2] + \ldots + [\alpha_r]$.

- **Corollary** Let $n_1, n_2, \ldots, n_r$ be integers such that $n = n_1 + n_2 + \ldots + n_r$. Then, if p is a non-zero integer,
$$\left[\frac{n}{p}\right] \geq \left[\frac{n_1}{p}\right] + \left[\frac{n_2}{p}\right] + \ldots + \left[\frac{n_r}{p}\right].$$

Proof Since $n/p = n_1/p + n_2/p + \ldots + n_r/p$, this result is an instance of the above theorem.

- **Theorem 4-5** If n is a positive integer and p a prime, the exponent[1] of the highest power of p that divides $n!$ is
$$\left[\frac{n}{p}\right] + \left[\frac{n}{p^2}\right] + \ldots + \left[\frac{n}{p^r}\right], \quad \text{where } \left[\frac{n}{p^r}\right] \geq 1 \text{ and } \left[\frac{n}{p^{r+1}}\right] = 0.$$

For example, since $[19/3] + [19/9] = 8$, 3^8 divides $19!$ but 3^9 does not.

First Proof In the set of consecutive integers $1, 2, 3, \ldots, 1 \cdot p, p + 1, \ldots, 2p, \ldots, 3p, \ldots, p \cdot p, \ldots, n$, let the last integer of the set that is divisible by p be kp. Then $kp \leq n < (k+1)p$, and $k \leq n/p < k+1$. Hence $k = [n/p]$. Thus, there are precisely $[n/p]$ integers in this set which are multiples of p, all other integers of the set being relatively prime to p. That is, each integer of the set which is a multiple of p is counted once in the total $[n/p]$. By an analogous consideration, if tp^2 is the largest multiple of p^2 not exceeding n, then $tp^2 \leq n < (t+1)p^2$; thus $t \leq n/p^2 < t+1$ and $t = [n/p^2]$. Therefore, each of the numbers of the original set which is a multiple of p^2 (and, of course, also of p) is counted once in the total $[n/p]$ and also counted once in the total $[n/p^2]$. Likewise, each number of the original set which is a multiple of p^3 is counted once in the total $[n/p]$, once in the total $[n/p^2]$, and also once in the total $[n/p^3]$. If $p^r \leq n < p^{r+1}$, $[n/p^r] > 0$ and $[n/p^{r+1}] = 0$. Since the number of multiples of $p, p^2, \ldots, p^r$ in the original set is, respectively, $[n/p], [n/p^2], \ldots, [n/p^r]$, the exponent of the highest power of the prime p dividing $n!$ is

$$\left[\frac{n}{p}\right] + \left[\frac{n}{p^2}\right] + \cdots + \left[\frac{n}{p^r}\right].$$

Second Proof Define $H_p(y)$ as the exponent of the highest power of the prime p dividing y. For example, $H_3(7) = 0$, $H_3(12) = 1$, and $H_3(6!) = 2$.

In the set of consecutive integers $1, 2, \ldots, 1 \cdot p, p+1, \ldots, 2p, \ldots, 3p, \ldots, p \cdot p, \ldots, n$, the last integer of the set divisible by p is $[n/p] p$. Hence, the integers of the set divisible by p are $1 \cdot p, 2p, \ldots, [n/p] p$. Hence,

$$H_p(n!) = H_p\left(p \cdot 2p \cdots \left[\frac{n}{p}\right]p\right) = H_p\left(1 \cdot 2 \cdots \left[\frac{n}{p}\right] \cdot p^{[n/p]}\right)$$
$$= \left[\frac{n}{p}\right] + H_p\left(1 \cdot 2 \cdots \left[\frac{n}{p}\right]\right).$$

The last integer of the set $1, 2, \ldots, [n/p]$ which is a multiple of p is

$$\left[\frac{[n/p]}{p}\right]p = \left[\frac{n}{p^2}\right]p.$$

Hence, proceeding as before, we see that the integers $1 \cdot p, 2p, \ldots, [n/p^2]p$ are the only integers of the set $1, 2, \ldots, [n/p]$ having the factor p. Therefore,

$$H_p\left(1 \cdot 2 \cdots \left[\frac{n}{p}\right]\right) = H_p\left(p \cdot 2p \cdots \left[\frac{n}{p^2}\right]p\right)$$
$$= H_p\left(1 \cdot 2 \cdots \left[\frac{n}{p^2}\right] \cdot p^{[n/p^2]}\right)$$
$$= \left[\frac{n}{p^2}\right] + H_p\left(1 \cdot 2 \cdots \left[\frac{n}{p^2}\right]\right).$$

The Greatest Integer Function

Likewise,

$$H_p\left(1 \cdot 2 \cdot \ldots \left[\frac{n}{p^2}\right]\right) = \left[\frac{n}{p^3}\right] + H_p\left(1 \cdot 2 \cdot \ldots \left[\frac{n}{p^3}\right]\right).$$

Proceeding in like fashion, we eventually reach the stage where $p^r \leq n < p^{r+1}$; thus $[n/p^r] \geq 1$ and $[n/p^{r+1}] = 0$.
Consequently,

$$H_p(n!) = \left[\frac{n}{p}\right] + \left[\frac{n}{p^2}\right] + \ldots + \left[\frac{n}{p^r}\right].$$

For example, the exponent of the highest power of 7 which divides 347! is

$$H_7(347!) = [347/7] + [347/7^2] + [347/7^3] = 49 + 7 + 1 = 57.$$

- **Theorem 4-6** If the positive integer m is written in the scale of a prime p so that $m = a_r p^r + a_{r-1} p^{r-1} + \cdots + a_1 p + a_0$, then the exponent of the highest power of the prime p dividing $m!$ is

$$\frac{m - \sum_{i=0}^{r} a_i}{p - 1}.$$

Proof By the preceding theorem, $H_p(m!)$

$$= \left[\frac{m}{p}\right] + \ldots + \left[\frac{m}{p^r}\right]$$

$$= a_r p^{r-1} + \ldots + a_2 p + a_1$$
$$\quad + a_r p^{r-2} + \ldots + a_2$$
$$\quad \vdots \qquad \vdots \qquad \vdots$$
$$\quad + a_r p + a_{r-1}$$
$$\quad + a_r,$$

$$= a_r \frac{p^r - 1}{p - 1} + a_{r-1} \frac{p^{r-1} - 1}{p - 1} + \ldots + a_2 \frac{p^2 - 1}{p - 1} + a_1,$$

$$= \frac{a_r p^r + a_{r-1} p^{r-1} + \cdots + a_2 p^2 + a_1 p + a_0 - (a_r + a_{r-1} + \cdots + a_2 + a_1 + a_0)}{p - 1}$$

$$= \frac{m - \sum_{i=0}^{r} a_i}{p - 1}.$$

For example, writing 347 in the scale of 7, we have

$$347 = 1 \cdot 7^3 + 0 \cdot 7^2 + 0 \cdot 7 + 4 = (1004)_7 \text{ and } 1 + 0 + 0 + 4 = 5.$$

Thus, $H_7(347!) = (347 - 5)/(7 - 1) = 57.$

The following useful theorem is a well-known result, since the expression denotes the number of ordered arrangements of n things of which n_1 are alike of one kind, n_2 of a second kind, ..., and n_r of an rth kind. Our proof, however, will depend upon the arithmetical properties just established.

- **Thereom 4-7** If $n_1, n_2, \ldots, n_r$ are positive integers such that $n = n_1 + n_2 + \ldots + n_r$, then $n!/(n_1! n_2! \ldots n_r!)$ is an integer.[2]

Proof We shall establish this result by showing that for every prime p dividing the denominator, the exponent of the highest power of this prime contained in the numerator is at least as great as the exponent of the highest power of this prime contained in the denominator. Thus, since the entire denominator would then divide into the numerator, the desired result would be established.

If the denominator is 1, then the theorem is obviously true. Let the denominator, then, be greater than one. If p is an *arbitrary* prime factor of the denominator, the exponent of the highest power of p dividing the numerator is

$$\left[\frac{n}{p}\right] + \left[\frac{n}{p^2}\right] + \left[\frac{n}{p^3}\right] + \cdots.$$

On the other hand, the exponent of the highest power of p dividing the factors in the denominator, $n_1!, n_2!, \ldots,$ and $n_r!$, is the sum, respectively, of each of the following r rows:

$$\left[\frac{n_1}{p}\right] + \left[\frac{n_1}{p^2}\right] + \left[\frac{n_1}{p^3}\right] + \cdots ;$$

$$\left[\frac{n_2}{p}\right] + \left[\frac{n_2}{p^2}\right] + \left[\frac{n_2}{p^3}\right] + \cdots ;$$

$$\vdots \qquad \vdots \qquad \vdots$$

$$\left[\frac{n_r}{p}\right] + \left[\frac{n_r}{p^2}\right] + \left[\frac{n_r}{p^3}\right] + \cdots.$$

But, by a preceding theorem (Theorem 4-4, Cor.),

$$\left[\frac{n}{p^i}\right] \geq \left[\frac{n_1}{p^i}\right] + \left[\frac{n_2}{p^i}\right] + \cdots + \left[\frac{n_r}{p^i}\right].$$

Hence, summing by columns, we find that the exponent of the highest power of the prime p contained in the numerator is greater than or equal to the exponent of the highest power of this same prime p contained in the denominator.

- **Corollary** The product of n consecutive positive integers is divisible by $n!$.

Proof We note that

$$\frac{(k+1)\ldots(k+n)}{n!} = \frac{k!(k+1)(k+2)\ldots(k+n)}{k!\,n!} = \frac{(k+n)!}{k!\,n!}.$$

As a consequence of the above theorem, this last expression is an integer.

4-2 Problems[3]

1. Show that if α is an arbitrary real quantity, $[\alpha] + [-\alpha] = -1$ or 0.

2. Show that $[\alpha] - 2[\alpha/2] = 0$ or 1.

3. Show that $-[-\alpha]$ is the least integer not less than α.

4. (a) Show that $[\alpha + \tfrac{1}{2}]$ is the integer nearest to the real quantity α; and if there are two integers equally near to α, this is the larger of the two.

   (b) Show that $-[-\alpha + \tfrac{1}{2}]$ is the integer nearest to α; and if there are two integers equally near to α, this is the smaller of the two.

5. Show that, for positive integers m and n,

$$\frac{(2m)!\,(3n)!}{(m!)^2(n!)^3}$$

is an integer.

6. Show that, for positive integers m and n, $(mn)!$ is divisible by both $(m!)^n$ and $(n!)^m$, and hence by their least common multiple.

7. For real quantities α and β, show that $[2\alpha] + [2\beta] \geq [\alpha] + [\beta] + [\alpha + \beta]$.

8. (a) For real quantities α and β, show that $[4\alpha] + [4\beta] \geq 2[\alpha] + 2[\beta] + 2[\alpha + \beta]$.

   (b) Similarly, show that $[4\alpha] + [4\beta] \geq 3[\alpha] + 3[\beta] + [\alpha + \beta]$.

9. For real α and β, show that $[3\alpha] + [3\beta] \geq [\alpha] + [\beta] + 2[\alpha + \beta]$.

10. For positive integers m and n show that

$$\frac{(2m)!\,(2n)!}{m!\,n!\,(m+n)!}$$

is an integer.

11. For real quantities α and β, prove that $[4\alpha] + [4\beta] \geq [\alpha] + [\beta] + [2\alpha + \beta] + [\alpha + 2\beta]$.

12. If m and n are relatively prime positive integers, show that $(m + n - 1)!/(m!\,n!)$ is an integer.

13. (a) If m and n are positive integers, show that

$$\frac{(4m)!\,(4n)!}{(m!)^2(n!)^2\{(m+n)!\}^2}$$

is an integer.

   (b) For positive integers m and n show that

$$\frac{(4m)!\,(4n)!}{(m!)^3(n!)^3(m+n)!}$$

is an integer.

14. If the integer n is greater than one and relatively prime to 6, prove that $(2n - 4)!/\{n!(n - 2)!\}$ is an integer.

15. Prove that, for positive integers m and n,
$$\frac{(3m)!\,(3n)!}{m!\,n!\{(m + n)!\}^2}$$
is an integer.

16. If m and n are positive integers, show that
$$\frac{(4m)!\,(4n)!}{m!\,n!(2m + n)!(m + 2n)!}$$
is an integer.

17. If the integer n is positive or zero, show that $[(1 + \sqrt{3}\,)^n]$ is even or odd according as n is odd or even, respectively.

18. If an odd integer n greater than 3 is relatively prime to 5, show that
$$\frac{(3n - 6)!}{(n!)^2(n - 4)!}$$
is an integer.

19. (a) Show that the exponent of the highest power of 7 which divides $(7^n - 3)!$ is $(7^n - 6n - 1)/6$.

(b) Show that the exponent of the highest power of 5 contained in $(5^t - 4)!$ is $(5^t - 4t - 1)/4$.

20. (a) Determine the number of zeros with which $101!$ terminates.

(b) Find the exponent of the highest power of 12 which divides $289!$.

21. For a prime p find an expression for the exponent of the highest power of p dividing $(p^n - 1)!$.

22. Where m and a are positive integers, prove that the integer $(2m)!/(m!)^2$ is divisible by every prime p such that $1 \leq m < p^a \leq 2m$.

23. If m and n are positive integers, show that the integer $(mn)!/(n!)^m$ is divisible by $m!$. (See Problem 6.)

24. Prove that, if $m \geq 5$, $4^m/m < (2m)!/(m!)^2 < 4^{m-1}$.

4-3 Generalization of Euler's ϕ-Function

We have obtained properties of Euler's ϕ-function, $\phi(n)$, defined as the number of positive integers relatively prime to the positive integer n and not greater than n. A function $f(n)$ defined for positive integral values of n is

Generalization of Euler's ϕ-Function

called an *arithmetical function*. An arithmetical function $f(n)$ such that $f(1) = 1$ and $f(mn) = f(m)f(n)$ when $(m, n) = 1$ is called a *multiplicative function*. An arithmetical function $f(n)$ such that $f(1) = 1$ and $f(mn) = f(m)f(n)$ without the above restriction on m and n is said to be *completely multiplicative* or *totally multiplicative*. Euler's ϕ-function is an example of an arithmetical function that is multiplicative. Several generalizations of Euler's ϕ-function have been given.

One interesting generalization, given by E. Lucas,[4] is defined as follows. Let $e_1, e_2, \cdots, e_k$ be a set of arbitrary integers. Define $\psi(n; e_1, e_2, \ldots, e_k)$, written simply as $\psi(n)$, as the number of integers h selected from the set $1, 2, \ldots, n$ such that each of

$$h + e_1, h + e_2, \ldots, h + e_k$$

is relatively prime to n.

For example, if $e_1 = 0$, $e_2 = 1$ for $k = 2$ and $n = 15$, we have $\psi(15)$ as the number of integers h, $1 \leq h \leq 15$, for which each number of the pair $h + 0$, $h + 1$ is relatively prime to 15. Since there are only three such pairs (namely: 1, 2; 7, 8; 13, 14), $\psi(15) = 3$. If the set of e's selected had been $e_1 = 0$, $e_2 = 1$, $e_3 = 2$, then, of course, $\psi(15) = 0$. If the set of e's selected had been $e_1 = e_2 = \ldots = e_k = 0$, we see that the function $\psi(n)$ would then have become Euler's ϕ-function.

In the following theorem, we shall prove $\psi(n)$ to be a multiplicative function.

- **Theorem 4-8** If $(m, n) = 1$, $\psi(mn) = \psi(m)\psi(n)$.

  Proof Select two integers r and s which satisfy the following conditions:

  $$r \equiv 1 \pmod{m}, \quad r \equiv 0 \pmod{n},$$
  $$s \equiv 0 \pmod{m}, \quad s \equiv 1 \pmod{n}.$$

  Then as x ranges over the complete set of residues $1, 2, \ldots, m$, modulo m, and as y ranges independently over the complete set of residues $1, 2, \ldots, n$, modulo n, the mn numbers

  $$z \equiv rx + sy \pmod{mn}$$

  range over a complete set of residues, modulo mn. For, if $rx' + sy' \equiv rx'' + sy'' \pmod{mn}$, then $r(x' - x'') \equiv s(y'' - y') \pmod{mn}$; that is, $r(x' - x'') \equiv s(y'' - y') \pmod{m}$ and $r(x' - x'') \equiv s(y'' - y') \pmod{n}$. Consequently, $x' \equiv x'' \pmod{m}$ and $y'' \equiv y' \pmod{n}$; and the mn values of z form a complete set of residues, modulo mn.

  Hence, for each e_i, $i = 1, 2, \ldots, k$, there exists a pair of integers x_i and y_i such that

  $$e_i \equiv rx_i + sy_i \pmod{mn};$$

that is,
$$e_i \equiv 1 \cdot x_i \pmod{m} \quad \text{and} \quad e_i \equiv 1 \cdot y_i \pmod{n}.$$

We now get
$$z + e_i \equiv r(x + x_i) + s(y + y_i) \pmod{mn}.$$

We know that $z + e_i$ is relatively prime to mn if and only if it is relatively prime to both m and n. Now $z + e_i$ is relatively prime to m if and only if $x + x_i$ is relatively prime to m; and $z + e_i$ is relatively prime to n if and only if $y + y_i$ is relatively prime to n. This amounts to saying that $x + e_i$ is to be relatively prime to m and $y + e_i$ is to be relatively prime to n. This, however, occurs for all $i = 1, 2, \ldots, k$ simultaneously for all $\psi(m)$ values of x of the set $1, 2, \ldots, m$ and for all $\psi(n)$ values of y of the set $1, 2, \ldots, n$. This gives $\psi(m)\psi(n)$ as the number of permissible values of z for which the $z + e_i$ ($i = 1, 2, \ldots, k$) are relatively prime to mn. This number, however, we have denoted by $\psi(mn)$. Hence we have established the fact that, for a given set of e's, $\psi(mn) = \psi(m)\psi(n)$.

Consider n to be a power of a prime $n = p^a$, where $a \geq 1$. Let the integers from 1 to n be arranged in p^{a-1} rows of p integers each:

$$\begin{array}{ccccc}
1 & 2 & 3 & \cdots & p-1 \quad p \\
p+1 & p+2 & p+3 & \cdots & 2p-1 \quad 2p \\
\vdots & \vdots & \vdots & & \vdots \quad \vdots \\
(p^{a-1}-1)p+1 & (p^{a-1}-1)p+2 & (p^{a-1}-1)p+3 & \cdots & p^a-1 \quad p^a.
\end{array}$$

Let t denote the number of distinct residues, modulo p, among the $e_1, e_2, \ldots, e_k$ and let $r_1, r_2, \ldots, r_t$ be their least non-negative (incongruent) residues, modulo p. Then in the first row of the above array, there are $p - t$ integers h, incongruent modulo p to the $-r_1, -r_2, \ldots, -r_t$, such that $h + r_1, h + r_2, \ldots, h + r_t$ are relatively prime to p (and therefore to p^a). Likewise, each number in a column headed by one of these $p - t$ integers h would likewise provide an h such that $h + r_1, \ldots, h + r_t$ are each relatively prime to p. Thus, we see that $\psi(p^a) = p^{a-1}(p - t)$ and, in particular, $\psi(p) = p - t$.

- **Theorem 4-9** If the canonical form of n as a product of powers of distinct primes is $n = p_1^{a_1} p_2^{a_2} \ldots p_r^{a_r}$ and if t_i ($i = 1, 2, \ldots, r$) denotes the number of integers among $e_1, e_2, \ldots, e_k$ which are incongruent modulo p_i, then

$$\psi(n) = \frac{n}{p_1 p_2 \cdots p_r}(p_1 - t_1)(p_2 - t_2) \cdots (p_r - t_r)$$
$$= n\left(1 - \frac{t_1}{p_1}\right)\left(1 - \frac{t_2}{p_2}\right) \cdots \left(1 - \frac{t_r}{p_r}\right).$$

Proof Since $\psi(n)$ is a multiplicative function,

$$\psi(n) = \psi(p_1^{a_1} \cdot (p_2^{a_2} \ldots p_r^{a_r})) = \psi(p_1^{a_1}) \cdot \psi(p_2^{a_2} \cdot (p_3^{a_3} \ldots p_r^{a_r}))$$
$$= \psi(p_1^{a_1})\psi(p_2^{a_2})\psi(p_3^{a_3} \ldots p_r^{a_r})$$
$$\ldots\ldots\ldots\ldots\ldots\ldots$$
$$= \psi(p_1^{a_1})\psi(p_2^{a_2}) \ldots \psi(p_r^{a_r})$$
$$= p_1^{a_1-1}(p_1 - t_1)p_2^{a_2-1}(p_2 - t_2) \ldots p_r^{a_r-1}(p_r - t_r)$$
$$= \frac{n}{p_1 p_2 \ldots p_r}(p_1 - t_1)(p_2 - t_2) \ldots (p_r - t_r)$$
$$= n\left(1 - \frac{t_1}{p_1}\right)\left(1 - \frac{t_2}{p_2}\right) \ldots \left(1 - \frac{t_r}{p_r}\right).$$

Example Find the number of terms of the set $1 \cdot 2, 2 \cdot 3, \ldots, n(n+1)$ which are relatively prime to n.

Solution Take $e_1 = 0$, $e_2 = 1$. Then the number of terms of the given set relatively prime to n would be the number of integers h, $1 \leq h \leq n$, such that $h + 0$ and $h + 1$ are both relatively prime to n; that is, $\psi(n)$. For example, if $n = 3^2 \cdot 5$, then $\psi(n) = \psi(3^2)\psi(5) = 3 \cdot (3-2) \cdot (5-2) = 9$. If n were even, $\psi(n) = 0$ since each term of the given set is even.

4-4 Functions $\tau(n)$ and $\sigma(n)$

We shall now consider two arithmetical functions of considerable interest and importance in the theory of numbers. We define $\tau(n)$ as the number of (distinct) positive integral divisors of the positive integer n. For example, if $n = 12$, the positive divisors are $1, 2, 3, 4, 6, 12$; thus, $\tau(12) = 6$. Similarly, $\tau(1) = 1$, $\tau(2) = \tau(3) = 2$, and $\tau(4) = 3$. The function $\sigma(n)$ is defined as the sum of all the positive integral divisors of the positive integer n. For example, $\sigma(12) = 28$; $\sigma(1) = 1$, $\sigma(2) = 3$, $\sigma(3) = 4$, $\sigma(4) = 7$.

Let us now assume that $n > 1$ and that $n = p_1^{a_1} p_2^{a_2} \ldots p_r^{a_r}$ is the canonical form of n as a product of primes.

• **Theorem 4-10** For $n = p_1^{a_1} \ldots p_r^{a_r}$, $\tau(n) = (a_1 + 1) \ldots (a_r + 1)$.

Proof Every divisor of n must be of the form $p_1^{c_1} \ldots p_r^{c_r}$, where $0 \leq c_1 \leq a_1$, $0 \leq c_2 \leq a_2, \ldots, 0 \leq c_r \leq a_r$. The divisor 1 is obtained if $c_1 = c_2 = \ldots = c_r = 0$ and the number n itself is obtained if $c_1 = a_1, \ldots, c_r = a_r$. Since the number of possibilities for each c_i is $a_i + 1$, we see that the total number of distinct positive divisors of n is $(a_1 + 1) \cdot (a_2 + 1) \cdot \ldots (a_r + 1)$.

- **Theorem 4-11** For $n = p_1^{a_1} p_2^{a_2} \cdots p_r^{a_r}$,

$$\sigma(n) = \frac{p_1^{a_1+1} - 1}{p_1 - 1} \cdot \frac{p_2^{a_2+1} - 1}{p_2 - 1} \cdot \ldots \cdot \frac{p_r^{a_r+1} - 1}{p_r - 1}.$$

Proof Every term in the expansion of

$$(1 + p_1 + p_1^2 + \ldots + p_1^{a_1})(1 + p_2 + p_2^2 + \ldots + p_2^{a_2}) \cdots$$
$$\cdot (1 + p_r + p_r^2 + \ldots + p_r^{a_r})$$

is obviously a divisor of n; moreover, every divisor of n, including one and n, is uniquely a term in this expansion. Hence, $\sigma(n)$ is given by this expansion, which equals

$$\frac{p_1^{a_1+1} - 1}{p_1 - 1} \cdot \frac{p_2^{a_2+1} - 1}{p_2 - 1} \cdot \ldots \cdot \frac{p_r^{a_r+1} - 1}{p_r - 1}.$$

- **Theorem 4-12** Both $\tau(n)$ and $\sigma(n)$ are multiplicative functions.

Proof Assuming $(m, n) = 1$, let $m = p_1^{a_1} \ldots p_r^{a_r}$, $n = q_1^{b_1} \ldots q_s^{b_s}$ be the canonical forms for m and n, respectively. Hence the primes p_1, $p_2, \ldots, p_r, q_1, q_2, \ldots, q_s$ are all distinct. Every divisor of mn has a unique representation as the product of a divisor of m and a divisor of n; conversely, the product of a divisor of m and a divisor of n is a divisor of mn. Thus $\tau(mn) = \tau(m)\tau(n) = (a_1 + 1) \ldots (a_r + 1)(b_1 + 1) \ldots$
$\cdot (b_s + 1)$. Moreover, since $mn = p_1^{a_1} \ldots p_r^{a_r} q_1^{b_1} \ldots q_s^{b_s}$, $\sigma(mn) = (1 + p_1 + \ldots + p_1^{a_1}) \cdots (1 + p_r + \ldots + p_r^{a_r})(1 + q_1 + \ldots + q_1^{b_1}) \cdots (1 + q_s + \ldots + q_s^{b_s}) = \sigma(m)\sigma(n)$.

For an arithmetical function $f(n)$, let $\sum_{d \mid n} f(d)$ denote the sum of the terms $f(d)$ as d ranges over all positive integral divisors of the positive integer n. That is, if $d_1, d_2, \ldots, d_k$ are all the positive divisors of n (including 1 and n), then

$$\sum_{d \mid n} f(d) \text{ means } f(d_1) + f(d_2) + \ldots + f(d_k).$$

Moreover, $\sum_{d \mid n} 1$ means that we are to add together as many 1's as there are distinct positive divisors of n; namely,

$$\sum_{d \mid n} 1 = \tau(n).$$

Also,

$$\sum_{d \mid n} d = \sum_{d \mid n} \frac{n}{d} = \sigma(n).$$

The sum

$$\sum_{\substack{i,j=1 \\ i<j}}^{n} f(x_i, y_j)$$

Functions $\tau(n)$ and $\sigma(n)$

means the sum of all terms $f(x_i, y_j)$, where i and j range independently over the set $1, 2, \ldots, n$ subject to the restriction that $i < j$. Moreover, if α is a real quantity ≥ 1, $\sum_{j=1}^{\alpha} f_j$ means

$$\sum_{j=1}^{[\alpha]} f_j = f_1 + f_2 + \cdots + f_{[\alpha]}.$$

For $m > 1$ and $m' > 1$, let $d_1, d_2, \ldots, d_r$ and $d'_1, d'_2, \ldots, d'_s$ be the positive divisors of m and m', respectively. We then have the following obvious result.

- **Theorem 4-13** If d ranges over all positive divisors of m and if d' ranges over all positive divisors of m', then for arithmetical functions $f(n)$ and $g(n)$,

$$\sum_{\substack{d \mid m \\ d' \mid m'}} f(d)g(d') = \left(\sum_{d \mid m} f(d)\right) \cdot \left(\sum_{d' \mid m'} g(d')\right).$$

The following result[5] will have useful applications.

- **Theorem 4-14** For arithmetical functions $F(n)$ and $f(n)$, let

$$F(n) = \sum_{d \mid n} f(d).$$

Then, for a real quantity $\alpha \geq 1$,

$$\sum_{n=1}^{\alpha} F(n) = \sum_{k=1}^{\alpha} f(k) \cdot \left[\frac{\alpha}{k}\right].$$

Proof Now,

$$\sum_{n=1}^{\alpha} F(n) = \sum_{n=1}^{\alpha} \sum_{d \mid n} f(d).$$

In the sum $\sum_{d \mid n} f(d)$ the term $f(k)$ appears if and only if k is a divisor of n. But, since every positive integer n has itself as a divisor, the sum in the right member of the last equation obviously includes (at least once) the term $f(k)$ for each k such that $1 \leq k \leq [\alpha]$. Now, for a specific integer k, k is a divisor exclusively of the following positive integers ($\leq [\alpha]$):

$$1 \cdot k, 2 \cdot k, \ldots, \left[\frac{\alpha}{k}\right] \cdot k.$$

Hence, $f(k)$ is counted $[\alpha/k]$ times in this sum. Consequently,

$$\sum_{n=1}^{[\alpha]} F(n) = \sum_{k=1}^{[\alpha]} f(k) \cdot \left[\frac{\alpha}{k}\right].$$

4-5 Problems

1. (a) Find an expression for the number of terms of the sequence $1 \cdot 2 \cdot 3, 2 \cdot 3 \cdot 4, \ldots, n(n+1)(n+2)$ which are relatively prime to n.
   (b) Find explicitly the number of such terms if $n = 539$.
   (c) Find, similarly, the number of such terms if $n = 231$.

2. Determine the number of integers of the set $1 \cdot 2 \cdot 3, 2 \cdot 3 \cdot 4, 3 \cdot 4 \cdot 5, \ldots, 8575 \cdot 8576 \cdot 8577$ which are relatively prime to 35.

3. Determine the number of integers of the set $1 \cdot 4 \cdot 7, 2 \cdot 5 \cdot 8, \ldots, 3025 \cdot 3028 \cdot 3031$ which are relatively prime to 3025.

4. If the positive integers m and n are relatively prime, let $P = 1$; if $(m, n) > 1$, let P denote the product of all the distinct primes dividing (m, n). Prove that
$$\psi(mn) = \frac{P\psi(m)\psi(n)}{\psi(P)}.$$

5. Where $\phi(n)$ denotes the Euler ϕ-function, find all integral solutions n of the given equation:
   (a) $\phi(n) = \phi(2n)$;
   (b) $\phi(2n) = \phi(3n)$;
   (c) $\phi(4n) = \phi(5n)$;
   (d) $\phi(5n) = \phi(6n)$.

6. If $\sigma_k(n)$ denotes the sum of the kth powers ($k \geq 0$) of the positive divisors of the positive integer n, prove that, for a real quantity $\alpha \geq 1$,
$$\sum_{n=1}^{[\alpha]} \sigma_k(n) = \sum_{m=1}^{[\alpha]} m^k \left[\frac{\alpha}{m}\right].$$

7. For a real quantity $\alpha \geq 1$, prove that:
   (a) $\sum_{n=1}^{[\alpha]} \tau(n) = \sum_{s=1}^{[\alpha]} \left[\frac{\alpha}{s}\right];$
   (b) $\sum_{n=1}^{[\alpha]} \sigma(n) = \sum_{s=1}^{[\alpha]} s \cdot \left[\frac{\alpha}{s}\right].$

8. Show that the number of irreducible fractions not exceeding one, whose numerator and denominator are positive integers not greater than the real quantity $\alpha \geq 1$, is
$$T(\alpha) = \sum_{k=1}^{[\alpha]} \phi(k).$$

9. For a real value $\alpha \geq 1$, prove the following results:
   (a) $\sum_{k=1}^{[\alpha]} \sum_{i=1}^{[\alpha/k]} \phi(i) = \sum_{r=1}^{[\alpha]} \left[\frac{\alpha}{r}\right] \phi(r);$
   (b) $\sum_{k=1}^{[\alpha]} \sum_{i=1}^{[\alpha/k]} \phi(i) = \frac{[\alpha]^2 + [\alpha]}{2}.$

4-6 Perfect Numbers

The Greeks were interested in the problem of determining positive integers n such that the sum of its positive divisors is $2n$. Such an integer was called a *perfect number*. If we understand a *proper divisor* (or *aliquot divisor*) of n to be a positive divisor less than n itself, we could say that a perfect number is one the sum of whose proper divisors is equal to the number itself. The two smallest perfect numbers are 6 and 28. As has been pointed out by L. E. Dickson, perfect numbers have, in fact, engaged the attention of arithmeticians of every century of the Christian era. A number is regarded as *deficient* or *abundant* according as the sum of its proper divisors is, respectively, less than or greater than the number itself. For example, 8 is deficient and 12 is abundant. A *multiply perfect number* is one the sum of whose proper divisors is equal to a multiple of the number (greater than the number itself). For example, 120 and 672 are multiply perfect numbers. Two numbers are called *amicable* if each equals the sum of the proper divisors of the other. That is, m and n are amicable if and only if $\sigma(m) = \sigma(n) = m + n$. The followers of Pythagoras were acquainted with the pair 220 and 284.

In 1968, only 23 perfect numbers were known, all of them being even. One of the oldest unsolved problems of the theory of numbers is concerned with the existence of *odd* perfect numbers. A number of necessary conditions have to be satisfied by a possible odd perfect number. In any event, if an odd perfect number exists, it must be exceedingly large,[6] not less than 100 quintillion (that is, 10^{20}), and must have at least six different prime factors, at least one of which must be not less than 61. Moreover, an odd perfect number must be of the form $12m + 1$ or $36m + 9$. We shall now establish the following theorem given by Euclid.

- **Theorem 4-15** If $2^p - 1$ is a prime, then $2^{p-1}(2^p - 1)$ is a perfect number.

   Proof Since 2^{p-1} and $2^p - 1$ are relatively prime and since $2^p - 1$ and 1 are the only divisors of $2^p - 1$, then, $\sigma(n)$ being a multiplicative function, the sum of *all* positive divisors of $2^{p-1}(2^p - 1)$ is:

   $$(1 + 2 + 2^2 + \ldots + 2^{p-1})(1 + 2^p - 1) = \frac{2^p - 1}{2 - 1} \cdot 2^p$$
   $$= 2 \cdot 2^{p-1}(2^p - 1).$$

   Hence, $2^{p-1}(2^p - 1)$ is perfect.

In view of the following theorem, all *even* perfect numbers are known to be of Euclid's type, as described in the preceding theorem.

- **Theorem 4-16** If q is a positive odd integer and if the even number $2^{n-1}q$ is perfect, then q is a prime of the form $q = 2^n - 1$.

Proof[7] Because 2^{n-1} is itself not perfect, the odd number q is at least 3. Since 2^{n-1} and q are relatively prime and since $2^{n-1}q$ is perfect, the sum of its positive divisors is $(1 + 2 + \ldots + 2^{n-1}) s = (2^n - 1) s$, where s is the sum of *all* positive divisors of q. Write $s = q + d$. Thus, $2^n q = (2^n - 1)(q + d)$; that is, $q = (2^n - 1)d$. Now, since $n \geq 2$, d is a divisor of q less than q. Hence, since the sum of all the (positive) divisors of q is $q + d$, d and q are the *only* possible divisors of q. Thus, $d = 1$ and q is a prime $2^n - 1$.

4-7 Möbius μ-Function

The arithmetic function $\mu(n)$ introduced by A. F. Möbius plays an important role in the theory of numbers, particularly in the theory of the asymptotic distribution of primes. Möbius employed the function in the reversion of an infinite series. This function may be defined as follows:
$\mu(1) = 1$; $\mu(n) = (-1)^r$ if n is the product of r distinct primes, where $r \geq 1$; $\mu(n) = 0$ if n is divisible by the square of a prime. For example, $\mu(2) = -1$, $\mu(3) = -1$, $\mu(6) = (-1)^2 = 1$, $\mu(4) = \mu(8) = \mu(9) = 0$, $\mu(10) = 1$, $\mu(12) = 0$.

As a first property, we shall establish the following result.

- **Theorem 4-17** If m and n are positive relatively prime integers, $\mu(mn) = \mu(m) \mu(n)$; in other words, $\mu(n)$ is a multiplicative function.

Proof First, if either m or n is divisible by the square of a prime, so is mn, and conversely. Hence, in this case, $\mu(mn) = 0 = \mu(m)\mu(n)$. Suppose, next, that neither m nor n is divisible by the square of a prime. If either $m = 1$ or $n = 1$, the result is obviously true. If $m > 1$ and $n > 1$, let the distinct prime factors of m and n be $p_1, p_2, \ldots, p_r$, and $q_1, q_2, \ldots, q_s$, respectively. Thus, $m = p_1 p_2 \cdots p_r$, $n = q_1 q_2 \cdots q_s$, $mn = p_1 p_2 \cdots p_r q_1 q_2 \cdots q_s$. Hence, $\mu(mn) = (-1)^{r+s} = (-1)^r (-1)^s = \mu(m)\mu(n)$.

We shall now establish the following interesting result.

- **Theorem 4-18** $\sum_{d|n} \mu(d) = 1$ if $n = 1$; $\sum_{d|n} \mu(d) = 0$ if $n > 1$.

Proof If $n = 1$, $d = 1$ is the only positive divisor of 1 and $\mu(1) = 1$. Next, let $n > 1$ and let its canonical decomposition into primes be $n = p_1^{a_1} p_2^{a_2} \cdots p_r^{a_r}$. Then, by the definition of $\mu(n)$,

Möbius µ-Function

$$\sum_{d|n} \mu(d) = \sum_{d|p_1 p_2 \cdots p_r} \mu(d) = \mu(1) + \sum_{i=1}^{r} \mu(p_i) + \sum_{\substack{i,j=1 \\ i<j}}^{r} \mu(p_i p_j)$$

$$+ \sum_{\substack{i,j,k=1 \\ i<j<k}}^{r} \mu(p_i p_j p_k) + \cdots + \mu(p_1 p_2 \cdots p_r)$$

$$= 1 + \frac{r}{1}(-1) + \frac{r(r-1)}{1 \cdot 2}(-1)^2 + \frac{r(r-1)(r-2)}{1 \cdot 2 \cdot 3}(-1)^3$$

$$+ \cdots + 1 \cdot (-1)^r$$

$$= (1-1)^r = 0.$$

We shall find it convenient to make use of the following notation. For a fixed positive integer n, $\sum_{kl|n} f(k)g(l)$ means the sum of all terms $f(k)g(l)$ as the product kl ranges over all divisors of n. This means, of course, that both k and l range over the positive divisors of n, subject to the restriction that kl also divides n.

By $\sum_{kl=n} f(k)g(l)$ is meant the sum of all terms $f(k)g(l)$ such that the product of the positive integers k and l is n.

We shall now derive the useful result known as *Möbius's Inversion Formula*.

- **Theorem 4-19** Let $f(n)$ be an arithmetical function such that $F(n) = \sum_{d|n} f(d)$. Then $f(n) = \sum_{d|n} \mu(d) F\left(\frac{n}{d}\right)$.

Proof First, we can readily see that the function $f(n)$ is uniquely determined by the function $F(n)$. For, from the equations $F(1) = f(1)$, $F(2) = f(2) + f(1)$, $F(3) = f(3) + f(1)$, $F(4) = f(4) + f(2) + f(1)$, ..., we can successively compute $f(1), f(2), f(3), f(4), \ldots$ in terms of the values of $F(1), F(2), F(3), F(4), \ldots$. We shall now not only prove that the function $f(n)$ is uniquely determined by the function $F(n)$, but derive a formula for this representation.

For every positive integer n/d, we have

$$F\left(\frac{n}{d}\right) = \sum_{m|\frac{n}{d}} f(m);$$

which yields

$$\mu(d) F\left(\frac{n}{d}\right) = \mu(d) \sum_{m|\frac{n}{d}} f(m)$$

and

$$\sum_{d|n} \mu(d) F\left(\frac{n}{d}\right) = \sum_{d|n} \mu(d) \sum_{m|\frac{n}{d}} f(m) = \sum_{d|n} \sum_{m|\frac{n}{d}} \mu(d) f(m).$$

It is obvious that here both d and m range over all positive divisors of n. Moreover, since, for a given d, $m|(n/d)$, we see that the sum

$$\sum_{d|n} \sum_{m|\frac{n}{d}} \mu(d)f(m)$$

is extended over *all* pairs d, m for which dm divides n. That is, this sum is $\sum_{dm|n} \mu(d)f(m)$, where the summation is extended over all positive integers d and m such that dm divides n. But this is the same as the sum extended over all pairs m, d such that $m|n$ and $d|(n/m)$; that is, it is the same as

$$\sum_{m|n} \sum_{d|\frac{n}{m}} \mu(d)f(m) = \sum_{m|n} f(m) \sum_{d|\frac{n}{m}} \mu(d).$$

However, $\sum_{d|(n/m)} \mu(d) = 0$ if $n/m > 1$, but equals 1 if $n/m = 1$ (that is, $m = n$). Hence, the sum is finally equal to $f(n)$. It is to be noted that $\sum_{d|n} \mu(d)F(n/d)$ is the same sum as $\sum_{d|n} \mu(n/d)F(d)$.

The converse of Theorem 4-19 is given in the following statement.

- **Theorem 4-20** If $F(n)$ is an arithmetical function such that for every positive integer n, $f(n) = \sum_{d|n} \mu(d)F(n/d)$, then $F(n) = \sum_{d|n} f(d)$.

Proof We shall show that the value of $\sum_{d|n} f(d)$ is $F(n)$. Now,

$$f(d) = \sum_{l|d} \mu(l)F\left(\frac{d}{l}\right) \quad \text{and} \quad \sum_{d|n} f(d) = \sum_{d|n} \sum_{l|d} \mu(l)F\left(\frac{d}{l}\right).$$

Set $k = d/l$, $n/d = n'$, which yields $n = dn' = kln'$. The last double sum becomes

$$\sum_{d|n} \sum_{kl=d} \mu(l)F(k) = \sum_{kl|n} \mu(l)F(k) = \sum_{k|n} F(k) \sum_{l|\frac{n}{k}} \mu(l).$$

For a given n and a fixed k,

$$\sum_{l|\frac{n}{k}} \mu(l) = 0$$

unless $k = n$. Therefore, the only value of k which makes a contribution to this last double sum is the value $k = n$. This gives us $F(n) \cdot 1 = F(n)$.

- **Theorem 4-21** If $f(n)$ and $F(n)$ are arithmetical functions such that $F(n) = \sum_{d|n} f(d)$, then $F(n)$ is a multiplicative function whenever $f(n)$ is.

Proof Let $(m, n) = 1$. Every divisor d of mn can be uniquely written

Möbius µ-Function

as $d_1 d_2$, where $d_1 | m$ and $d_2 | n$ and, of course, $(d_1, d_2) = 1$. Assuming that $f(n)$ is multiplicative, we have

$$F(mn) = \sum_{d | mn} f(d) = \sum_{\substack{d_1 | m \\ d_2 | n}} f(d_1 d_2) = \sum_{\substack{d_1 | m \\ d_2 | n}} f(d_1) f(d_2)$$

$$= \left(\sum_{d_1 | m} f(d_1) \right) \left(\sum_{d_2 | n} f(d_2) \right) = F(m) F(n).$$

We consider the converse in the following theorem.

- **Theorem 4-22** If $f(n)$ and $F(n)$ are arithmetical functions such that $F(n) = \sum_{d | n} f(d)$ and if $F(n)$ is multiplicative, then $f(n)$ is also.

    Proof Let $(m, n) = 1$. Now,

$$f(mn) = \sum_{d | mn} \mu(d) F\left(\frac{mn}{d} \right) = \sum_{d | mn} \mu\left(\frac{mn}{d} \right) F(d)$$

$$= \sum_{\substack{d_1 | m \\ d_2 | n}} \mu\left(\frac{m}{d_1} \cdot \frac{n}{d_2} \right) F(d_1 d_2)$$

$$= \sum_{\substack{d_1 | m \\ d_2 | n}} \mu\left(\frac{m}{d_1} \right) \mu\left(\frac{n}{d_2} \right) F(d_1) F(d_2)$$

$$= \sum_{\substack{d_1 | m \\ d_2 | n}} \mu\left(\frac{m}{d_1} \right) F(d_1) \cdot \mu\left(\frac{n}{d_2} \right) F(d_2)$$

$$= \left(\sum_{d_1 | m} \mu\left(\frac{m}{d_1} \right) F(d_1) \right) \left(\sum_{d_2 | n} \mu\left(\frac{n}{d_2} \right) F(d_2) \right)$$

$$= f(m) f(n).$$

Another inversion formula[8] of a somewhat different type is given in the following theorem.

- **Theorem 4-23** If $f(n)$ and $F(n)$ are arithmetical functions such that $F(n) = \sum_{m=1}^{n} f\left(\left[\frac{n}{m} \right] \right)$, then $f(n) = \sum_{d=1}^{n} \mu(d) F\left(\left[\frac{n}{d} \right] \right)$.

    Proof

$$\sum_{d=1}^{n} \mu(d) F\left(\left[\frac{n}{d} \right] \right) = \sum_{d=1}^{n} \mu(d) \sum_{m=1}^{[n/d]} f\left(\left[\frac{[n/d]}{m} \right] \right)$$

$$= \sum_{d=1}^{n} \mu(d) \sum_{m=1}^{[n/d]} f\left(\left[\frac{n}{dm} \right] \right).$$

Set $k = dm$. Since $k = dm \leq d[n/d]$, k ranges over the integers from 1 to n, and d divides k. Hence, we have the above expression equal to

$$\sum_{1 \leq k \leq n} \sum_{d | k} \mu(d) f\left(\left[\frac{n}{k} \right] \right) = \sum_{1 \leq k \leq n} f\left(\left[\frac{n}{k} \right] \right) \sum_{d | k} \mu(d).$$

Since $\sum_{d|k} \mu(d)$ is zero unless $k = 1$, the only value of k that makes a contribution to the expression is $k = 1$. Hence, the final result is $f([n/1]) = f(n)$.

As in the case of the Möbius inversion formula, this theorem has a converse.

- **Theorem 4-24** If $f(n)$ and $F(n)$ are arithmetical functions such that

$$f(n) = \sum_{d=1}^{n} \mu(d) F\left(\left[\frac{n}{d}\right]\right), \quad \text{then} \quad F(n) = \sum_{m=1}^{n} f\left(\left[\frac{n}{m}\right]\right).$$

Proof We desire to show that $F(n)$ is the value of

$$\sum_{m=1}^{n} f\left(\left[\frac{n}{m}\right]\right) = \sum_{m=1}^{n} \sum_{d=1}^{[n/m]} \mu(d) F\left(\left[\frac{[n/m]}{d}\right]\right)$$

$$= \sum_{m=1}^{n} \sum_{d=1}^{[n/m]} \mu(d) F\left(\left[\frac{n}{md}\right]\right)$$

$$= \sum_{m=1}^{n} \sum_{\substack{1 \le d \le [n/m] \\ k = md}} \mu(d) F\left(\left[\frac{n}{k}\right]\right)$$

$$= \sum_{1 \le k \le n} \sum_{d|k} \mu(d) F\left(\left[\frac{n}{k}\right]\right)$$

$$= \sum_{1 \le k \le n} F\left(\left[\frac{n}{k}\right]\right) \sum_{d|k} \mu(d)$$

$$= F\left(\left[\frac{n}{1}\right]\right) = F(n).$$

Example 1 Now consider an application of the Möbius inversion formula. Where $\phi(n)$ denotes the Euler ϕ-function, we know that $n = \sum_{d|n} \phi(d)$. Using the inversion formula, we obtain

$$\phi(n) = \sum_{d|n} \mu(d) \frac{n}{d} = n \sum_{d|n} \frac{\mu(d)}{d}.$$

In canonical form, let $n = p_1^{a_1} p_2^{a_2} \ldots p_r^{a_r} > 1$. Since $\mu(n)$ is a multiplicative function, $\mu(n)/n$ is also. Let

$$d = p_1^{b_1} p_2^{b_2} \ldots p_r^{b_r}, \ 0 \le b_1 \le a_1, \ 0 \le b_2 \le a_2, \ldots, \ 0 \le b_r \le a_r.$$

We note that $\mu(p_i^{b_i})$ equals 1 for $b_i = 0$, -1 for $b_i = 1$, and 0 for $b_i > 1$. Then

Möbius μ-Function

$$\phi(n) = n \sum_{d\mid n} \frac{\mu(d)}{d} = n \sum_{\substack{0\leq b_1\leq a_1 \\ \cdots \\ 0\leq b_r\leq a_r}} \frac{\mu(p_1^{b_1})\mu(p_2^{b_2})}{p_1^{b_1}\, p_2^{b_2}} \cdots \frac{\mu(p_r^{b_r})}{p_r^{b_r}}$$

$$= n\left(\sum_{0\leq b_1\leq a_1} \frac{\mu(p_1^{b_1})}{p_1^{b_1}}\right)\left(\sum_{0\leq b_2\leq a_2} \frac{\mu(p_2^{b_2})}{p_2^{b_2}}\right) \cdots \left(\sum_{0\leq b_r\leq a_r} \frac{\mu(p_r^{b_r})}{p_r^{b_r}}\right)$$

$$= n\left(1 - \frac{1}{p_1}\right)\left(1 - \frac{1}{p_2}\right) \cdots \left(1 - \frac{1}{p_r}\right),$$

the formula we had established previously (Theorem 3-7).

Example 2 Where $\tau(n)$ is the number of positive divisors of n, we have $\tau(n) = \sum_{d\mid n} 1$. Using the inversion formula with $F(n)$ and $f(n)$ replaced by $\tau(n)$ and 1, respectively, we have $1 = \sum_{d\mid n} \mu(n/d)\tau(d)$ valid for $n \geq 1$.

4-8 Liouville's Function $\lambda(n)$

We shall now consider another useful arithmetical function, first introduced by J. Liouville.[9] The function $\lambda(n)$ is defined as follows:

$$\lambda(1) = 1;$$
$$\lambda(n) = (-1)^{a_1+a_2+\cdots+a_r}$$

if the canonical decomposition into primes of n is $n = p_1^{a_1} p_2^{a_2} \cdots p_r^{a_r}$. We note that $\lambda(1) = 1$, $\lambda(2) = \lambda(3) = \lambda(5) = -1$, $\lambda(4) = 1$, $\lambda(6) = 1$, $\lambda(7) = \lambda(8) = -1$, $\lambda(9) = \lambda(10) = 1$. We note, too, that $\lambda(n)$ is obviously a multiplicative function; in fact, it is completely multiplicative. Its properties show analogies with those of the Möbius μ-function. We shall next establish the following important result.

- **Theorem 4-25** The sum, extended over all positive divisors of n, $\sum_{d\mid n} \lambda(d)$ equals 1 if n is a perfect square and equals 0 if n is not a perfect square.

Proof If $n = 1$, the theorem is obviously true. Let, then, $n > 1$ and $n = p_1^{a_1} p_2^{a_2} \cdots p_r^{a_r}$, in canonical form as a product of primes.

We shall derive a result somewhat more general than that needed for the theorem itself. If $s \geq 0$, we see that the product

$$\{1 - p_1^s + p_1^{2s} - \cdots + (-1)^{a_1} p_1^{a_1 s}\} \times \cdots$$
$$\times \{1 - p_r^s + p_r^{2s} - \cdots + (-1)^{a_r} p_r^{a_r s}\}$$
$$= \frac{1 - (-p_1^s)^{a_1+1}}{1 + p_1^s} \cdot \frac{1 - (-p_2^s)^{a_2+1}}{1 + p_2^s} \cdots \frac{1 - (-p_r^s)^{a_r+1}}{1 + p_r^s}$$
$$= \prod_{i=1}^{r} \frac{1 + (-1)^{a_i} p_i^{(a_i+1)s}}{1 + p_i^s}.$$

However, since each term in the expansion of the original product
$$\prod_{i=1}^{r}(1 - p_i^s + p_i^{2s} - \cdots + (-1)^{a_i}p_i^{a_is})$$
is of the form
$$(-1)^{b_1+b_2+\cdots+b_r}(p_1^{b_1}p_2^{b_2}\cdots p_r^{b_r})^s,$$
where $0 \leq b_1 \leq a_1$, $0 \leq b_2 \leq a_2$, ..., $0 \leq b_r \leq a_r$, we have
$$\prod_{i=1}^{r}\frac{1+(-1)^{a_i}p_i^{(a_i+1)s}}{1+p_i^s} = \prod_{i=1}^{r}(1 - p_i^s + p_i^{2s} - \cdots + (-1)^{a_i}p_i^{a_is})$$
$$= \sum_{d\mid n}\lambda(d)d^s.$$

For the particular value $s = 0$, we obtain the formula
$$\frac{1+(-1)^{a_1}}{2} \cdot \frac{1+(-1)^{a_2}}{2} \cdots \frac{1+(-1)^{a_r}}{2} = \sum_{d\mid n}\lambda(d).$$

If any value a_i is odd, the left member of this last equation is zero; if the value of each of the a's is even, the left member is 1. Since n is not a perfect square if one of the a's is odd, but is a perfect square if all a's are even, our theorem is established.

If $F(n) = \sum_{d\mid n}\lambda(d)$, we see that $F(n) = 1$ when and only when n is a perfect square; otherwise $F(n) = 0$.

Example 1 For a real value $\alpha \geq 1$, prove that
$$\sum_{m=1}^{[\alpha]}\lambda(m)\left[\frac{\alpha}{m}\right] = [\sqrt{\alpha}].$$

Proof Let $F(n) = \sum_{d\mid n}\lambda(d)$. Therefore, $F(n) = 1$ when n is a perfect square and 0 otherwise. By Theorem 4-14 we have
$$\sum_{n=1}^{[\alpha]}F(n) = \sum_{n=1}^{[\alpha]}\sum_{d\mid n}\lambda(d) = \sum_{m=1}^{[\alpha]}\lambda(m)\left[\frac{\alpha}{m}\right].$$

The sum $\sum_{n=1}^{[\alpha]}F(n)$ consists of as many terms equal to 1 as there are positive integers $\leq [\alpha]$ which are perfect squares, the other terms of the sum being zero. Where s^2 is the largest square $\leq \alpha$, the squares are $1^2, 2^2, \ldots, s^2$. Therefore the number of such squares is $[\sqrt{\alpha}]$.

Example 2 Prove that
$$\sum_{d\mid n}\frac{n}{d}\sigma(d) = \sum_{d\mid n}d\tau(d).$$

Proof This is one of several results given (in 1857) by J. Liouville.[10]
If we write $F(n) = \sum_{d\mid n}\frac{n}{d}\sigma(d)$ and $G(n) = \sum_{d\mid n}d\tau(d)$, we can readily

see that both $F(n)$ and $G(n)$ are multiplicative functions of n. If we are able to prove that $F(p^a) = G(p^a)$, where p^a is a power of a prime, then, since $F(n)$ and $G(n)$ are multiplicative functions, $F(n) = G(n)$ if, in canonical form, $n = p_1^{a_1} p_2^{a_2} \cdots p_r^{a_r} > 1$.

First, let $n = n_1 n_2$, $d = d_1 d_2$, where $(n_1, n_2) = 1$, and $d|n$, $d_1|n_1$, $d_2|n_2$. We note also that $\tau(n)$ and $\sigma(n)$ are multiplicative functions of n. Then

$$F(n) = F(n_1 n_2) = \sum_{d_1 d_2 | n_1 n_2} \frac{n_1 n_2}{d_1 d_2} \sigma(d_1 d_2)$$

$$= \sum_{\substack{d_1|n_1 \\ d_2|n_2}} \left(\frac{n_1}{d_1}\right) \sigma(d_1) \cdot \left(\frac{n_2}{d_2}\right) \sigma(d_2) = F(n_1) F(n_2).$$

Likewise,

$$G(n) = G(n_1 n_2) = \sum_{d_1 d_2 | n_1 n_2} d_1 d_2 \tau(d_1 d_2) = \sum_{\substack{d_1|n_1 \\ d_2|n_2}} d_1 \tau(d_1) \cdot d_2 \tau(d_2) = G(n_1) G(n_2).$$

Next, the result is obviously true for $n = 1$. Let $n = p^a > 1$, where p is a prime. Let $d = p^b$, $0 \leq b \leq a$, and $e = a - b \geq 0$. Then

$$F(p^a) = \sum_{0 \leq b \leq a} \frac{p^a}{p^b} \sigma(p^b) = \sum_{0 \leq e \leq a} p^e \sum_{e \leq f \leq a} p^{a-f} = \sum_{e=0}^{a} \sum_{f=e}^{a} p^{a+e-f}$$

$$= \sum_{e=0}^{a}(p^a + p^{a-1} + \ldots + p^e) = \sum_{e=0}^{a}(p^e + p^{e+1} + \ldots + p^{e+a-e})$$

$$= 1 + p + p^2 + \ldots + p^{a-1} + p^a$$
$$ + p + p^2 + \ldots + p^{a-1} + p^a$$
$$ + p^2 + \ldots + p^{a-1} + p^a$$
$$\cdots$$
$$ + p^{a-1} + p^a$$
$$\phantom{= 1 + p + p^2 + \ldots + p^{a-1}} + p^a$$

$$= 1 + 2p + 3p^2 + \ldots + a p^{a-1} + (a+1) p^a.$$

Now, $$G(p^a) = \sum_{p^b | p^a} p^b \tau(p^b) = \sum_{b=0}^{a} p^b \tau(p^b)$$
$$= 1 + p \cdot 2 + p^2 \cdot 3 + \ldots + p^{a-1} \cdot a + p^a \cdot (a+1)$$
$$= F(p^a).$$

Hence, if $n = p_1^{a_1} p_2^{a_2} \ldots p_r^{a_r}$, since $F(p_i^{a_i}) = G(p_i^{a_i})$, we have

$$F(n) = F(p_1^{a_1} p_2^{a_2} \ldots p_r^{a_r}) = G(p_1^{a_1} p_2^{a_2} \ldots p_r^{a_r}) = G(n)$$

4-9 Problems

1. Prove that a power of a prime cannot be a perfect number.

2. Prove that the sum of the reciprocals of the divisors of a perfect number is 2.

3. If n is a positive integer and if α is any real quantity, prove that
$$[\alpha] + \left[\alpha + \frac{1}{n}\right] + \left[\alpha + \frac{2}{n}\right] + \ldots + \left[\alpha + \frac{n-1}{n}\right] = [n\alpha].$$

4. For a positive integer n, prove that
$$n = \sum_{m=1}^{n} \mu(m) \sum_{k=1}^{[n/m]} \tau(k).$$

5. If α is a real quantity ≥ 1 and if n is a positive integer, show that:

(a) $\sum_{n=1}^{[\alpha]} \sum_{d|n} \mu(d) = 1;$

(b) $\sum_{k=1}^{n} \sum_{d|k} \mu(d) = \sum_{k=1}^{n} \mu(k)\left[\frac{n}{k}\right];$

(c) $\sum_{k=1}^{n} \mu(k)\left[\frac{n}{k}\right] = 1.$

6. If n is a positive integer having r distinct prime factors $p_1, p_2, \ldots, p_r$, prove the following:

(a) $\sum_{d|n} \mu(d)\tau(d) = (-1)^r;$

(b) $\sum_{d|n} \mu(d)\sigma(d) = (-1)^r p_1 p_2 \ldots p_r.$

7. If the canonical decomposition of n is $n = p_1^{a_1} p_2^{a_2} \ldots p_r^{a_r}$, prove the following:

(a) $\sum_{d|n} \mu(d)\psi(d) = \prod_{i=1}^{r}(t_i + 1 - p_i);$

(b) $\sum_{d|n} \mu(d)\phi(d) = \prod_{i=1}^{r}(2 - p_i).$

8. If $n = p_1^{a_1} p_2^{a_2} \ldots p_r^{a_r}$, where the p's are distinct primes, prove that
$$\sum_{d|n} \mu(d)\lambda(d) = 2^r.$$

9. If the canonical decomposition of n is $n = p_1^{a_1} p_2^{a_2} \ldots p_r^{a_r}$, prove that
$$\sum_{d|n} \mu(d)\lambda\left(\frac{n}{d}\right) = (-1)^{a_1+a_2+\cdots+a_r} 2^r.$$

10. Show that
$$\sum_{k=1}^{n} k \sum_{d|k} \lambda(d) = \frac{[\sqrt{n}]([\sqrt{n}]+1)(2[\sqrt{n}]+1)}{6}.$$

11. If $n = p_1^{a_1} p_2^{a_2} \cdots p_r^{a_r}$ (p's being distinct primes), then prove that
$$1 > \frac{n}{\sigma(n)} > \left(1 - \frac{1}{p_1}\right)\left(1 - \frac{1}{p_2}\right) \cdots \left(1 - \frac{1}{p_r}\right).$$

12. Prove that an *odd* perfect number must have more than two distinct prime factors.

Liouville's Function $\lambda(n)$

13. For a positive integer n, show that
$$\sum_{d|n} \phi(d)\tau\left(\frac{n}{d}\right) = \sigma(n).$$

14. For a positive integer n, show that
$$\sum_{d|n} \sigma(d) = \sum_{d|n} \frac{n}{d}\tau(d).$$

15. For a positive integer n, show that
$$\sum_{d|n} \phi\left(\frac{n}{d}\right)\sigma(d) = n\tau(n).$$

16. If n is a positive integer, prove that
$$\sum_{d|n} d\sigma(d) = \sum_{d|n} \left(\frac{n}{d}\right)^2 \sigma(d).$$

17. If n is a positive integer, prove that
$$\sum_{d|n} \sigma(d)\sigma\left(\frac{n}{d}\right) = \sum_{d|n} d\tau(d)\tau\left(\frac{n}{d}\right).$$

(Hint: For n a power of a prime, p^a, try induction on a.)

4-10 Recurrence Formulae

We shall here briefly consider recurring sequences and formulae associated with them. The infinite sequence
$$x_1, x_2, \ldots, x_n, \ldots$$
is called a *recurring sequence* if, from a certain point on, every term can be represented as a linear combination of preceding terms of the sequence. We shall not consider the general case, but merely special cases of those sequences for which
$$x_n = a_1 x_{n-1} + \cdots + a_k x_{n-k}, \; 1 \leq k \leq n-1,$$
the k and a's being fixed. This relation is a special type of an arithmetical function, known as a *recurrence formula*. We shall, in fact, restrict ourselves to integral values throughout. If the values of $x_1, x_2, \ldots, x_{n-1}$ are known, the values of the x_i, where $i \geq n$, can be determined by suitably employing the above recurrence formula.

Let us consider the recurring sequence[11] defined by $x_0 = 0$, $x_1 = 1$, and, where a and c are relatively prime, $x_r = ax_{r-1} + cx_{r-2}$ for $r \geq 2$. That is, $x_0 = 0$, $x_1 = 1$, $x_2 = a$, $x_3 = a^2 + c$, $x_4 = a^3 + 2ac$, $x_5 = a^4 + 3a^2c + c^2$, $x_6 = a^5 + 4a^3c + 3ac^2$, $x_7 = a^6 + 5a^4c + 6a^2c^2 + c^3, \ldots$.

First, we note the following relation between binomial coefficients:

$$\binom{n-k-2}{k} + \binom{n-k-2}{k-1} = \binom{n-k-1}{k}.$$

Next, we can prove,[12] by induction on n, that for odd n, we have x_n as the sum of $(n+1)/2$ terms:

$$x_n = a^{n-1} + \binom{n-2}{1} a^{n-3} c + \binom{n-3}{2} a^{n-5} c^2 + \cdots$$

$$+ \binom{n - \frac{n-3}{2}}{\frac{n-5}{2}} a^4 c^{(n-5)/2}$$

$$+ \binom{n - \frac{n-1}{2}}{\frac{n-3}{2}} a^2 c^{(n-3)/2} + \left(\frac{n}{2}a\right)^0 c^{(n-1)/2}.$$

For even n, we have x_n as the sum of $n/2$ terms:

$$x_n = a^{n-1} + \binom{n-2}{1} a^{n-3} c + \binom{n-3}{2} a^{n-5} c^2 + \cdots$$

$$+ \binom{n - \frac{n-4}{2}}{\frac{n-6}{2}} a^5 c^{(n-6)/2} + \binom{n - \frac{n-2}{2}}{\frac{n-4}{2}} a^3 c^{(n-4)/2}$$

$$+ \left(\frac{n}{2}a\right)^1 c^{(n-2)/2}.$$

We shall now obtain the x_i expressed in a different form. Let a and c be integers such that $b = a^2 + 4c \neq 0$. Then the equation $x^2 - ax - c = 0$ has distinct solutions, α and β, such that $\alpha + \beta = a$ and $\alpha\beta = -c$. Moreover, if for $n \geq 0$, $y_n = (\alpha^n - \beta^n)/(\alpha - \beta)$, we see that

$$ay_{n+1} + cy_n = (\alpha + \beta)\frac{\alpha^{n+1} - \beta^{n+1}}{\alpha - \beta} - \alpha\beta \frac{\alpha^n - \beta^n}{\alpha - \beta}$$

$$= \frac{\alpha^{n+2} - \beta^{n+2}}{\alpha - \beta}$$

$$= y_{n+2}.$$

Also, since $y_0 = 0$, $y_1 = 1$, $y_2 = \alpha + \beta = a$, we see that $y_n = x_n$ for $n = 0, 1, 2$. Where $k \geq 1$, assume that $y_n = x_n$ for $n = 0, 1, \ldots, k$. Then, where $k \geq 1$,

$$y_{k+1} = ay_k + cy_{k-1} = ax_k + cx_{k-1} = x_{k+1}.$$

Hence, by complete mathematical induction, $y_n = x_n$ for all $n \geq 0$.

- **Theorem 4-26** For m a non-negative integer and n a positive integer,
$$x_{m+n} = cx_m x_{n-1} + x_{m+1} x_n.$$

First Proof For $n = 1$ and 2, the statement is obviously true. Assume the result true for $n = 1, 2, \ldots, k, k \geq 2$. Then, using the result for x_{m+k} and x_{m+k-1}, we have

$$\begin{aligned} x_{m+k+1} &= cx_{m+k-1} + ax_{m+k} \\ &= c(cx_m x_{k-2} + x_{m+1} x_{k-1}) + a(cx_m x_{k-1} + x_{m+1} x_k) \\ &= cx_m(ax_{k-1} + cx_{k-2}) + x_{m+1}(ax_k + cx_{k-1}) \\ &= cx_m x_k + x_{m+1} x_{k+1}. \end{aligned}$$

Hence, since the result is now true for $n = k + 1$, the induction is complete.

Second Proof Employing the α and β, we see that the right member of the proposed relation is

$$-\alpha\beta \frac{\alpha^m - \beta^m}{\alpha - \beta} \cdot \frac{\alpha^{n-1} - \beta^{n-1}}{\alpha - \beta} + \frac{\alpha^{m+1} - \beta^{m+1}}{\alpha - \beta} \cdot \frac{\alpha^n - \beta^n}{\alpha - \beta}$$
$$= \frac{(\alpha - \beta)\alpha^{m+n} - (\alpha - \beta)\beta^{m+n}}{(\alpha - \beta)^2} = \frac{\alpha^{m+n} - \beta^{m+n}}{\alpha - \beta} = x_{m+n}.$$

For the case $\alpha = \beta$, the theorem remains valid.[13]

- **Theorem 4-27** For $n \geq 1$, $(x_n, c) = 1$.

Proof For $n = 1$ and 2, the statement is obviously true. Assume the statement to be true for $n = 1, 2, \ldots, k$, where $k \geq 2$. Let $(x_{k+1}, c) = d \geq 1$. Since $d|c$ and $d|x_{k+1}$, we see from $x_{k+1} = ax_k + cx_{k-1}$ that $d|ax_k$. Since $(a, c) = 1$, $d|x_k$ and $d|c$. Hence d divides $(x_k, c) = 1$. Therefore, $d = 1$ and our induction is complete.

- **Theorem 4-28** If $x_m \neq 0$, for a positive integer k, x_m divides x_{km}; in other words, if $s|t$ and $x_s \neq 0$, then $x_s|x_t$.

Proof We shall use induction on k. The result is obvious if $k = 1$. Assume the result true for k. We desire to prove the result true for $k + 1$. Now, by Theorem 4-26, $x_{(k+1)m} = x_{km+m} = cx_{km} x_{m-1} + x_{km+1} \cdot x_m$. Since x_m divides each of the two terms of the sum on the right, x_m divides $x_{(k+1)m}$. Hence, by induction, the theorem is established.

- **Corollary** If the positive integer d divides both the positive integers p and q, then, provided $x_d \neq 0$, x_d divides both x_p and x_q.

- **Theorem 4-29** For a positive or zero integer p, x_p and x_{p+1} are relatively prime.

Proof Since $x_0 = 0$, $x_1 = 1$, and $x_2 = a$, the theorem is obviously true if $p = 0$ or 1. Let $p \geq 1$.

Assume the truth of the theorem for $p = k \geq 1$. We shall then prove the theorem for $p = k + 1$. Let $(x_{k+1}, x_{k+2}) = d \geq 1$. We know that $x_{k+2} = ax_{k+1} + cx_k$ and $x_{k+1} = ax_k + cx_{k-1}$. Since d divides x_{k+1} and x_{k+2}, $d|cx_k$. If $d|c$, then $d|ax_k$; hence $d|x_k$. Thus, in any case, d must divide x_k. Since d now divides both x_k and x_{k+1}, it divides $(x_k, x_{k+1}) = 1$. Hence $d = 1$, and the theorem follows by induction on p.

- **Theorem 4-30** If p and q are positive integers and if $d = (p, q)$, then, when $x_d \neq 0$, $(x_p, x_q) = |x_d|$.

Proof[14] Since the statement of the theorem is symmetric in p and q, let $q \geq p \geq 1$. Since $x_1 = 1$, the theorem is obviously true when $p = 1$. Hence the theorem is true for $q = 1$. We shall give a proof by complete mathematical induction on q.

Assume the theorem to be true for $q = 1, 2, \ldots, k$, that is, where $1 \leq l \leq k$, the theorem is assumed true when $q = l$ and $p = 1, 2, \ldots, l$. Let $q = k + 1$. Then, when $p = k + 1$, $(p, q) = (k + 1, k + 1) = k + 1$. Hence, since $(x_p, x_q) = (x_{k+1}, x_{k+1}) = |x_{k+1}|$, the theorem holds for $q = p = k + 1$. Next, let $1 \leq p \leq k$, $q = k + 1$, and $(k + 1, p) = d_1$, where $0 < k + 1 - p < k + 1$. Consequently, $(k + 1 - p, p) = d_1$. By the induction hypothesis,

$$(x_{k+1-p}, x_p) = |x_{d_1}|.$$

Now, by Theorem 4-26,

$$x_{k+1} = x_{(k+1-p)+p} = cx_{k+1-p}x_{p-1} + x_{k+1-p+1}x_p.$$

Then

$$(x_p, x_{k+1}) = (x_p, cx_{k+1-p}x_{p-1}).$$

Since, however, by Theorems 4-29 and 4-27, $(x_p, x_{p-1}) = (x_p, c) = 1$, we have

$$(x_p, cx_{k+1-p}x_{p-1}) = (x_p, x_{k+1-p}) = |x_{d_1}|.$$

Hence, since

$$(x_{k+1}, x_p) = |x_{d_1}|,$$

where $d_1 = (k + 1, p)$, the theorem is true for $q = k + 1$, and the induction on q is complete.

- **Corollary 1** $(x_p, x_q) = 1$ if and only if $|x_d| = 1$, where $d = (p, q)$.

Proof Assume that $(x_p, x_q) = 1$. It is readily seen (from Theorem 4-26) by induction on k that if $x_t = 0$, then $x_{kt} = 0$ for all $t \geq 0$ and all $k \geq 0$. Now, if x_d were zero, we would have $x_p = x_q = 0$. This contradicts the statement that $(x_p, x_q) = 1$. Hence $x_d \neq 0$, and by Theorem 4-30, $|x_d| = (x_p, x_q) = 1$. Conversely, if $|x_d| = 1$, $(x_p, x_q) = |x_d| = 1$.

- **Corollary 2** If $(p, q) = 1$, then $(x_p, x_q) = 1$, but not conversely (cf. $a = -c = -1$, where $x_2 = -1$, $x_4 = -3$).

4-11 Fibonacci's and Lucas' Sequences

We shall now consider the Fibonacci sequence, a special case of the preceding recurring sequence of Siebeck obtained by setting $a = c = 1$. We define $u_0 = 0$, $u_1 = 1$, $u_n = u_{n-1} + u_{n-2}$ for $n \geq 2$. The Fibonacci sequence then is given by: $u_0 = 0$, $u_1 = 1$, $u_2 = 1$, $u_3 = 2$, $u_4 = 3$, $u_5 = 5$, $u_6 = 8$, $u_7 = 13$, $u_8 = 21$, $u_9 = 34$, ... The numbers 1, 1, 2, 3, 5, 8, ..., are called *Fibonacci numbers*.[15] Some of the properties of this sequence are already known from the results given above for the more general case. Many of the properties of the Fibonacci sequence are readily established by mathematical induction.

If we were to define $y_0 = 2$, $y_1 = a$ and $y_n = ay_{n-1} + cy_{n-2}$ for $n \geq 2$, we would have the sequence

$$y_0 = 2, y_1 = a, y_2 = a^2 + 2c, y_3 = a^3 + 3ac, \ldots$$

In this case, it is readily seen that, where α and β are the solutions of $x^2 - ax - c = 0$, the y_n can be expressed in the form[16]

$$y_n = \alpha^n + \beta^n, n \geq 0.$$

If we place $a = c = 1$ and write v_i for y_i, we get the Lucas sequence $v_0 = 2$, $v_1 = 1$, $v_2 = 3$, $v_3 = 4$, $v_4 = 7$, $v_5 = 11, \ldots$, and the *Lucas numbers* v_i ($i \geq 1$).

4-12 Problems

1. For the recurring sequence given by $x_0 = 0$, $x_1 = 1$, $x_r = x_{r-1} - x_{r-2}$ for $r \geq 2$, prove the following:
   (a) $x_l = -x_{l-3}$ for $l \geq 3$;
   (b) $x_l = (-1)^{(l-a)/3} x_a$ if $l = 3t + a$;
   (c) No two consecutive x_i's can be zero.

2. Give an independent proof that two consecutive Fibonacci numbers are relatively prime.

3. For an arbitrary integer c, define $w_0 = 1$, $w_1 = w_0 + c$, $w_r = w_0 w_1 \cdot w_2 \ldots w_{r-1} + c$ for $r \geq 2$.
   (a) Show $(w_i, c) = 1$ for $i \geq 1$.
   (b) Show $(w_r, w_s) = 1$ for $r \neq s$.
   (c) Using the result in part (b), obtain another proof of the infinitude of primes.

4. For the Fibonacci sequence defined by $u_0 = 0$, $u_1 = 1$ and $u_n = u_{n-1} + u_{n-2}$ when $n \geq 2$, prove:
   (a) $u_{n+2} = 1 + \sum_{i=0}^{n} u_i$;
   (b) $u_{2m+1} = 1 + \sum_{i=0}^{m} u_{2i}$;
   (c) $u_{2m} = \sum_{i=2}^{2m+1} (-1)^{i+1} u_i$.

5. For the Fibonacci sequence, prove the following:
   (a) $u_{2m+2} = \sum_{i=0}^{m} u_{2i+1}$;
   (b) $\sum_{k=1}^{n} u_{2k} = \sum_{k=0}^{n} (n-k) u_{2k+1}$;
   (c) $\sum_{i=1}^{n} u_i^2 = u_n u_{n+1}$.

6. Prove that the nth Fibonacci number u_n is less than $(\frac{5}{3})^{n-1}$ when $n > 1$.

7. For the Fibonacci sequence given above, when n is positive, prove the following:
   (a) $2 \sum_{i=1}^{n+1} u_i u_{i-1} + 2 = u_{n+1}^2 + u_{n+2}^2 - \sum_{i=1}^{n} u_{i+1}^2$;
   (b) $2 \sum_{i=1}^{n} u_{i+1} u_i + 1 = u_{n+3}^2 - 3 u_{n+2} u_{n+1}$.
   (c) For positive integers m and s, give an independent proof that u_s divides u_{ms}.

8. For the same Fibonacci sequence with n positive, prove the following relations:
   (a) $2(u_n^2 + u_{n+1}^2) = u_{n-1}^2 + u_{n+2}^2$;
   (b) $u_m u_{m+n} - u_{m-1} u_{m+n+1} = (-1)^{m-1} u_{n+1}$ for positive values of m and n;
   (c) $u_{n+1} u_n - 3 u_n u_{n-1} + u_{n-1} u_{n-2} = (-1)^{n+1}$ for $n \geq 2$.

9. For the same Fibonacci sequence with positive n, establish the following relations:
   (a) $u_{n+1}^2 - 3 u_n^2 + u_{n-1}^2 = 2(-1)^n$;
   (b) $u_{n+3}^2 - u_{n-1}^2 = 3(u_{n+2}^2 - u_n^2)$;
   (c) $u_{n+3}^2 - u_{n+2}^2 - 4 u_{n+1}^2 - u_n^2 + u_{n-1}^2 = 0$.

10. For the Fibonacci sequence with $n \geq 1$, prove the following relations:
 (a) $u_n^2 - u_{n+1}u_{n-1} = (-1)^{n+1}$;
 (b) $u_{n+1}^3 - 4u_n^3 - u_{n-1}^3 = 3(-1)^n u_n$;
 (c) $u_{m+n} - (u_{n-1} + u_{n+1})u_m + (-1)^n u_{m-n} = 0$
   for $m \geq n$.

11. For the Fibonacci sequence $u_0 = 0$, $u_1 = 1$, $u_n = u_{n-1} + u_{n-2}$ when $n \geq 2$, prove the following relations:
 (a) $u_n^2 + u_{n+1}^2 = u_{2n+1}$;
 (b) $u_n^2 - u_{n+l}u_{n-l} = (-1)^{n+l}u_l^2$.

12. For the Fibonacci sequence $u_0 = 0$, $u_1 = 1$, $u_n = u_{n-1} + u_{n-2}$ when $n \geq 2$, and the Lucas sequence $v_0 = 2$, $v_1 = 1$, $v_n = v_{n-1} + v_{n-2}$ when $n \geq 2$, prove the following results for positive values of n:
 (a) $v_n = u_{n-1} + u_{n+1}$;
 (b) $v_n = \dfrac{u_{2n}}{u_n}$;
 (c) $(u_n, v_n) = 1$ or 2;
 (d) $v_{4n} = v_{2n}^2 - 2$;
 (e) $v_{4n+2} = v_{2n+1}^2 + 2$.

13. For the same Fibonacci sequence $u_i (i \geq 0)$ and the Lucas sequence $v_i (i \geq 0)$ as above, establish the following results:
 (a) $v_n^2 - v_{n-1}v_{n+1} = (-1)^n 5$;
 (b) $u_n v_{n+k} - u_{n+k} v_n = 2(-1)^{n+1} u_k$ for $k \geq 0$.

14. Where $(a, c) = 1$ and $a^2 + 4c \neq 0$, consider the previous sequences of x's and y's defined as follows:

$$x_0 = 0, \ x_1 = 1, \ x_r = ax_{r-1} + cx_{r-2} \text{ for } r \geq 2;$$
$$y_0 = 2, \ y_1 = a, \ y_r = ay_{r-1} + cy_{r-2} \text{ for } r \geq 2.$$

For these sequences, prove the following statements:
 (a) If x_n is odd, $(x_n, y_n) = 1$;
 (b) If x_n is even, $(x_n, y_n) = 2$.

Primitive Roots and Indices   5

5-1 Belonging to an Exponent

Let integers a and m be relatively prime. Then, assuming $m > 1$, we know that $a^{\phi(m)} \equiv 1 \pmod{m}$.

DEFINITION *If, when $(a, m) = 1$, e is the least positive integer such that $a^e \equiv 1 \pmod{m}$, we say that a belongs to exponent e, modulo m, or that the positive number e is the exponent to which a belongs, modulo m.*

For example, if $m = 14$, 3 belongs to exponent 6 and 13 belongs to exponent 2, modulo 14.

- **Theorem 5-1** *If $(a, m) = 1$, the exponent to which a belongs, modulo m, is a divisor of $\phi(m)$.*

Proof Let a belong to exponent e, modulo m, and let $\phi(m) = eq + r$, $0 \leqq r < e$.

Then $1 \equiv a^{\phi(m)} \equiv a^{eq+r} \equiv (a^e)^q a^r \equiv a^r \pmod{m}$. By the definition of e, r cannot be positive. Hence $r = 0$ and the theorem follows.

Example 1 If $m = 12$ and $a = 7$, then, since $7 \not\equiv 1 \pmod{12}$ and $7^2 \equiv 1 \pmod{12}$, 7 belongs to exponent 2, modulo 12. Moreover, since $\phi(12) = 4$, 2 divides $\phi(12)$.

Example 2 For an arbitrary divisor e of $\phi(m)$, there is not always a number a belonging to this e as exponent. For instance, if $m = 36$, $\phi(36) = 12$. There is, however, no number belonging to exponent 4 or to exponent 12, modulo 36. If, though, the modulus m is a prime, we shall see that there is a number a belonging to an arbitrary divisor e of $\phi(m)$ as exponent.

- **Theorem 5-2** If a belongs to exponent e, modulo m, then
$$a^s \equiv a^t \pmod{m}$$
if and only if $s \equiv t \pmod{e}$.

  Proof Let $a^s \equiv a^t \pmod{m}$. Then, assuming $s \geq t$, we have $a^{s-t} a^t \equiv a^t \pmod{m}$; thus, $a^{s-t} \equiv 1 \pmod m$. Set $s - t = eq + r$, where $0 \leq r < e$. Then $1 \equiv a^{s-t} \equiv a^{eq+r} \equiv (a^e)^q a^r \equiv a^r \pmod{m}$. Therefore, since r cannot be positive, $r = 0$ and $s \equiv t \pmod{e}$.

  Conversely, if, where $s \geq t \geq 0$,
$$s \equiv t \pmod{e},$$
then
$$s - t = qe,$$
and
$$a^{s-t} \equiv (a^e)^q \equiv 1 \pmod{m}.$$
Therefore, multiplying both members by a^t, we obtain
$$a^s \equiv a^t \pmod{m}.$$

- **Corollary** $a^s \equiv 1 \pmod{m}$ if and only if $e|s$.

  Proof Take $t = 0$ in the above theorem.

Note that Theorem 5-1 is a special case of the present theorem when $s = \phi(m)$ and $t = 0$.

- **Theorem 5-3** If a belongs to exponent e and if b belongs to exponent f, modulo m, then ab belongs to exponent ef, modulo m, provided that e and f are relatively prime.

Belonging to an Exponent

Proof Let ab belong to exponent g, modulo m. Then
$$1 \equiv (ab)^{ge} \equiv (a^e)^g b^{ge} \equiv b^{ge} \pmod{m}.$$
Hence, by Theorem 5-2, Corollary, $f|ge$. Since $(e, f) = 1$, $f|g$. Similarly,
$$1 \equiv (ab)^{gf} \equiv a^{gf}(b^f)^g \equiv a^{gf} \pmod{m}.$$
Thus, $e|gf$ and $e|g$. Therefore, $ef|g$ and $g \geq ef$. But
$$(ab)^{ef} \equiv (a^e)^f (b^f)^e \equiv 1 \pmod{m}.$$
Hence, $g|ef$ and $g \leq ef$. Consequently $g = ef$.

Example Since 2 belongs to exponent 3, and 6 belongs to exponent 2, modulo 7, 12 (that is, 5) belongs to exponent 6, modulo 7.

- **Theorem 5-4** *If a belongs to exponent e, modulo m, then $a, a^2, a^3, \ldots, a^e$ are incongruent, modulo m.*

Proof If $a^s \equiv a^t \pmod{m}$, where $1 \leq s \leq e$ and $1 \leq t \leq e$, then by Theorem 5-2, $e|(s-t)$. This is impossible if $s \neq t$. Hence, the e numbers given in the theorem are incongruent, modulo m.

- **Theorem 5-5** *If p is a prime, there are precisely $\phi(e)$ incongruent numbers, modulo p, which belong to an arbitrary divisor e of $p-1$ as exponent, modulo p.*

For example, if $p = 13$, then $e = 1, 2, 3, 4, 6$, or 12. The number 1 belongs to exponent 1; the number 12 to exponent 2; the numbers 3 and 9 to exponent 3; the numbers 5 and 8 to exponent 4; the numbers 4 and 10 to exponent 6; the numbers 2, 6, 7, 11 to exponent 12.

Proof If $e = 1$, there is only the number 1 belonging to exponent 1, modulo p.

Next, let $e > 1$ be a power of a prime: $e = q^r$, where $r \geq 1$. Then by Theorem 3-20, Corollary 2,

$$(5\text{-}1) \qquad x^{q^r} \equiv 1 \pmod{p}$$

has q^r distinct solutions, modulo p. Let the solution x_i belong to the exponent e_i. Now by Theorem 5-2, Corollary, $e_i|q^r$, and, therefore, e_i is a power of q. If $e_i = q^s < q^r$, then x_i, being a solution of $x^{q^s} \equiv 1 \pmod{p}$, is also a solution of $(x^{q^s})^{q^{r-s-1}} \equiv 1 \pmod{p}$; that is, it is a solution of $x^{q^{r-1}} \equiv 1 \pmod{p}$. There being q^{r-1} distinct solutions of this last congruence, all of which are obviously solutions of (5-1), there are $q^r - q^{r-1} = \phi(q^r)$ solutions of (5-1) which are not solutions of $x^{q^{r-1}} \equiv 1 \pmod{p}$. Hence, the number of (incongruent) solutions of (5-1) belonging to exponent $e = q^r$, modulo p, is $\phi(q^r)$.

Next, consider an arbitrary divisor e (greater than unity) of $p - 1$ and write e in the canonical form

$$e = q_1^{b_1} \ldots q_k^{b_k},$$

as the product of powers of distinct primes. By Theorem 5-3 and the first case just considered, we see that there exists an integer a belonging to exponent e. Moreover, by Theorem 5-4, the e integers $a, a^2, \ldots, a^e$ are incongruent, modulo p. Hence, since $x^e \equiv 1 \pmod{p}$ has exactly e incongruent solutions, modulo p, these integers give *all* the solutions of the congruence

$$x^e \equiv 1 \pmod{p}.$$

We shall now show that a solution a^t of this congruence belongs to exponent e if and only if $(t, e) = 1$. For, if $(t, e) = 1$ and if a^t belongs to exponent g, then $e | tg$ and $e | g$; but, since $(a^t)^e \equiv (a^e)^t \equiv 1 \pmod{p}$, $g | e$; whence $g = e$.

If $(t, e) = d > 1$, then $(a^t)^{e/d} \equiv (a^e)^{t/d} \equiv 1 \pmod{p}$. Consequently, a^t belongs to an exponent $\leq e/d < e$. Therefore, if a^t belongs to exponent e, modulo p, $(t, e) = 1$. Hence Theorem 5-5 is established.

- **Theorem 5-6** For a positive integer t, if the highest power of the odd prime p dividing $a^t - 1$ is p^n, where $n \geq 1$, then the highest power of p dividing $a^{tp} - 1$ is p^{n+1}.

Proof First, write $a^t = 1 + hp^n$, where $(h, p) = 1$. Then, using the binomial theorem, we obtain

$$\begin{aligned}
a^{tp} &= (1 + hp^n)^p \\
&= 1 + hp^{n+1} + \frac{p-1}{2} h^2 p^{2n+1} + \frac{p(p-1)(p-2)}{1 \cdot 2 \cdot 3} h^3 p^{3n} + \cdots \\
&\quad + h^p p^{pn},
\end{aligned}$$

the coefficients being integers.

Since $2n + 1 > n + 1$ and $kn > n + 1$ for $3 \leq k \leq p$, $a^{tp} = 1 + p^{n+1}(h + Mp)$ for a suitable integer M. Consequently, p^{n+1} is the highest power of p dividing $a^{tp} - 1$.

The converse of Theorem 5-6 is also true.

- **Theorem 5-7** If, for positive integers t and n, the highest power of the odd prime p dividing $a^{tp} - 1$ is p^{n+1}, then the highest power of p dividing $a^t - 1$ is p^n.

Proof First, we see that p does divide $a^t - 1$. For, let $a^t \equiv b \pmod{p}$, where $1 \leq b < p$. Then $1 \equiv a^{tp} \equiv b^p \pmod{p}$; but, by Fermat's theorem, $b^p \equiv b \pmod{p}$. Hence $b \equiv 1 \pmod{p}$ and $a^t \equiv 1 \pmod{p}$.

Belonging to an Exponent

If the highest power of p dividing $a^t - 1$ is p^s, $s \geq 1$, then, from Theorem 5-6 above, the highest power of p dividing $a^{tp} - 1$ is p^{s+1}. But we are given that this highest power of p is p^{n+1}. Hence, $s = n$ and the result is established.

Theorem 5-8 If, for positive integers n, t and k, p^{n+k} is the highest power of the odd prime p dividing $a^{tp^k} - 1$, then p^n is the highest power of p dividing $a^t - 1$.

Proof For $k = 1$, the result is true by Theorem 5-7. Assuming the truth of the theorem for a fixed value k, we desire to establish the result for $k + 1$. Now, if p^{n+k+1} is the highest power of p dividing $a^{tp^{k+1}} - 1 = a^{(tp^k)p} - 1$, we know, by Theorem 5-7, that the highest power of p dividing $a^{tp^k} - 1$ is p^{n+k}. Hence, by our assumption of the truth of the theorem for k, we see that the highest power of p dividing $a^t - 1$ is p^n. Hence our theorem is established by induction on k.

Theorem 5-9 If, for positive integers t, k, and n, the highest power of the odd prime p dividing $a^t - 1$ is p^n, then the highest power of p dividing $a^{tp^k} - 1$ is p^{n+k}.

Proof Since, by Theorem 3-4,

$$(a^t)^{p^k} \equiv 1 \pmod{p^{n+k}},$$

the highest power of p dividing $a^{tp^k} - 1$ is p^{n+s}, where $s \geq k$. Then, by Theorem 5-8, the highest power of p dividing $a^t - 1$ is p^{n+s-k}. This was, however, given as p^n. Hence $s = k$, and the result follows.

Example 1 If p is an odd prime and if a and $-a$ belong to exponents e and f, respectively, modulo p, prove that $e = 2f$ when f is odd, $f = 2e$ when e is odd, and $e = f$ when both e and f are even.

Proof We have $a^e \equiv 1 \pmod{p}$ and $(-a)^f \equiv 1 \pmod{p}$; thus $a^f \equiv (-1)^f \pmod{p}$. Consequently, since $a^{2f} \equiv 1 \pmod{p}$, e divides $2f$.

Let f be odd. If e were to be odd, we would have $e|f$ and $(-a)^e \equiv (-1)^e a^e \equiv -1 \pmod{p}$. Hence, since $(-a)^{2e} \equiv 1 \pmod{p}$, $f|2e$ and $f|e$. Consequently, $e = f$. This contradicts the result that $a^f \equiv (-1)^f \pmod{p}$. We cannot have e and f both odd; consequently, e must be even when f is odd. Therefore, if $e = 2e_0$, we have $2e_0|2f$ and $e_0|f$. Since $0 \equiv a^{2e_0} - 1 \equiv (a^{e_0} + 1)(a^{e_0} - 1) \pmod{p}$, $a^{e_0} \equiv -1 \pmod{p}$; from this we have $(-a)^{e_0} \equiv (-1)^{e_0} a^{e_0} \equiv (-1)^{e_0+1} \pmod{p}$. Since e_0 is odd, $(-a)^{e_0} \equiv 1 \pmod{p}$ and $f|e_0$. Therefore, $e_0 = f$ and $e = 2f$.

Next, let f be even; that is, $f = 2f_0$. Consequently, $(-a)^{2f_0} \equiv a^{2f_0} \equiv 1 \pmod{p}$ and $e|2f_0$. If e were odd, $e|f_0$ and $(-a)^e \equiv (-1)^e a^e \equiv -1 \pmod{p}$; thus $(-a)^{2e} \equiv 1 \pmod{p}$, $f|2e$, and $f_0|e$. Consequently, since $e|f_0$ and $f_0|e$, $e = f_0$ and $f = 2e$. If, on the other hand, e is even, $(-a)^e \equiv a^e \equiv 1 \pmod{p}$ and also $1 \equiv (-a)^f \equiv a^f \pmod{p}$. Hence, since $f|e$ and $e|f$, $e = f$.

Example 2 Let a and b be relatively prime and, where $m > 1$, let ab and m be relatively prime. Let a^k be the least positive power of a congruent, modulo m, to a positive power of b. Prove that k divides the exponent to which a belongs, modulo m.

Proof Since $a^{\phi(m)} \equiv 1 \equiv b^{\phi(m)} \pmod{m}$, there is a positive power of a congruent to a positive power of b. Let a and b belong, respectively, to exponents e and f, modulo m; moreover, let

(5-2) $$a^k \equiv b^l \pmod{m},$$

where both k and l are positive. Divide e and f by k and l, respectively:

$$e = kq_1 + r_1, \ 0 \leq r_1 < k; \ f = lq_2 + r_2, \ 0 \leq r_2 < l.$$

It is obvious that $1 \leq k \leq e$; moreover, assume that $1 \leq l \leq f$. Raising both sides of (5-2) to the power q_1 and then multiplying through by a^{r_1}, we have

(5-3) $$a^{kq_1} a^{r_1} \equiv b^{lq_1} a^{r_1} \pmod{m}.$$

Hence,

$$1 \equiv a^e \equiv a^{kq_1 + r_1} \equiv b^{lq_1} a^{r_1} \pmod{m}.$$

Assume that $r_1 > 0$. Since $1 \equiv b^{lq_2 + r_2} \pmod{m}$,

(5-4) $$b^{lq_2 + r_2} \equiv b^{lq_1} a^{r_1} \pmod{m}.$$

Now,

$$b^{lq_2 + r_2} \equiv b^{lq_1} \cdot b^{l(q_2 - q_1) + r_2} \equiv b^{lq_1} \cdot a^{r_1} \pmod{m}.$$

If $q_2 > q_1$, we have, on dividing each side by b^{lq_1}, $b^{l(q_2 - q_1) + r_2} \equiv a^{r_1} \pmod{m}$. Since $0 < r_1 < k$, this contradicts the definition of k.

If $q_2 = q_1$, then, on dividing each side of (5-4) by b^{lq_1}, we obtain

$$b^{r_2} \equiv a^{r_1} \pmod{m}.$$

By the definition of k, we must have $r_2 = 0$ and $1 \equiv a^{r_1} \pmod{m}$. Hence $e|r_1$. The positive number e cannot be less than r_1, since q_1 (and also q_2 and f) would then be negative. Hence $e = r_1$. However, $e = r_1 < k$; nevertheless, $k \leq e = r_1$. Again we have a contradiction.

Finally, if $q_2 < q_1$, we obtain from (5-2)

$$a^{kq_1} \equiv 1 \cdot b^{lq_1} \pmod{m},$$
$$a^{kq_1} \equiv a^{kq_1 + r_1} \cdot b^{lq_1} \pmod{m}.$$

Dividing through by a^{kq_1}, we obtain $1 \equiv a^{r_1} b^{lq_1} \pmod{m}$. Take s sufficiently large so that $sf = s(lq_2 + r_2) > lq_1$. Since $1 \equiv b^{fs} \equiv b^{(lq_2+r_2)s}$ (mod m), we have

$$b^{(lq_2+r_2)s-lq_1+lq_1} \equiv a^{r_1}b^{lq_1} \pmod{m}.$$

Dividing through by b^{lq_1}, we obtain

$$b^{(lq_2+r_2)s-lq_1} \equiv a^{r_1} \pmod{m}.$$

Since $0 < r_1 < k$ and the exponent of b is positive, we again have a contradiction. Hence $r_1 = 0$, and $k|e$.

5-2 Problems

1. Find the exponents to which 11 and 19 belong, modulo 36.

2. If a belongs to exponent uv, modulo m, prove that a^u belongs to exponent v, modulo m.

3. If a belongs to exponent e modulo m, show that a^s, $s \geq 1$, belongs to exponent e/d, where $d = (s, e)$.

4. (a) If a belongs to an even exponent $2k$, modulo an odd prime p, show that $a^k \equiv -1 \pmod{p}$.

   (b) Give an example of a composite modulus for which the statement in part (a) is untrue.

5. For a prime p exceeding 3, show that there is an integer a, where $1 < a < p - 1$, such that some positive power of a is congruent to -1, modulo p.

6. Let $m > 2$, and let some positive power of a be congruent to -1, modulo m. If a^s is the least such positive power, prove that a belongs to exponent $2s$, modulo m.

7. Let p be an odd prime.

   (a) If a belongs to exponent 3, modulo p, prove that $a + 1$ belongs to exponent 6, modulo p.

   (b) Conversely, if $a + 1$ belongs to the exponent 6, modulo p, prove that a belongs to exponent 3, modulo p.

8. Let $k = (m_1, m_2)$, and let a belong to exponents e_1 and e_2, modulo m_1 and modulo m_2, respectively. Prove that, if $d = (e_1, e_2)$, a belongs to exponent e_1e_2/d, modulo m_1m_2/k.

9. Prove that all odd prime factors of the number $n^2 + 1$ are of the form $4M + 1$.

10. Prove that all odd prime factors of the number $n^4 + 1$ are of the form $8M + 1$.

11. Show that there exists an infinitude of primes of each of the forms $4M + 1$ and $8M + 1$.

12. If p is an odd prime, prove that the prime factors of $a^{p-1} + a^{p-2} + \ldots + a^2 + a + 1$ are either p or prime numbers of the form $2pM + 1$. (Hint: Note that when q is an odd prime, the congruence $a^p \equiv 1 \pmod{q}$ implies that a belongs to exponent 1 or to exponent p, modulo q.)

13. If p is an odd prime, prove that all prime factors of $2^p - 1$ are of the form $2pM + 1$.

14. Where p is an odd prime, show that there is an infinitude of primes of the form $2pM + 1$.

5-3 Primitive Roots

We have seen that every number relatively prime to m belongs to some exponent, modulo m.

DEFINITION *If $(a, m) = 1$ and if a belongs to exponent $\phi(m)$, modulo m, then a is called a primitive root of m.*

From Example 2 at the beginning of Section 5-1, we see that 36 does not have a primitive root. Consequently, not every positive integer has a primitive root. By Theorem 5-5, however, with $e = p - 1$, we see that a prime does have a primitive root. We shall now determine integers which do not have a primitive root.

Consider $m = 2^r$, where $r \geq 1$. For an odd integer a, $a^2 \equiv 1 \pmod{2^3}$. Thus, by Theorem 3-4, $(a^2)^{2^{r-3}} \equiv a^{2^{r-2}} \equiv 1 \pmod{2^r}$ when $r \geq 3$. Hence, when $r \geq 3$,

$$a^{\frac{1}{2}\{\phi(2^r)\}} \equiv 1 \pmod{2^r},$$

and consequently, there are no primitive roots of 2^r. The integers 2 and 2^2, however, have the primitive roots 1 and 3, respectively.

Next,[1] let $m = p_1^{a_1} \ldots p_r^{a_r}$ be the canonical form of m. Now, $\phi(p_i^{a_i}) = p_i^{a_i-1}(p_i - 1)$. This is even if p_i is an odd prime or if $p_i = 2$ and $a_i \geq 2$. In other words, $\phi(p_i^{a_i})$ is even if $p_i^{a_i} > 2$. Let l denote the least common multiple of $\phi(p_1^{a_1}), \ldots, \phi(p_r^{a_r})$. By Euler's theorem,

$$a^{\phi(p_i^{a_i})} \equiv 1 \pmod{p_i^{a_i}};$$

and

$$a^l \equiv 1 \pmod{p_i^{a_i}}, \text{ for } i = 1, 2, \ldots, r.$$

Hence, $a^l \equiv 1 \pmod{m}$. Consequently, if any two of the $p_i^{a_i}$ exceed 2, the least

Primitive Roots

common multiple of $\phi(p_1^{a_1}), \ldots, \phi(p_r^{a_r})$ is less than their product $\phi(m)$. Thus, in this case, there can be no primitive roots of m. If no two of the $p_i^{a_i}$ exceed 2, then m must be either a power of a prime p or the double of a power of an odd prime. Hence, m cannot have a possible primitive root unless it is 2, 2^2, p^n, or $2p^n$, where $n \geq 1$ and p is an odd prime.

- **Theorem 5-10** There exist $\phi(p-1)$ primitive roots of a prime p.

    Proof This result follows from Theorem 5-5, with $e = p - 1$.

    For example, for $p = 7$, $\phi(6) = 2$. The primitive roots of 7 are 3 and 5.

- **Theorem 5-11** Let p be an odd prime[2] and let a belong to exponent e, modulo p. Then, if p^l, where $0 < l \leq n$, is the highest power of p dividing $a^e - 1$, a belongs to the exponent ep^{n-l}, modulo p^n.

    Proof Now, from $a^e \equiv 1 \pmod{p^l}$, we obtain, by Theorem 3-4,
    $$a^{ep^{n-l}} \equiv 1 \pmod{p^n}.$$
    Assume that $0 < l < n$. Hence, the exponent to which a belongs, modulo p^n, must be a multiple of e and a divisor of ep^{n-l}. Suppose that a belongs to exponent ep^k, where $0 \leq k < n - l$, modulo p^n. We shall prove this supposition to be false, and consequently, that a belongs to exponent ep^{n-l}, modulo p^n.

    Since $a^{ep^k} - 1$ is divisible by p^n, it follows by Theorem 5-8 that $a^e - 1$ is divisible by p^{n-k}. But, since $n - k > l$, we have a contradiction of the requirement that p^l be the highest power of p dividing $a^e - 1$. Hence, we must have $k = n - l$.

    Considering the case $l = n$, let a belong to exponent $d \leq e$, modulo p^n. Then, since $a^d \equiv 1 \pmod{p}$, $d \geq e$. Consequently, a belongs to exponent e, modulo p^n.

Consider now a converse of the above theorem.

- **Theorem 5-12** Let p be an odd prime and let a belong to exponent e, modulo p. If a belongs to the exponent ep^{n-l}, modulo p^n, where $0 < l < n$, then $a^e - 1$ is divisible by p^l but not by p^{l+1}.

    Proof Now, p^n is the highest power of p dividing $a^{ep^{n-l}} - 1$. For, if p^{n+t} ($t > 0$) is the highest power of p dividing this expression, then, by Theorem 5-7, p^{n+t-1} is the highest power of p dividing $a^{ep^{n-l-1}} - 1$. That is, a belongs to an exponent $\leq ep^{n-l-1}$, modulo p^n. This is a contradiction. Hence, by Theorem 5-8, $p^{n-(n-l)} = p^l$ is the highest power of p dividing $a^e - 1$.

If a belongs to exponent e modulo an odd prime p, we have $a^e \equiv 1 \pmod{p}$. There are instances where we have at the same time $a^e \equiv 1 \pmod{p^2}$

for $1 < a < p$. For example, 3 belongs to exponent 5 (mod 11) and also $3^5 \equiv 1$ (mod 11^2); moreover, 9 belongs to exponent 5 (mod 11) and $9^5 \equiv 1$ (mod 11^2); 252 is found[3] to belong to exponent 12, modulo the prime 997, but also $252^{12} \equiv 1$ (mod 997^2). Moreover, 2 belongs[4] to exponent 364, modulo the prime 1093, and $2^{364} \equiv 1$ (mod 1093^2). This means, of course, that $2^{p-1} \equiv 1$ (mod p^2) holds for the prime $p = 1093$. It has been established[5] that the only primes not exceeding 200183 for which this last congruence is true are $p = 1093$ and $p = 3511$. Consequently, these examples lend significance to the following theorem.

- **Theorem 5-13** If p is an odd prime and if e is an arbitrary divisor $(e > 1)$ of $p - 1$, then there exists a positive integer a belonging to exponent e, modulo p, such that p is the highest power of p dividing $a^e - 1$.

    Proof Let c be any one of the $\phi(e)$ numbers belonging to exponent e $(e > 1)$, modulo p. Assume that $1 < c < p$. Then $e | (p - 1)$ and $c^e = 1 + Mp^s$, where $s \geq 1$ and $(M, p) = 1$. Since $(ce, p) = 1$, there exists a unique solution x, modulo p, of the congruence

    $$ec^{e-1}x \equiv -Mp^{s-1} \pmod{p}.$$

    Let w denote any one of the $p - 1$ incongruent numbers, modulo p, which do not satisfy this last congruence. Then $a = c + wp$ satisfies $a^e \equiv 1$ (mod p) and $a^e \not\equiv 1$ (mod p^2). For, by the binomial theorem,

    $$a^e \equiv (c + wp)^e \equiv c^e + ec^{e-1}wp \pmod{p^2}$$
    $$\equiv 1 + (Mp^{s-1} + ec^{e-1}w)p \pmod{p^2}$$
    $$\not\equiv 1 \pmod{p^2}.$$

    Since $(ec^{e-1}M, p) = 1$, we see that when $s > 1$, $a = c + p$ and $a = c + Mp$ are numbers belonging to exponent e, modulo p, such that p^2 does not divide $a^e - 1$.

- **Theorem 5-14** If p is an odd prime and n a positive integer, there exists a primitive root of p^n.

    Proof By Theorem 5-13 above we see that there exist integers a belonging to exponent e, modulo p, such that p^2 does not divide $a^e - 1$. Hence, using Theorem 5-11 with $e = p - 1$ and $l = 1$, we see that a is a primitive root of p and belongs to exponent $p^{n-1}(p - 1) = \phi(p^n)$, modulo p^n. Hence, this a is a primitive root of p^n.

- **Theorem 5-15** For an odd prime p and a positive integer n, a primitive root r of p^n is a primitive root of p.

    Proof If r belongs to exponent e, modulo p, then $r^{ep^{n-1}} \equiv 1$ (mod p^n); but r belongs to exponent $(p - 1)p^{n-1}$, modulo p^n. Hence

$(p-1)p^{n-1}|ep^{n-1}$ and, since $e \leq p-1$, $e = p-1$. That is, r is a primitive root of p.

We now state an important theorem on primitive roots.

- **Theorem 5-16** There exist primitive roots of m if and only if m is 2, 4, p^n, or $2p^n$, where p is an odd prime and n a positive integer.

   Proof We have already seen that a number not of one of these forms cannot have a primitive root. Previously we saw that 2 and 4 have primitive roots and, by Theorem 5-14, that p^n has a primitive root. Next, let r be a primitive root of p^n. Let g denote that one of the two numbers r and $r + p^n$ which is odd. Then, if g belongs to exponent e modulo $2p^n$, e divides $\phi(2p^n) = \phi(p^n) = p^{n-1}(p-1)$. But, since g is a primitive root of p^n and since $g^e \equiv 1 \pmod{p^n}$, $\phi(p^n) = p^{n-1}(p-1)$ divides e. Hence, $e = p^{n-1}(p-1)$; and g, therefore, is a primitive root of $2p^n$.

5-4 Obtaining Primitive Roots

The problem of finding primitive roots of an odd prime p is reduced, in a greater or lesser degree, to trial and error. Gauss[6] proposed a method of obtaining a sequence of integers belonging to higher and higher exponents. Hence, employing his method, one eventually attains to the exponent $p - 1$, and thereby obtains a primitive root. Knowing one primitive root r, we obtain all $\phi(p-1)$ of them by taking the least positive residues of r^k, where $1 \leq k \leq p-1$ and $(k, p-1) = 1$.

Following Gauss's method, let us take any number a, greater than one and less than p: in practice, 2 being the smallest such number, it is usually the most suitable number with which to start. Let a belong to the exponent e (a divisor of $p-1$), modulo p. Consider the least positive residues of the positive powers of a. If $e = p - 1$, a is then a primitive root of p; if $e < p - 1$, consider next a number b, where $1 < b < p$, not congruent to a positive power of a. Let b belong to exponent f, modulo p. If $f = p - 1$, b is a primitive root. If $f < p - 1$, we consider the two possibilities: first, $f \not\equiv 0 \pmod{e}$; second $f \equiv 0 \pmod{e}$.

Consider the first case, $f \not\equiv 0 \pmod{e}$. Let m be the least common multiple of e and f. We shall see that m can now be written $m = e_1 f_1$, where $e_1 | e$, and $f_1 | f$, and $(e_1, f_1) = 1$. For, let q^k be the highest power of an arbitrary prime q dividing m. Then q^k must be a divisor of one or both of e and f. If q^k divides e but not f, let it be a factor of e_1; if q^k divides f but not e, let it be a factor of f_1; if it divides both e and f, let it be taken arbitrarily as a factor of *either* e_1 or f_1. Hence, m can be written as the product of the relatively prime integers e_1 and f_1. Writing $e = e_1 e_2$ and $f = f_1 f_2$, we see that a^{e_2} and b^{f_2} belong to the exponents e_1 and f_1, respectively, modulo p. Hence, $a^{e_2} b^{f_2}$ belongs, modulo p, to the exponent $e_1 f_1 = m$. Since e has a factor not dividing f, $m > f$. Moreover, since $m \geq e$, we shall see that $m > e$. For, when $m = e$, obviously $f|e$. Therefore, in the case where $f = e = m$, we would have a contradiction of the condition $f \not\equiv 0 \pmod{e}$. Next, in the case where $f|e$ and $f < e$, all the incongruent solutions $b, b^2, \ldots, b^f$ of

(5-5) $$x^f \equiv 1 \pmod{p}$$

are solutions also of

(5-6) $$x^e \equiv 1 \pmod{p}.$$

But, all solutions of (5-6) are the e incongruent numbers $a, a^2, \ldots, a^e$. Hence, b is congruent, modulo p, to some power of a. This contradicts the definition of b. Consequently, it follows that $m > e$. Thus, we have found a number $a^{e_2} b^{f_2}$ belonging, modulo p, to a higher exponent than that to which either a or b belongs.

Consider next the case $f \equiv 0 \pmod{e}$. Here $f \geq e > 1$. Now, (5-5) has precisely the f incongruent solutions

$$b, b^2, \ldots, b^f.$$

But, each of the e incongruent solutions of (5-6), (namely, $a, a^2, \ldots, a^e$) is a solution of (5-5). If $f = e$, then b, being a solution of (5-6), is a power of a. This contradicts our condition on the choice of b. Hence $f > e$; and we have found an integer b belonging to a higher exponent than that to which a belongs.

Proceeding in this way, we can obtain a sequence of integers belonging to higher and higher exponents. Eventually, we shall obtain an integer belonging to the maximum exponent $p - 1$, modulo p. Certain expedients can often be employed to shorten the work of arriving at a primitive root of p.

Example   Let $p = 71$. We desire to obtain a primitive root of 71.

By trial, we find that 2 belongs, modulo 71, to the exponent 35. The least positive residues, modulo 71, of the successive powers of 2 are:

```
2,  4,  8, 16, 32, 64, 57, 43, 15, 30
60, 49, 27, 54, 37,  3,  6, 12, 24, 48
25, 50, 29, 58, 45, 19, 38,  5, 10, 20
40,  9, 18, 36,  1.
```

We see that the smallest positive number not in this sequence is 7. Using

$b = 7$, we find by trial that 7 belongs to exponent 70, and therefore 7 is a primitive root of 71.

The work could be shortened by noting that $70 \equiv -1 \pmod{71}$ and 70 belongs to exponent 2. Since 2 belongs to exponent 35 and $(2, 35) = 1$, $(-1) \cdot 2$ (which is congruent to 69) belongs to exponent $2 \cdot 35 = 70$. Consequently, 69 is a primitive root of 71. We could also have selected a number among the positive residues of the powers of 2 which is near the value 71, such as $64 \equiv 2^6 \pmod{71}$. Since $(6, 35) = 1$, 64 belongs to exponent 35. Hence $(-1) \cdot 64$ belongs to exponent $2 \cdot 35$. Hence, since $-64 \equiv 7 \pmod{71}$, 7 is a primitive root of 71. Moreover, since $58 \equiv 2^{24} \pmod{71}$ and $(24, 35) = 1$, $(-1) \cdot 58$ belongs to exponent $2 \cdot 35 = 70$. Hence 13 is also a primitive root of 71. Also, since $43 \equiv 2^8 \pmod{71}$ and $(8, 35) = 1$, $(-1) \cdot 43$ belongs to exponent $2 \cdot 35$. Consequently, 28 is a primitive root of 71.

5-5 Sum of Numbers Belonging to an Exponent

• **Theorem 5-17** If p is an odd prime, the sum[7] of all (incongruent) numbers belonging to exponent d ($d > 1$), modulo p, is divisible by p when d is divisible by the square of a prime, but is congruent to $(-1)^s$, modulo p, if d is the product of s distinct primes.

Example For $p = 13$, the numbers belonging to exponent $d = 3$ are 3 and 9, those belonging to exponent $d = 4$ are 5 and 8, those belonging to exponent $d = 6$ are 4 and 10, and those belonging to exponent $d = 12$ are the primitive roots 2, 6, 7, 11. Moreover, $3 + 9 \equiv 12 \equiv (-1)^1 \pmod{13}$; $5 + 8 \equiv 13 \equiv 0 \pmod{13}$; $4 + 10 \equiv 14 \equiv (-1)^2 \pmod{13}$; $2 + 6 + 7 + 11 \equiv 26 \equiv 0 \pmod{13}$.

Proof Where the q's are distinct primes and the b's are positive integers, let $d = q_1^{b_1} \ldots q_s^{b_s}$ be a divisor of $\phi(p) = p-1$, and let r belong to exponent d, modulo p. Moreover, write $d = q_i^{b_i} d_i$, $i = 1, 2, \ldots, s$. By the Chinese Remainder Theorem, determine non-negative integers f_i so that

$$f_i \equiv 1 \pmod{q_i^{b_i}}$$

(5-7) and

$$f_i \equiv 0 \pmod{d_i}.$$

Hence, $(f_i, q_i^{b_i}) = 1$ and $f_i = M_i d_i$ for $i = 1, 2, \ldots, s$. Now $f_1 + f_2 + \ldots + f_s \equiv 1 \pmod{q_i^{b_i}}$, for $i = 1, 2, \ldots, s$. Hence, $f_1 + f_2 + \ldots + f_s \equiv 1 \pmod{d}$. We now see that r^{f_i} belongs to exponent $q_i^{b_i}$, modulo p. For, supposing r^{f_i} belongs to exponent g_i, modulo p, we see that

$$r^{f_i g_i} \equiv 1 \pmod{p}$$

and
$$d|f_i g_i.$$
That is, $q_i^{b_i} d_i | M_i d_i g_i$; thus $q_i^{b_i} | g_i$, for $i = 1, 2, \ldots, s$. But, since $d | f_i q_i^{b_i}$, $(r^{f_i})^{q_i^{b_i}} \equiv 1 \pmod{p}$ and $g_i | q_i^{b_i}$. Hence, $g_i = q_i^{b_i}$ and r^{f_i} belongs to exponent $q_i^{b_i}$, modulo p, for $i = 1, 2, \ldots, s$. Consequently, by Theorem 5-3, $r^{f_1} r^{f_2} \cdots r^{f_s} = r^{f_1 + f_2 + \cdots + f_s}$ belongs to exponent d, modulo p.

Moreover, since r belongs to exponent d, modulo p, and since $f_1 + f_2 + \cdots + f_s \equiv 1 \pmod{d}$, then, by Theorem 5-2,
$$r^{f_1 + f_2 + \cdots + f_s} \equiv r \pmod{p}.$$
Consequently, if r is a number belonging to exponent $d = q_1^{b_1} \cdots q_s^{b_s}$, modulo p, it is congruent (uniquely, as we shall see) to the product of s numbers $r_1 = r^{f_1}, r_2 = r^{f_2}, \ldots, r_s = r^{f_s}$, modulo p, where r_i belongs, modulo p, to the exponent $q_i^{b_i}$, $i = 1, 2, \ldots, s$. Conversely, if $a_1, \ldots, a_s$ belong, respectively, to exponents $q_1^{b_1}, \ldots, q_s^{b_s}$, modulo p, then $r \equiv a_1 a_2 \cdots a_s \pmod{p}$ is a number belonging to exponent d, modulo p. Moreover, by (5-7),
$$r^{f_i} \equiv a_1^{f_i} a_2^{f_i} \cdots a_{i-1}^{f_i} a_i^{f_i} a_{i+1}^{f_i} \cdots a_s^{f_i}$$
$$\equiv a_i \pmod{p}.$$
For, since $d_j | f_j$ and, where $j \neq i$, $q_j^{b_j} | d_i$ and $d_i | f_i$, we see that, by Theorem 5-2,
$$a_i^{f_i} \equiv a_i^1 \pmod{p}$$
and
$$a_j^{f_i} \equiv a_j^0 \equiv 1 \pmod{p}$$
for $j = 1, 2, \ldots, i-1, i+1, \ldots, s$. Hence, every number r belonging to exponent d is uniquely expressible in the form $r \equiv r^{f_1} r^{f_2} \cdots r^{f_s} \pmod{p}$, where $r_i = r^{f_i}$ belongs to exponent $q_i^{b_i}$, $i = 1, 2, \ldots, s$.

Next, for $i = 1, 2, \ldots, s$, let $r_i^{(j)}$, $j = 1, 2, \ldots, \phi_i$, where ϕ_i means $\phi(q_i^{b_i})$, be all the numbers belonging to exponent $q_i^{b_i}$, modulo p. Then, the sum of all numbers belonging to exponent d, modulo p, will be given by the expression

(5-8)
$$(r_1^{(1)} + r_1^{(2)} + \ldots + r_1^{(\phi_1)})(r_2^{(1)} + r_2^{(2)} + \ldots + r_2^{(\phi_2)}) \cdots$$
$$\cdot (r_s^{(1)} + r_s^{(2)} + \ldots + r_s^{(\phi_s)})$$
$$\equiv \sum_{\substack{1 \leq l_1 \leq \phi_1 \\ \vdots \\ 1 \leq l_s \leq \phi_s}} r_1^{(l_1)} r_2^{(l_2)} \cdots r_s^{(l_s)} \pmod{p}$$
$$\equiv \sum_{i=1}^{\phi(d)} R_i \pmod{p},$$
where the R_i belong to exponent d, modulo p.

Sum of Numbers Belonging to an Exponent **117**

Next, if r_i belongs to the exponent q_i, then the remaining numbers belonging to this exponent, modulo p, will be

$$r_i^2, r_i^3, \ldots, r_i^{q_i-1}.$$

But

$$(1 + r_i + r_i^2 + \ldots + r_i^{q_i-1})(r_i - 1) \equiv r_i^{q_i} - 1 \equiv 0 \pmod{p}.$$

Since r_i belongs to exponent $q_i > 1$, $r_i \not\equiv 1 \pmod{p}$. Therefore, $r_i + r_i^2 + \ldots + r_i^{q_i-1} \equiv -1 \pmod{p}$. If all the b's are unity ($b_1 = b_2 = \ldots = b_s = 1$), then by (5-8) the sum of all numbers belonging to the exponent $d = q_1 q_2 \ldots q_s$ is congruent (mod p) to $(-1)^s$. If, however, r_i belongs to exponent $q_i^{b_i}$ ($b_i \geq 2$), modulo p, the remaining numbers belonging to this exponent will be found if, from the set

$$r_i^2, r_i^3, \ldots, r_i^{q_i^{b_i}-1}, r_i^{q_i^{b_i}}$$

we delete

$$r_i^{q_i}, r_i^{2q_i}, \ldots, r_i^{(q_i^{b_i-1}-1)q_i}, r_i^{q_i^{b_i}}.$$

The sum of all numbers belonging to the exponent $q_i^{b_i}$, modulo p, will, therefore, be congruent to

(5-9) $$1 + r_i + r_i^2 + \ldots + r_i^{q_i^{b_i}-1}$$
$$- (1 + r_i^{q_i} + r_i^{2q_i} + \ldots + r_i^{(q_i^{b_i-1}-1)q_i}),$$

modulo p. But, since

$$r_i^{q_i^{b_i}} - 1 \equiv (r_i - 1)(1 + r_i + r_i^2 + \ldots + r_i^{q_i^{b_i}-1}) \pmod{p}$$

and $r_i - 1 \not\equiv 0 \pmod{p}$,

$$1 + r_i + r_i^2 + \ldots + r_i^{q_i^{b_i}-1} \equiv 0 \pmod{p}.$$

Likewise, since

$$(r_i^{q_i})^{q_i^{b_i-1}} - 1 \equiv (r_i^{q_i} - 1) \cdot (1 + r_i^{q_i} + r_i^{2q_i} + \ldots + r_i^{(q_i^{b_i-1}-1)q_i}) \pmod{p}$$

and

$$r_i^{q_i} - 1 \not\equiv 0 \pmod{p},$$

then

$$1 + r_i^{q_i} + r_i^{2q_i} + \cdots + r_i^{(q_i^{b_i-1}-1)q_i} \equiv 0 \pmod{p}.$$

Hence, by (5-9), the sum of all numbers belonging to exponent $q_i^{b_i}$ is congruent to zero, modulo p. Thus, in view of (5-8), if merely one of the exponents b_i is greater than unity, the sum of all numbers belonging to exponent d is congruent to zero, modulo p.

5-6 Further Consideration of Primitive Roots of p^n

We have already seen (Theorem 5-13) that for every odd prime p and for every positive divisor $e(e > 1)$ of $p - 1$, there is an integer a belonging to exponent e, modulo p, such that $a^e - 1 \not\equiv 0 \pmod{p^2}$. From this it immediately follows (with $e = p - 1$) that every odd prime p has a primitive root r such that $r^{p-1} - 1 \not\equiv 0 \pmod{p^2}$.

- **Theorem 5-18** If p is an odd prime and n a positive integer, there exists a number belonging to the exponent $p - 1$, modulo p^n.

    Proof Let a be an integer belonging to exponent $p - 1$, modulo p, such that $a^{p-1} - 1 \not\equiv 0 \pmod{p^2}$. Then, by Theorem 3-4, $a^{(p-1)p^{n-1}} \equiv 1 \pmod{p^n}$, and by Theorem 5-9, p^n is the highest power of p for which this last congruence is true.

    Suppose $a^{p^{n-1}}$ belongs to exponent d, modulo p^n, where $0 < d < p - 1$. Let p^{n+s}, $s \geq 0$, be the highest power of p for which

    $$a^{dp^{n-1}} \equiv 1 \pmod{p^{n+s}}.$$

    Then by Theorem 5-8, p^{s+1} is the highest power of p (as modulus) for which $a^d \equiv 1 \pmod{p^{s+1}}$. Since a belongs to exponent $p - 1$, modulo p, this is impossible when $0 < d < p - 1$. Hence $d = p - 1$. Consequently, $a^{p^{n-1}}$ is a number belonging to exponent $p - 1$, modulo p^n, $n \geq 1$.

- **Theorem 5-19** For an odd prime p and a positive integer n, there exists a number b belonging to the exponent p^{n-1}, mod p^n.

    Proof By Theorem 5-13 there exists for an arbitrary positive divisor e ($e > 1$) of $p - 1$, an integer a belonging to exponent e such that p is the highest power of p dividing $a^e - 1$; that is, $a^e \equiv 1 \pmod{p}$ and $a^e \not\equiv 1 \pmod{p^2}$. Then, by Theorem 5-9, $a^{ep^{n-1}} \equiv 1 \pmod{p^n}$ and $a^{ep^{n-1}} \not\equiv 1 \pmod{p^{n+1}}$. Let a^e belong to exponent t, modulo p^n. Then, since $(a^e)^{p^{n-1}} \equiv 1 \pmod{p^n}$, $t | p^{n-1}$. Let $t = p^k$, where $0 \leq k \leq n - 1$. Then since $a^{ep^k} \equiv 1 \pmod{p^n}$, we see that

    $$(a^{ep^k})^{p^{n-1-k}} \equiv 1 \pmod{p^{n+(n-1-k)}};$$

    that is,

    $$a^{ep^{n-1}} \equiv 1 \pmod{p^{n+n-1-k}}.$$

    But we know that $n + n - 1 - k \leq n$; that is, $k \geq n - 1$. Hence $k = n - 1$, and a^e belongs to exponent p^{n-1}, modulo p^n.

Further Consideration of Primitive Roots of p^n

We now have a further proof of the existence of a primitive root of p^n.

- **Theorem 5-20** If p is an odd prime and n a positive integer, then there exists a primitive root of p^n.

    Proof Noting that $p - 1$ and p^{n-1} are relatively prime and that there exists, by Theorem 5-18, an integer a belonging to exponent $p - 1$, modulo p^n, and by Theorem 5-19 an integer b belonging to exponent p^{n-1}, modulo p^n, we see by Theorem 5-3 that ab belongs to exponent $p^{n-1}(p - 1) = \phi(p^n)$, modulo p^n. That is, p^n has a primitive root, namely ab.

We shall now give another (independent) proof[8] of the existence of a primitive root of p^n.

- **Theorem 5-21** If p is an odd prime, there exists a primitive root of p^n, where $n \geq 2$.

    Proof Let R be a primitive root of p. Then, by Fermat's Theorem, $R^p - R = Kp$. Now, where t is any integer such that $K - t \not\equiv 0 \pmod{p}$, $r = R + pt$ being a primitive root of p such that $r \equiv R \pmod{p}$, we have, by Theorem 3-4, $r^p \equiv R^p \pmod{p^2}$. Consequently,

    $$r^p - r \equiv R^p - R - pt \pmod{p^2} \equiv (K - t)p \pmod{p^2}.$$

    Hence, $r^{p-1} - 1 \not\equiv 0 \pmod{p^2}$; that is, $r^{p-1} = 1 + Mp$, where $(M, p) = 1$.

    By the binomial theorem,

    (5-10) $$(1 + Mp^j)^p \equiv 1 + Mp^{j+1} \pmod{p^{j+2}},$$

    where $j \geq 1$. The case $j = 1$ shows that

    (5-11) $$(r^{p-1})^{p^s} \equiv 1 + Mp^{s+1} \pmod{p^{s+2}}$$

    is valid when $s = 1$. Assuming the truth of (5-11) for a fixed value of $s \geq 1$, we desire to show that (5-11) is true for $s + 1$; that is, we desire to show that

    $$(r^{p-1})^{p^{s+1}} \equiv 1 + Mp^{s+2} \pmod{p^{s+3}}.$$

    Now, we have from (5-11), by Theorem 3-4,

    (5-12) $$(r^{p-1})^{p^{s+1}} \equiv (1 + Mp^{s+1})^p \pmod{p^{s+3}}.$$

    Applying (5-10) for $j = s + 1$, we have

    $$(1 + Mp^{s+1})^p \equiv 1 + Mp^{s+2} \pmod{p^{s+3}}.$$

    Thus, from (5-12) we get

    $$(r^{p-1})^{p^{s+1}} \equiv 1 + Mp^{s+2} \pmod{p^{s+3}}.$$

Hence, by induction on s, (5-11) is valid for every positive integer s. Taking $s = n - 2$ in (5-11), we obtain

(5-13)
$$r^{(p-1)p^{n-2}} \equiv 1 + Mp^{n-1} \pmod{p^n}$$
$$\not\equiv 1 \pmod{p^n}.$$

Let r belong to exponent e, modulo p^n. Then $e \mid p^{n-1}(p-1)$. But since $r^e \equiv 1 \pmod{p}$, e is divisible by $(p-1)$. If $e < \phi(p^n)$, let $e = p^t(p-1)$, $0 \leq t \leq n-2$. Then

$$r^{p^t(p-1)} \equiv 1 \pmod{p^n},$$

and

$$1 \equiv (r^{p^t(p-1)})^{p^{n-t-2}} \equiv r^{p^{n-2}(p-1)} \pmod{p^n}.$$

This statement contradicts congruence (5-13). Therefore, since r must belong to exponent $\phi(p^n)$, modulo p^n, r is a primitive root of p^n.

5-7 Problems[9]

1. (a) Find the exponents to which 10 and 26 belong, modulo 37.
   (b) Find the exponents to which 6 and 31 belong, modulo 37.
   (c) Find all numbers belonging to exponent 12, modulo 37.
   (d) Find a number belonging to exponent 2, modulo 37.
   (e) Prove that $7, 9, 12, 16, 33$, and 34 all belong to exponent 9, modulo 37.
   (f) Determine all primitive roots of 37.

2. (a) Determine all primitive roots of the prime 13.
   (b) Determine all primitive roots of 26.

3. (a) Determine all primitive roots of the prime 7.
   (b) Determine all the primitive roots of 7^2, incongruent modulo 49.

4. (a) Find all primitive roots of 23.
   (b) Find all primitive roots of 46.

5. (a) Find all primitive roots of 41.
   (b) Find all primitive roots of 82.

6. (a) Show that the product of an even number of primitive roots of an odd prime p is not a primitive root of p.
   (b) Prove that the product of an odd number of such primitive roots is either a primitive root of the odd prime p or belongs to an exponent not dividing $(p-1)/2$.

7. (a) Prove that a primitive root r of the odd prime p such that $r^{p-1} \not\equiv 1 \pmod{p^2}$ is a primitive root of p^n, $n \geq 2$.

(b) Prove that, for an odd prime p, a primitive root of p^2 is a primitive root of p^n, $n \geq 2$.

8. Show that there are $(p-1)\phi(p-1)$ primitive roots of p^n, $n \geq 2$, incongruent modulo p^2.

9. Show that, if p is an odd prime and n a positive integer, there are $\phi\{\phi(p^n)\}$ primitive roots of p^n, incongruent modulo p^n.

10. Let p be a prime greater than 3. Show that the product of all primitive roots of p is congruent to 1, modulo p.

11. Prove that the product of all (incongruent) numbers belonging to an exponent d ($d > 2$) is congruent to 1, modulo p, an odd prime.

12. Let p and q be distinct odd primes and m an integer such that $m \equiv a \pmod{p}$ and $m \equiv b \pmod{q}$. Then, if a belongs to exponent e, mod p, and if b belongs to exponent f, mod q, with d the greatest common divisor of e and f, show that m belongs to the exponent ef/d modulo pq.

13. Let p be an odd prime and let d be a divisor not less than 3 of $p-1$. Show that, if d is the product of an even number of distinct primes, then the sum of all (incongruent) numbers belonging to exponent d, modulo p, is congruent to their product, modulo p.

14. If r is a primitive root ($0 < r < p$) of the odd prime p and if $r' = r^{p-2} + p$, then show that r' is a primitive root of p, and also that at least one of the numbers r and r' is a primitive root of p^n (for $n \geq 1$).

15. If r is a primitive root of m, prove that r^s belongs, modulo m, to the exponent $\phi(m)/d$, where $d = (s, \phi(m))$.

16. Let m be an odd number > 1 having a primitive root; and let $\phi(m) = q_1^{b_1} \cdots q_s^{b_s}$, where the q's are distinct primes. Prove that r is a primitive root of m if and only if $x = r$ is a solution of no one of the congruences

$$x^{\phi(m)/q_1} \equiv 1 \pmod{m}, x^{\phi(m)/q_2} \equiv 1 \pmod{m}, \ldots, x^{\phi(m)/q_s} \equiv 1 \pmod{m}.$$

17. For an odd prime p, prove that there exist primitive roots r and r_1 such that $r^2 + r_1 \equiv 0 \pmod{p}$ when and only when p is of the form $4M + 3$. Also, show that r and r_1 are distinct primitive roots when $p > 3$.

18. Given that 2 is a primitive root of each of the primes 53 and 61, find:

(a) all positive numbers less than 53 belonging to exponent 4, modulo 53;

(b) all positive numbers less than 61 belonging to exponent 4, modulo 61.

19. Given that 5 is a primitive root of 103, find all positive numbers less than 103 belonging to exponent 6, modulo 103.

20. Prove that in a possible odd perfect number only one of its prime factors occurs to an odd power; all the others occur to even powers.

21. Show that an odd perfect number n, if it exists, must contain one prime factor p such that, if the highest power of p occurring in n is p^a, both p and a are congruent to 1, modulo 4; all other prime factors must occur to an even power.

22. Prove that every possible odd perfect number having three distinct prime factors must have two of its prime factors 3 and 5.

23. Prove that there exists no odd perfect number having exactly three distinct prime factors.

24. For the following, determine the type of decimal fraction and, if a repeating decimal, the length of the period:

(a) $\frac{7}{30}$; (b) $\frac{3}{40}$; (c) $\frac{3}{35}$; (d) $\frac{3}{26}$.

25. Find the length of the period in the repeating part of the decimal representation of the following:

(a) $\frac{1}{7}$; (b) $\frac{5}{84}$; (c) $\frac{2}{13}$; (d) $\frac{1}{34}$.

5-8 Indices

Let m be a positive integer having a primitive root r. Then, the $\phi(m)$ numbers, $r, r^2, \ldots, r^{\phi(m)}$, being incongruent, modulo m, form a reduced set of residues, modulo m. Hence, if N is relatively prime to m, N is congruent, modulo m, to one of these powers of r.

DEFINITION *Let r be a primitive root of m and let N be relatively prime to m. The integer t, where $1 \leq t \leq \phi(m)$, such that $N \equiv r^t \pmod{m}$, is called the index of N to base r, modulo m; in symbols, we have*

$$N \equiv r^{\mathrm{Ind}_r N} \pmod{m}.$$

For example, since 2 is a primitive root of 9, we have $1 \equiv 2^6 \pmod 9$, $2 \equiv 2^1 \pmod 9$, $4 \equiv 2^2 \pmod 9$, $5 \equiv 2^5 \pmod 9$, $7 \equiv 2^4 \pmod 9$, $8 \equiv 2^3 \pmod 9$. Consequently, the indices to base 2 of 1, 2, 4, 5, 7, 8, modulo 9, are, respectively, 6, 1, 2, 5, 4, and 3.

As we shall see, in theory and application, indices show analogy to logarithms. This fact will be evident from the results contained in the following theorem.

• **Theorem 5-22** *If m has a primitive root r and if the product MN is relatively prime to m, we have the following results:*

Indices

(i) $\text{Ind}_r M \equiv \text{Ind}_r N \pmod{\phi(m)}$ if and only if $M \equiv N \pmod{m}$;
(ii) $\text{Ind}_r MN \equiv \text{Ind}_r M + \text{Ind}_r N \pmod{\phi(m)}$;
(iii) $\text{Ind}_r M^k \equiv k \, \text{Ind}_r M \pmod{\phi(m)}$ when k is a positive integer.

Proof Since $M \equiv r^{\text{Ind}_r M} \pmod{m}$ and $N \equiv r^{\text{Ind}_r N} \pmod{m}$, $M \equiv N \pmod{m}$ is equivalent to $r^{\text{Ind}_r M} \equiv r^{\text{Ind}_r N} \pmod{m}$. This last congruence, by Theorem 5-2, with $e = \phi(m)$, is true if and only if $\text{Ind}_r M \equiv \text{Ind}_r N \pmod{\phi(m)}$. Thus, (i) is established. Moreover,

$$MN \equiv r^{\text{Ind}_r M} r^{\text{Ind}_r N} \pmod{m}$$
$$\equiv r^{\text{Ind}_r M + \text{Ind}_r N} \pmod{m};$$

and by definition,

$$MN \equiv r^{\text{Ind}_r MN} \pmod{m}.$$

Hence,

$$r^{\text{Ind}_r MN} \equiv r^{\text{Ind}_r M + \text{Ind}_r N} \pmod{m}.$$

This last result follows when and only when

$$\text{Ind}_r MN \equiv \text{Ind}_r M + \text{Ind}_r N \pmod{\phi(m)}.$$

Consequently, (ii) is true. Moreover, $M^k \equiv r^{\text{Ind}_r M^k} \pmod{m}$ and $M^k \equiv (r^{\text{Ind}_r M})^k \equiv r^{k \, \text{Ind}_r M} \pmod{m}$. Therefore, $r^{\text{Ind}_r M^k} \equiv r^{k \, \text{Ind}_r M} \pmod{m}$. Consequently, it follows that $\text{Ind}_r M \equiv k \, \text{Ind}_r M^k \pmod{\phi(m)}$. That is, (iii) is valid.

From the definition of index, we have also the obvious results:

$$\text{Ind}_r 1 \equiv 0 \pmod{\phi(m)} \quad \text{and} \quad \text{Ind}_r r \equiv 1 \pmod{\phi(m)}.$$

- **Corollary** If r is a primitive root of an odd prime p and if $(MN, p) = 1$, we have the following results:

(i) $\text{Ind}_r M \equiv \text{Ind}_r N \pmod{p-1}$ if and only if $M \equiv N \pmod{p}$;
(ii) $\text{Ind}_r MN \equiv \text{Ind}_r M + \text{Ind}_r N \pmod{p-1}$;
(iii) $\text{Ind}_r M^k \equiv k \, \text{Ind}_r M \pmod{p-1}$ when k is a positive integer.

If we use a different primitive root, we naturally obtain a different value for the index of N, modulo m. The following theorem will enable us to change from one base to another. If the primitive root used is evident, we frequently omit writing the base and simply write $\text{Ind } N$ instead of $\text{Ind}_r N$.

- **Theorem 5-23** If r and s are distinct primitive roots of m, then $\text{Ind}_r N \equiv (\text{Ind}_s N)(\text{Ind}_r s) \pmod{\phi(m)}$. Moreover, $(\text{Ind}_r s)(\text{Ind}_s r) \equiv 1 \pmod{\phi(m)}$.

Proof Now,

$$N \equiv r^{\text{Ind}_r N} \equiv s^{\text{Ind}_s N} \pmod{m}.$$

Then, since $s \equiv r^{\text{Ind}_r s} \pmod{m}$,
$$N \equiv s^{\text{Ind}_s N} \equiv (r^{\text{Ind}_r s})^{\text{Ind}_s N} \pmod{m}.$$

Thus
$$r^{\text{Ind}_r N} \equiv r^{(\text{Ind}_r s)(\text{Ind}_s N)} \pmod{m}.$$

This last congruence is true when and only when
$$\text{Ind}_r N \equiv (\text{Ind}_s N)(\text{Ind}_r s) \pmod{\phi(m)}.$$

Moreover, when $N = r$, we have
$$(\text{Ind}_s r)(\text{Ind}_r s) \equiv 1 \pmod{\phi(m)}.$$

- **Corollary** If r and s are distinct primitive roots of the odd prime p,
$$\text{Ind}_r N \equiv (\text{Ind}_s N)(\text{Ind}_r s) \pmod{p - 1}.$$
Moreover, $(\text{Ind}_r s)(\text{Ind}_s r) \equiv 1 \pmod{p - 1}.$

For example, 2 and 6 are both primitive roots of the prime 13. We construct the *companion tables* for the primitive root 2. Here I denotes Ind N, modulo 13.

I	1	2	3	4	5	6	7	8	9	10	11	12
N	2	4	8	3	6	12	11	9	5	10	7	1

N	1	2	3	4	5	6	7	8	9	10	11	12
I	12	1	4	2	9	5	11	3	8	10	7	6

Since $2 \equiv 6^5 \pmod{13}$, $\text{Ind}_6 2 = 5$. Hence, if $(N, 13) = 1$, $\text{Ind}_6 N \equiv (\text{Ind}_2 N) \cdot (\text{Ind}_6 2) \pmod{12}$; thus $\text{Ind}_6 N \equiv 5 (\text{Ind}_2 N) \pmod{12}$. Using the above table and this last congruence, we can easily compute $\text{Ind}_6 N$ for $N = 1, 2, \ldots, 12$, modulo 13.

We can readily solve binomial congruences modulo an odd prime p if we have the companion tables for a primitive root of p. We shall explain the method using the following illustrative examples.

Example 1 Solve the congruence
$$7x^3 \equiv 3 \pmod{13}.$$
Since 2 is a primitive root of 13, we may use the above table. Now, $7 \equiv 2^{\text{Ind}_2 7}$, $x^3 \equiv 2^{\text{Ind}_2 x^3}$, and $3 \equiv 2^{\text{Ind}_2 3} \pmod{13}$. Hence, we are seeking all values of x such that $2^{\text{Ind}_2 7} 2^{\text{Ind}_2 x^3} \equiv 2^{\text{Ind}_2 3} \pmod{13}$; that is, such that $\text{Ind}_2 7 + \text{Ind}_2 x^3 \equiv \text{Ind}_2 3 \pmod{12}$. That is, $11 + 3(\text{Ind}_2 x) \equiv 4 \pmod{12}$. Therefore, $3(\text{Ind}_2 x) \equiv 5 \pmod{12}$. In this case, there is *no* solution because the linear congruence $3w \equiv 5 \pmod{12}$ is not solvable.

Indices

Example 2 Solve the congruence

(5-14) $\qquad 7x^3 \equiv 4 \pmod{13}$.

Since 2 is a primitive root of 13, we obtain from the preceding table of indices: $7 \equiv 2^{\mathrm{Ind}_2 7}$, $x^3 \equiv 2^{\mathrm{Ind}_2 x^3}$ and $4 \equiv 2^{\mathrm{Ind}_2 4} \pmod{13}$. Hence, we desire to obtain all values of x such that $2^{\mathrm{Ind}_2 7} 2^{\mathrm{Ind}_2 x^3} \equiv 2^{\mathrm{Ind}_2 4} \pmod{13}$; that is, $\mathrm{Ind}_2 7 + \mathrm{Ind}_2 x^3 \equiv \mathrm{Ind}_2 4 \pmod{12}$, and $\mathrm{Ind}_2 7 + 3\,\mathrm{Ind}_2 x \equiv \mathrm{Ind}_2 4 \pmod{12}$. Hence, $11 + 3\,(\mathrm{Ind}_2 x) \equiv 2 \pmod{12}$; consequently, $3\,(\mathrm{Ind}_2 x) \equiv 3 \pmod{12}$ and $\mathrm{Ind}_2 x \equiv 1 \pmod 4$. Therefore, since $\mathrm{Ind}_2 x = 1, 5, 9$, the solutions of (5-14) are 2, 6, 5, modulo 13.

Example 3 Solve the congruence

(5-15) $\qquad 7x^3 \equiv 4 \pmod{169}$.

First, we note that the solutions of $7x^3 \equiv 4 \pmod{13}$ are $x \equiv 2, 5, 6 \pmod{13}$. Let $f(x) = 7x^3 - 4$ and let $f(x_0) \equiv 0 \pmod{13}$, and write $f(x_0) = M \cdot 13$. Referring to page 65, we wish to determine k so that $kf'(x_0) \equiv -M \pmod{13}$. As a consequence, $x_0 + k \cdot 13$ will be a solution of (5-15). Using $x_0 = 2$ we desire to find k so that $2 + k \cdot 13$ is a solution of (5-15) (see page 65). Now, $f(2) = 4 \cdot 13$ and $f'(2) = 84$. We wish to determine k so that $k \cdot 84 \equiv -4 \pmod{13}$; that is, $2k \equiv 3 \pmod{13}$. Hence,

$$2^{\mathrm{Ind}_2 2 + \mathrm{Ind}_2 k} \equiv 2^{\mathrm{Ind}_2 3} \pmod{13}.$$

Therefore, using the preceding table of indices, we obtain

$$1 + \mathrm{Ind}_2 k \equiv 4 \pmod{12};$$

consequently,

$$\mathrm{Ind}_2 k \equiv 3 \pmod{12}.$$

Therefore, $k \equiv 8 \pmod{13}$.

Hence, $2 + 8 \cdot 13 = 106$ is a solution of (5-15).

Next, taking $x_0 = 5$, we see that $f(5) = 67 \cdot 13$ and $f'(5) = 525$. We wish to determine k so that

$$525k \equiv -67 \pmod{13};$$

that is,

$$5k \equiv 11 \pmod{13}.$$

Since

$$\mathrm{Ind}_2 5 + \mathrm{Ind}_2 k \equiv \mathrm{Ind}_2 11 \pmod{12},$$

we have

$$9 + \mathrm{Ind}_2 k \equiv 7 \pmod{12},$$

and

$$\mathrm{Ind}_2 k \equiv 10 \pmod{12}.$$

Then $k \equiv 10 \pmod{13}$. Hence $5 + 10 \cdot 13 = 135$ is a solution of (5-15).

Finally, taking $x_0 = 6$, we note that $f(6) = 116 \cdot 13$ and $f'(6) = 756$. We desire k such that $k \cdot 756 \equiv -116 \pmod{13}$; that is, $2k \equiv 1 \pmod{13}$. Hence, $k \equiv 7 \pmod{13}$, and $6 + 7 \cdot 13 = 97$ is a solution of (5-15). Therefore, the solutions of (5-15) are 97, 106, 135.

Example 4 Solve the congruence

(5-16) $\qquad 7x^3 \equiv 4 \pmod{2197}$, where $2197 = 13^3$.

Solution Let $f(x) = 7x^3 - 4$, and assume, from Example 3, that the solutions of $f(x) \equiv 0 \pmod{13^2}$ are 97, 106, 135. Where $x = x_0$ is a solution modulo 13^2, we desire a value of k such that $f(x_0) + k \cdot 13^2 f'(x_0)$ be divisible by 13^3. Writing $f(x_0) = M \cdot 13^2$, we wish to obtain k so that $k \cdot f'(x_0) \equiv -M \pmod{13}$. Then it follows that $x_0 + k \cdot 13^2$ is a solution of (5-16).

First, for $x_0 = 97$, we have $f(97) = 6388707 = 37803 \cdot 13^2$ and $f'(97) = 197589$. Hence, $197589\, k \equiv -37803 \pmod{13}$; that is, $2k \equiv -12 \equiv 1 \pmod{13}$; thus $k \equiv 7 \pmod{13}$. Thus, $97 + 7 \cdot 13^2 = 1280$ is a solution of (5-16).

Next, for $x_0 = 106$, $f(106) = 8337108 = 49332 \cdot 13^2$ and $f'(106) = 235956$. Then $235956k \equiv -49332 \pmod{13}$; that is, $6k \equiv -10 \equiv 3 \pmod{13}$, and $k \equiv 7 \pmod{13}$. Hence, $106 + 7 \cdot 13^2 = 1289$ is a solution of (5-16).

Finally, for $x_0 = 135$, $f(135) = 17222621 = 101909 \cdot 13^2$, $f'(135) = 382725$. Then $382725k \equiv -101909 \pmod{13}$; that is, $5k \equiv -2 \equiv 11 \pmod{13}$. Therefore, $k \equiv 10 \pmod{13}$. Hence, $135 + 10 \cdot 13^2 = 1825$ is a solution of (5-16). Consequently, the only solutions of (5-16) are 1280, 1289, and 1825.

If, where $f(x) = 7x^3 - 4$, the congruence $f(x) \equiv 0 \pmod{13^4}$ were proposed for solution, the computation would, naturally, become heavier. We would use the fact that $x_0 = 1280, 1289, 1825$ are the solutions of $f(x) \equiv 0 \pmod{13^3}$ and set $f(x_0) = M \cdot 13^3$. We would then seek for each x_0 a value of k for which $kf'(x_0) \equiv -M \pmod{13}$. Hence, a solution of $f(x) \equiv 0 \pmod{13^4}$ would then be $x_0 + k \cdot 13^3$. The solutions of $f(x) \equiv 0 \pmod{13^4}$ are, in fact, 3477, 10613, and 14471.

5-9 Problems

1. (a) Make companion tables of indices for a primitive root of 17.
   (b) Using tables from part (a), solve $11x^5 \equiv 6 \pmod{17}$.
   (c) Completely solve $11x^5 \equiv 6 \pmod{578}$.

2. (a) Using a primitive root of 17 greater than 4, make companion tables of indices for this primitive root.

(b) Using the tables from part (a), solve $11x^5 \equiv 6 \pmod{17}$, and compare the result with that of Problem 1 (b), where presumably a different primitive root was used.

(c) Completely solve $9x^8 \equiv 8 \pmod{17}$.

3. (a) Using a table of indices for a primitive root of 17, solve $5x^6 \equiv 3 \pmod{17}$.

(b) Completely solve the congruence $5x^6 \equiv 3 \pmod{289}$.

(c) Completely solve the congruence $5x^6 \equiv 3 \pmod{578}$.

4. (a) Determine all values of l, where $1 \leq l \leq 17$, for which $5x^6 \equiv l \pmod{17}$ has a solution.

(b) Using a table of indices, solve the congruence $5x^6 \equiv 6 \pmod{34}$.

(c) Using indices, solve $5x^6 \equiv 6 \pmod{289}$.

(d) Using indices, solve $5x^6 \equiv 6 \pmod{578}$.

5. Using tables of indices, solve:
(a) $11x^3 \equiv 3 \pmod{13}$;
(b) $11x^3 \equiv 3 \pmod{29}$;
(c) $11x^3 \equiv 3 \pmod{841}$, where $841 = 29^2$;
(d) $11x^3 \equiv 3 \pmod{377}$.

6. (a) Where $1 \leq b \leq 28$, determine for what values of b the congruence $8x^7 \equiv b \pmod{29}$ is solvable.

(b) Where $1 \leq a \leq 12$, determine for what values of a the congruence $ax^4 \equiv b \pmod{13}$ is solvable when $b = 2, 5$, or 6.

7. If p is an odd prime, n a positive integer, and a and b positive integers not exceeding $p - 1$, obtain a necessary and sufficient condition for the congruence $ax^n \equiv b \pmod{p}$ to be solvable.

8. Let p be an odd prime, and let $m = p^n$, $n \geq 1$. For an arbitrary primitive root r of m, prove that the index of a number a belonging to exponent e, modulo m, is divisible by $\phi(m)/e$. In addition, prove that for a suitably chosen primitive root of m, the index of a will be precisely $\phi(m)/e$.

Quadratic Congruences   6

6-1 A Quadratic Congruence

A quadratic congruence in one unknown is a second degree congruence of the form

(6-1) $$ax^2 + bx + c \equiv 0 \pmod{m}.$$

We have already seen (page 63) that the problem of solving a congruence of degree n eventually depends upon the solution of congruences of degree n or less, with respect to prime moduli. Consequently, we shall consider a quadratic congruence modulo a prime p:

(6-2) $$ax^2 + bx + c \equiv 0 \pmod{p},$$

where $(a, p) = 1$.

In case the modulus is the prime 2, we have the following possible quadratic congruences:

$$x^2 + 0 \cdot x + 0 \equiv 0 \pmod{2}, \quad x^2 + 0 \cdot x + 1 \equiv 0 \pmod{2},$$
$$x^2 + 1 \cdot x + 0 \equiv 0 \pmod{2}, \quad x^2 + 1 \cdot x + 1 \equiv 0 \pmod{2}.$$

Each of these congruences is solvable except the last one.

Let us assume, then, that in the quadratic congruence (6-2), p is an odd prime. For a small value of p, the quadratic congruence (6-2) is readily solved by trial. However, the determination of the solvability of such a congruence for a large prime p requires other methods. Since $(4a, p) = 1$, we obtain a congruence equivalent to (6-2) by multiplying through by $4a$, transposing $4ac$ to the right and adding b^2 to each member:

$$4a^2x^2 + 4abx + b^2 \equiv b^2 - 4ac \pmod{p};$$

that is

(6-3) $$X^2 \equiv D \pmod{p}.$$

where $X = 2ax + b$ and $D = b^2 - 4ac$. If $x \equiv x_0 \pmod{p}$ is a solution of (6-2), then $X \equiv 2ax_0 + b \pmod{p}$ is a solution of (6-3). Conversely, if $X \equiv X_0$ is a solution of (6-3), then, since $(2a, p) = 1$, there exists a solution $x \equiv x_0 \pmod{p}$ not only of $2ax \equiv X_0 - b \pmod{p}$ but also of congruence (6-2). Hence, the solution of congruence (6-2) depends ultimately upon the solution of a congruence of the form

$$X^2 \equiv D \pmod{p}.$$

If $D \equiv 0 \pmod{p}$, congruence (6-3) has only the trivial solution $X \equiv 0 \pmod{p}$. We shall now consider the case where $(D, p) = 1$.

6-2 Quadratic Residue and Quadratic Nonresidue

DEFINITION[1] *Let $m > 1$ and $(D, m) = 1$. Then D is a quadratic residue of m if $x^2 \equiv D \pmod{m}$ has a solution; D is a quadratic nonresidue of m if $x^2 \equiv D \pmod{m}$ has no solution.*

For example, 3 is a quadratic residue of 22 since $x^2 \equiv 3 \pmod{22}$ has a solution. However, 3 is a quadratic nonresidue of 28, since $x^2 \equiv 3 \pmod{28}$ has no solution. If $(a, m) = 1$, a^2 is obviously a quadratic residue of m since $x^2 \equiv a^2 \pmod{m}$ has a solution.

There is, as we shall see, an intimate relation between primitive roots and quadratic residues of an odd prime.

• **Theorem 6-1** The quadratic residues of an odd prime p coincide with the residues, modulo p, of the even powers of a primitive root r of p; the quadratic nonresidues coincide with the residues of the odd powers of r.

Proof If r is a primitive root of p, then $r, r^2, \ldots, r^{p-1}$, being incongruent modulo p, provide a reduced set of residues, modulo p. The even powers of r are obviously quadratic residues of p. Conversely, if d is a quadratic residue of p, there is a number x such that $x^2 \equiv d \pmod{p}$; that is, since $x \equiv r^{\operatorname{Ind} x} \pmod{p}$, $d \equiv r^{2\operatorname{Ind} x} \pmod{p}$. Hence, a quadratic residue of p is congruent to an even power of r. Now $r^s \equiv r^t \pmod{p}$ if and only if the even number $p - 1$ divides $s - t$. Hence, no odd power of a primitive root is congruent to an even power, modulo p. Thus all odd powers of r are quadratic nonresidues of p.

- **Theorem 6-2** There are precisely $(p - 1)/2$ incongruent quadratic residues and $(p - 1)/2$ incongruent quadratic nonresidues of an odd prime p.

    Proof Since $r, r^2, \ldots, r^{p-1}$ form a reduced set of residues, modulo p, they yield all the quadratic residues and all the quadratic nonresidues of p: the even powers are quadratic residues and the odd powers quadratic nonresidues, modulo p. The number of each is $(p - 1)/2$. In fact, the quadratic residues of p are the residues of the $(p - 1)/2$ incongruent numbers $1^2, 2^2, \ldots, \{(p - 1)/2\}^2$. (See Section 6-3, Problem 7.)

- **Theorem 6-3** The product of two quadratic residues or two quadratic nonresidues of an odd prime p is a quadratic residue of p. The product of a quadratic residue and a quadratic nonresidue of p is a quadratic nonresidue of p.

    Proof Let R and N denote a quadratic residue and a quadratic nonresidue, respectively, of the odd prime p. Then R is congruent to an even power of a primitive root r and N congruent to an odd power of r, modulo p. Hence the product of two quadratic residues R_1 and R_2 or the product of two quadratic nonresidues N_1 and N_2, is congruent to an even power of r and hence is a quadratic residue of p. On the other hand, the product of a quadratic residue R_3 and a quadratic nonresidue N_3 is congruent to an odd power of r and hence is a quadratic nonresidue of p.

    Since a number relatively prime to the odd prime p is either a quadratic residue or a quadratic nonresidue of p, we shall now give an important and useful result.

- **Theorem 6-4** A quadratic residue R of the odd prime p satisfies the condition $R^{\frac{p-1}{2}} \equiv 1 \pmod{p}$, a quadratic nonresidue N of p satisfies the condition $N^{\frac{p-1}{2}} \equiv -1 \pmod{p}$.

    Proof If r is a primitive root of p, $(r^{(p-1)/2} - 1) \cdot (r^{(p-1)/2} + 1) \equiv 0 \pmod{p}$, thus $r^{(p-1)/2} \equiv -1 \pmod{p}$. Then, since $R \equiv r^{2k} \pmod{p}$, $R^{(p-1)/2} \equiv r^{(p-1)k} \equiv 1 \pmod{p}$; moreover, since $N \equiv r^{2l+1} \pmod{p}$, $N^{(p-1)/2} \equiv (r^{2l+1})^{(p-1)/2} \equiv (r^{(p-1)/2})^{2l+1} \equiv -1 \pmod{p}$.

6-3 Problems

1. Determine whether $2x^2 + x + 3 \equiv 0 \pmod{65}$ is solvable.

2. Using a primitive root of 17, list all quadratic residues R of 17 such that $0 < R < 17$.

3. Determine whether $2x^2 + x + 3 \equiv 0 \pmod{221}$ has a solution.

4. Show that if p and q are distinct odd primes and if $(c, q) = 1$, then $px^2 + 2qx + c \equiv 0 \pmod{p^n q}$, where $n \geq 1$, is solvable if and only if $-cp$ is a quadratic residue of q.

5. If p and q are distinct odd primes such that $(q, c) = (p, 3) = 1$, and if n is positive, prove that the congruence $pqx^3 + px^2 + 6qx + c \equiv 0 \pmod{p^n q}$ is solvable if and only if $-cp$ is a quadratic residue of q.

6. Show that the product of all the (incongruent) quadratic residues and the quadratic nonresidues of an odd prime p when increased by 1 is divisible by p.

7. Show that the integers $1^2, 2^2, \ldots, \{(p-1)/2\}^2$ yield all the (incongruent) quadratic residues of the odd prime p.

8. If r is a primitive root of an odd prime p, prove that $r^{(p^2-1)/4}$ is congruent, modulo p, to the product of the quadratic residues of p, and that $r^{\{(p-1)/2\}^2}$ is congruent, modulo p, to the product of its quadratic nonresidues.

9. If p is a prime exceeding 3, give an independent proof that p divides the sum of its quadratic nonresidues.

10. (a) Let D be relatively prime to the odd prime p. If D is a quadratic residue of p, prove that it is a quadratic residue of p^n, where $n \geq 1$.

    (b) If the canonical decomposition of the *odd* integer m into primes is $m = p_1^{a_1} \ldots p_r^{a_r}$, prove that a quadratic residue of m is a quadratic residue of $p_1^{b_1} p_2^{b_2} \ldots p_r^{b_r}$, where $b_1, b_2, \ldots b_r$ are non-negative integers with at least one of them positive.

11. If p is a prime exceeding 5, show that the sum of the squares of its quadratic nonresidues is divisible by p.

12. If a and b are positive integers, show that $4ab - a - b$ cannot be a perfect square.

6-4 Euler's Criterion

Euler's Criterion is usually stated as in the following theorem.

- **Theorem 6-5** If p is an odd prime and if $(D, p) = 1$, then the congruence $x^2 \equiv D \pmod{p}$ is solvable if and only if $D^{(p-1)/2} \equiv 1 \pmod{p}$. (cf. Section 6-2, Theorem 6-4.)

Proof Now, since $D^{p-1} - 1 \equiv (D^{(p-1)/2} - 1)(D^{(p-1)/2} + 1) \equiv 0 \pmod{p}$, it follows that either $D^{(p-1)/2} \equiv 1 \pmod{p}$ or $D^{(p-1)/2} \equiv -1 \pmod{p}$; but not both can be true. Hence, by the preceding Theorem 6-4, D is a quadratic residue of p if and only if $D^{(p-1)/2} \equiv 1 \pmod{p}$.

The next theorem gives a somewhat more general statement.

- **Theorem 6-6** Let m be an integer > 1 which has a primitive root r, and let D be relatively prime to m. If n is a positive integer and if $d = (n, \phi(m))$, then

$$x^n \equiv D \pmod{m}$$

is solvable if and only if

$$D^{\frac{\phi(m)}{d}} \equiv 1 \pmod{m}.$$

Proof Assume that $x^n \equiv D \pmod{m}$ has a solution x.

Since

$$D \equiv r^{\operatorname{Ind} D} \pmod{m} \quad \text{and} \quad x \equiv r^{\operatorname{Ind} x} \pmod{m},$$
$$(r^{\operatorname{Ind} x})^n \equiv x^n \equiv r^{\operatorname{Ind} D} \pmod{m}.$$

Consequently,

$$n (\operatorname{Ind} x) \equiv \operatorname{Ind} D \pmod{\phi(m)}$$

Hence, since the linear congruence $ny \equiv \operatorname{Ind} D \pmod{\phi(m)}$ has a solution $y = \operatorname{Ind} x$, $d | \operatorname{Ind} D$. Then

$$D^{\phi(m)/d} \equiv (r^{\operatorname{Ind} D})^{\phi(m)/d} \equiv (r^{\phi(m)})^{(\operatorname{Ind} D)/d} \equiv 1 \pmod{m}$$

Next, let $D^{\phi(m)/d} \equiv 1 \pmod{m}$.
Then

$$1 \equiv D^{\phi(m)/d} \equiv (r^{\operatorname{Ind} D})^{\phi(m)/d} \pmod{m}.$$

Since r is a primitive root of m, the integer $\operatorname{Ind} D(\phi(m)/d)$ is a multiple of $\phi(m)$. Therefore, $(\operatorname{Ind} D)/d$ is an integer; that is, $d | \operatorname{Ind} D$. Hence, $nw \equiv \operatorname{Ind} D \pmod{\phi(m)}$ has d solutions w, where $1 \leq w \leq \phi(m)$. Thus, there are values of w satisfying

$$r^{nw} \equiv r^{\operatorname{Ind} D} \pmod{m}.$$

In other words,

$$(r^w)^n \equiv D \pmod{m}.$$

Consequently, $x^n \equiv D \pmod{m}$ has a solution $x \equiv r^w \pmod{m}$.

6-5 Legendre's Symbol

Legendre introduced a notation which has been found useful in the study of the quadratic character of a number with respect to an odd prime p.

DEFINITION *Let p be an odd prime, and let $(m, p) = 1$. We then define the symbol $\left(\dfrac{m}{p}\right)$ to be 1 if m is a quadratic residue of p and to be -1 if m is a quadratic nonresidue of p.*

For example, $\left(\dfrac{2}{7}\right) = 1$ since $x^2 \equiv 2 \pmod{7}$ is solvable, but $\left(\dfrac{2}{5}\right) = -1$ since $x^2 \equiv 2 \pmod{5}$ is not solvable.

In terms of Legendre's symbol, we can now restate the preceding results regarding quadratic residues and quadratic nonresidues.

- **Theorem 6-7** *Let p be an odd prime and let $(mn, p) = 1$.*

  (i) *If $m \equiv n \pmod{p}$, $\left(\dfrac{m}{p}\right) = \left(\dfrac{n}{p}\right)$.*

  (ii) $\left(\dfrac{mn}{p}\right) = \left(\dfrac{m}{p}\right)\left(\dfrac{n}{p}\right).$

  (iii) $\left(\dfrac{m}{p}\right) \equiv m^{(p-1)/2} \pmod{p}.$

  (iv) $\left(\dfrac{-1}{p}\right) = (-1)^{(p-1)/2}.$

Proof From the definition of the Legendre symbol, (i) and (ii) follow. By Euler's Criterion, (iii) follows. Let $m \equiv -1 \pmod{p}$ in (iii). Then -1 is a quadratic residue of p if and only if $\left(\dfrac{-1}{p}\right) = 1$ and $(-1)^{(p-1)/2} \equiv 1 \pmod{p}$; -1 is a quadratic nonresidue of p if and only if $\left(\dfrac{-1}{p}\right) = -1$ and $(-1)^{(p-1)/2} \equiv -1 \pmod{p}$. Hence, since $1 \not\equiv -1 \pmod{p}$, the result in (iv) follows. Hence, $\left(\dfrac{-1}{p}\right) = 1$ when $p = 4M + 1$, and $\left(\dfrac{-1}{p}\right) = -1$ when $p = 4M + 3$.

6-6 The Quadratic Reciprocity Law

We shall now prove an extremely useful and elegant theorem concerning quadratic residues. Although the result was known to Euler and Legendre, it was Gauss who first gave a complete proof. In fact, he gave seven proofs

The Quadratic Reciprocity Law

of this pivotal theorem of the elementary theory of numbers. Many additional proofs[2] have been given since. In the first proof presented here, we shall employ Gauss's Lemma and the geometric argument of Eisenstein.

- **Theorem 6-8** (Gauss's Lemma) Let p be an odd prime and let q be relatively prime to p. Moreover, let μ denote the number of least positive residues greater than $p/2$, modulo p, of the $(p-1)/2$ integers

(6-4) $$q, 2q, \ldots, \frac{p-1}{2}q.$$

Then
$$\left(\frac{q}{p}\right) = (-1)^\mu.$$

Proof Since $(p, q) = 1$, none of the $(p-1)/2$ integers (6-4) is congruent to zero and no two belong to the same residue class, modulo p. Consider, then, the least positive residues of (6-4) modulo p. Let $a_1, \ldots, a_\mu$ be the least positive residues $> p/2$ and $b_1, \ldots, b_\lambda$ the least positive residues $< p/2$. Hence $\mu + \lambda = (p-1)/2$.

Now, each of

(6-5) $$p - a_1, p - a_2, \ldots, p - a_\mu$$

is positive and $< p/2$. Moreover, these μ integers (6-5) are in different residue classes, modulo p. Suppose one of the b's is congruent, modulo p, to an integer in (6-5): say,

$$b_j \equiv p - a_i \pmod{p}.$$

Let a_i and b_j be the least positive residues, respectively, of αq and βq, modulo p. Then $0 \equiv a_i + b_j \equiv \alpha q + \beta q \equiv (\alpha + \beta)q \pmod{p}$. But, since α and β are both positive integers and $\leq (p-1)/2$, p does not divide $\alpha + \beta$. Consequently, the $(p-1)/2$ positive integers

(6-6) $$p - a_1, \ldots, p - a_\mu, b_1, \ldots, b_\lambda$$

are distinct numbers $\leq (p-1)/2$, and are, therefore, an arrangement of the integers

$$1, 2, \ldots, \frac{p-1}{2}.$$

Hence, the product of the integers (6-6) is

$$\left(\frac{p-1}{2}\right)! : (p-a_1)(p-a_2)\ldots(p-a_\mu)b_1\ldots b_\lambda \equiv \left(\frac{p-1}{2}\right)! \pmod{p}.$$

Thus,

(6-7) $$(-1)^\mu a_1 a_2 \ldots a_\mu b_1 b_2 \ldots b_\lambda \equiv \left(\frac{p-1}{2}\right)! \pmod{p}.$$

We have also

(6-8) $$q \cdot 2q \cdot \ldots \cdot \frac{p-1}{2} q \equiv a_1 a_2 \ldots a_\mu b_1 \ldots b_\lambda \pmod{p}.$$

Hence, from (6-8) and (6-7),

$$\left(\frac{p-1}{2}\right)! \, q^{(p-1)/2} \equiv (-1)^\mu \left(\frac{p-1}{2}\right)! \pmod{p};$$

from this it follows that

$$q^{(p-1)/2} \equiv (-1)^\mu \pmod{p}.$$

Since

$$\left(\frac{q}{p}\right) \equiv q^{(p-1)/2} \pmod{p}, \quad \left(\frac{q}{p}\right) = (-1)^\mu.$$

Thus Gauss's Lemma is established.

We note that, if we divide a positive multiple of q by p, we have

$$kq = pQ + r,$$

where

$$0 \leq r < p;$$

thus

$$\frac{kq}{p} = Q + \frac{r}{p}.$$

Consequently,

$$Q = \left[\frac{kq}{p}\right].$$

Following Gauss, we can then rewrite the numbers (6-4) as follows:

(6-9)
$$q = p\left[\frac{q}{p}\right] + r_1,$$
$$2q = p\left[\frac{2q}{p}\right] + r_2,$$
$$\vdots \quad \vdots \quad \vdots$$
$$\frac{p-1}{2} q = p\left[\frac{\{(p-1)/2\} \cdot q}{p}\right] + r_{(p-1)/2},$$

where the r_i satisfy the condition

$$0 \leq r_i < p, \; 1 \leq i \leq \frac{p-1}{2}.$$

The Quadratic Reciprocity Law

Using

$$P = 1 + 2 + \ldots + \frac{p-1}{2} = \frac{\{(p-1)/2\} \cdot \{(p+1)/2\}}{2} = \frac{p^2-1}{8},$$

$$A = a_1 + a_2 + \ldots + a_\mu,$$
$$B = b_1 + b_2 + \ldots b_\lambda,$$

and

$$M = \left[\frac{q}{p}\right] + \left[\frac{2q}{p}\right] + \ldots + \left[\frac{\{(p-1)/2\} \cdot q}{p}\right],$$

and recalling that $r_1 + r_2 + \ldots + r_{(p-1)/2} = A + B$, we find upon adding the equations (6-9) that

(6-10) $$Pq = pM + A + B.$$

Since $p - a_1, p - a_2, \ldots, p - a_\mu, b_1, b_2, \ldots, b_\lambda$ is an arrangement of the integers $1, 2, \ldots, (p-1)/2$,

(6-11) $$P = \mu p - A + B.$$

Subtracting (6-11) from (6-10), we obtain

$$P(q-1) = p(M - \mu) + 2A;$$

that is,

$$P(q-1) \equiv M - \mu \pmod{2}.$$

If q is an odd integer, it follows that $M \equiv \mu \pmod{2}$ and

(6-12) $$\left(\frac{q}{p}\right) = (-1)^M$$

If q is 2, $M = 0$ and $P \equiv -\mu \equiv \mu \pmod{2}$; consequently, we have

(6-13) $$\left(\frac{2}{p}\right) = (-1)^{(p^2-1)/8}$$

That is, $\left(\frac{2}{p}\right) = 1$ when $p = 8x \pm 1$, and $\left(\frac{2}{p}\right) = -1$ when $p = 8x \pm 3$.

We shall now state the Quadratic Reciprocity Law and, employing (6-12), proceed to its proof.

- **Theorem 6-9** (Quadratic Reciprocity Law). Let p and q be distinct odd primes. Then

$$\left(\frac{p}{q}\right)\left(\frac{q}{p}\right) = (-1)^{\frac{p-1}{2} \cdot \frac{q-1}{2}}.$$

Proof In view of (6-12), which has just been proved for an odd prime p relatively prime to the odd number q, we see, by interchanging the roles of p and q, that for the odd prime q relatively prime to the odd number p,

(6-14) $$\left(\frac{p}{q}\right) = (-1)^N,$$

where

$$N = \left[\frac{p}{q}\right] + \left[\frac{2p}{q}\right] + \ldots + \left[\frac{\{(q-1)/2\}p}{q}\right].$$

Consequently, when p and q are distinct odd primes, we obtain from (6-12) and (6-14):

$$\left(\frac{p}{q}\right)\left(\frac{q}{p}\right) = (-1)^{M+N}.$$

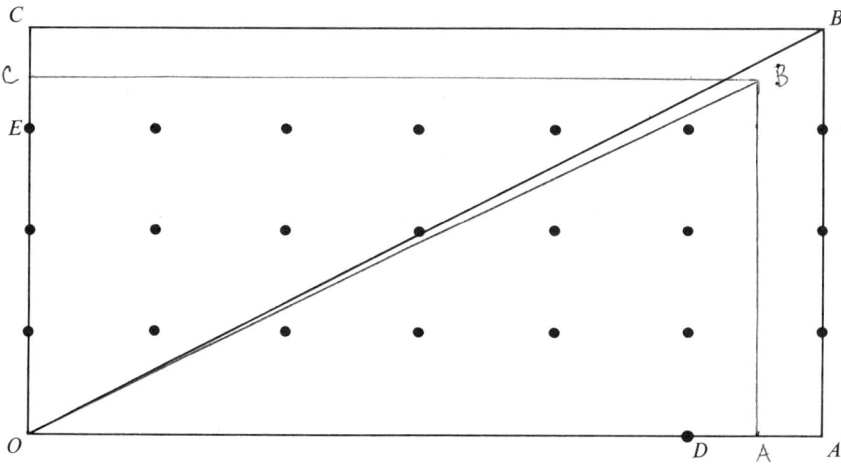

We shall now employ a simple geometrical argument of Eisenstein to evaluate the sum $M + N$ in terms of p and q. Using Cartesian coordinates, we shall call the point (x, y) a *lattice point*, if both x and y are integers. Consider the rectangle $OABC$, where A, B, and C have, respectively, the coordinates $(p/2, 0)$, $(p/2, q/2)$, $(0, q/2)$. Let D and E be the lattice points determined by $((p-1)/2, 0)$ and $(0, (q-1)/2)$, respectively. The diagram is drawn for $p = 11, q = 7$. The equation of the diagonal OB is $y = (q/p)x$. Since the slope q/p is in its lowest terms, there cannot be a lattice point on OB between O and B. Let k be a positive integer $< p/2$; that is, $1 \leq k \leq (p-1)/2$. The number of lattice points on the line $x = k$ above OA and below the diagonal OB is $[(qk)/p]$. Hence, the number of lattice points inside (but not on the boundary of) the triangle OAB is $[q/p] + [2q/p] + \ldots + [\{(p-1)/2\}q/p]$, that is, M.

Next, let l be a positive integer $< q/2$; that is, $1 \leq l \leq (q-1)/2$. The number of lattice points on the line $y = l$ to the right of OC and

The Quadratic Reciprocity Law

to the left of the diagonal OB (whose equation is $x = \frac{p}{q}y$), is $\left[\frac{pl}{q}\right]$.
Hence, the number of lattice points within the triangle OCB is

$$\left[\frac{p}{q}\right] + \left[\frac{2p}{q}\right] + \cdots + \left[\frac{\{(q-1)/2\} \cdot p}{q}\right];$$

that is, N.

Therefore, the number of lattice points inside (but not on the boundary of) the rectangle $OABC$ is $M + N$. But, the number of lattice points *within* this rectangle is also given by the length OD times the length OE; that is, by $\{(p-1)/2\} \cdot \{(q-1)/2\}$. Hence $M + N = \{(p-1)/2\} \cdot \{(q-1)/2\}$. Consequently, we have the final result:

$$\left(\frac{p}{q}\right)\left(\frac{q}{p}\right) = (-1)^{\{(p-1)/2\} \cdot \{(q-1)/2\}}.$$

Example 1 Determine whether the congruence $x^2 \equiv 53 \pmod{187}$ is solvable.

We note that $187 = 11 \cdot 17$. Hence, the given congruence is solvable when and only when each of the following congruences is solvable:

$$x^2 \equiv 53 \pmod{11}$$

and

$$x^2 \equiv 53 \pmod{17}.$$

Now, $x^2 \equiv 53 \equiv 9 \pmod{11}$ is obviously solvable for $x \equiv 3 \pmod{11}$. Moreover, since $\left(\frac{2}{17}\right) = (-1)^{(17^2-1)/8} = 1$, $x^2 \equiv 53 \equiv 2 \pmod{17}$ is also solvable. Hence, the given congruence $x^2 \equiv 53 \pmod{187}$ is solvable.

Example 2 Use the Quadratic Reciprocity Law to determine whether the congruence

(6-15) $\qquad 3x^2 + 7x - 43 \equiv 0 \pmod{391}$

is solvable.

First Method We note that $391 = 17 \cdot 23$. Multiplying the given congruence through by 12, we obtain the equivalent congruence

$$36x^2 + 84x - 516 \equiv 0 \pmod{391};$$

thus

$$(6x + 7)^2 \equiv 565 \pmod{391}.$$

This last congruence is solvable if and only if

(6-16) $$X^2 \equiv 174 \pmod{391}$$
is solvable. Now, (6-16) is solvable if and only if both
$$X^2 \equiv 174 \pmod{17} \quad \text{and} \quad X^2 \equiv 174 \pmod{23}$$
are solvable; that is, if and only if both $X^2 \equiv 4 \pmod{17}$ and $X^2 \equiv 13 \pmod{23}$ are solvable. The first of these last two congruences is obviously solvable since
$$\left(\frac{4}{17}\right) = \left(\frac{2}{17}\right)\left(\frac{2}{17}\right) = 1.$$
Since, by the Quadratic Reciprocity Law,
$$\left(\frac{13}{23}\right)\left(\frac{23}{13}\right) = (-1)^{[(13-1)/2]\cdot[(23-1)/2]} = 1,$$
$$\left(\frac{13}{23}\right) = \left(\frac{23}{13}\right) = \left(\frac{10}{13}\right) = \left(\frac{2}{13}\right)\left(\frac{5}{13}\right).$$
Now,
$$\left(\frac{2}{13}\right) = (-1)^{(13^2-1)/8} = -1.$$
By the Quadratic Reciprocity Law,
$$\left(\frac{5}{13}\right)\left(\frac{13}{5}\right) = 1.$$
Hence
$$\left(\frac{5}{13}\right) = \left(\frac{13}{5}\right) = \left(\frac{3}{5}\right).$$
Again, by the Quadratic Reciprocity Law, $\left(\frac{3}{5}\right)\left(\frac{5}{3}\right) = 1$; from this it follows that
$$\left(\frac{3}{5}\right) = \left(\frac{5}{3}\right) = \left(\frac{2}{3}\right) = (-1)^{(3^2-1)/8} = -1.$$
Therefore,
$$\left(\frac{13}{23}\right) = (-1)\cdot(-1) = 1.$$
Thus, the congruence $X^2 \equiv 13 \pmod{23}$ is solvable, and as a result, (6-15) is solvable.

Second Method Let us multiply (6-15) through by an integer k such that $3k \equiv 1 \pmod{391}$. We could use $k = -130$. Thus, (6-15) is equivalent to
$$-390x^2 - 910x + 5590 \equiv 0 \pmod{391};$$

that is,
$$x^2 - 128x + 64^2 \equiv 64^2 - 5590$$
$$\equiv 4096 - 5590$$
$$\equiv -1494 \equiv 70 \pmod{391};$$

or
$$(x - 64)^2 \equiv 70 \pmod{391}.$$

Hence (6-15) is solvable if and only if both $X^2 \equiv 70 \pmod{17}$ and $X^2 \equiv 70 \pmod{23}$ are solvable; that is, if both $X^2 \equiv 2 \pmod{17}$ and $X^2 \equiv 1 \pmod{23}$ are solvable. Since $\left(\frac{2}{17}\right) = 1$, we see that both of these are solvable.

Example 3 (a) Determine for what primes 3 is a quadratic residue.
(b) Determine for what primes 7 is a quadratic residue.

Solution (a) Since $x^2 \equiv 3 \pmod{2}$ is solvable, 3 is obviously a quadratic residue of 2. We desire, next, to find for what odd primes p $\left(\frac{3}{p}\right) = +1$.

By the Quadratic Reciprocity Law,
$$\left(\frac{3}{p}\right) = (-1)^{[(3-1)/2]\cdot[(p-1)/2]}\left(\frac{p}{3}\right).$$

In order that $\left(\frac{3}{p}\right)$ be 1, we must have either

(6-17) $$\frac{p-1}{2} = 2k \quad \text{and} \quad \left(\frac{p}{3}\right) = +1,$$

or

(6-18) $$\frac{p-1}{2} = 2l + 1 \quad \text{and} \quad \left(\frac{p}{3}\right) = -1.$$

For (6-17) we must have p of the form $4k + 1$ and $p = 3m + 1$; that is, $p \equiv 1 \pmod{4}$ and $p \equiv 1 \pmod{3}$; thus p is of the form $p = 12M + 1$. For (6-18) we must have $p \equiv 3 \pmod{4}$ and $p \equiv 2 \pmod{3}$; that is, p is of the form $p = 12M + 11$.

If p is of the form $12M + 5$ or $12M + 7$, then 3 is a quadratic non-residue of p. Consequently, 3 is a quadratic residue exclusively of the prime 2 and those primes of the form $12M \pm 1$.

(b) We desire to determine p so that $x^2 \equiv 7 \pmod{p}$ is solvable. Now, 7 is obviously a quadratic residue of $p = 2$. We next desire $\left(\frac{7}{p}\right) = 1$ where p is an odd prime (distinct from 7). Now, since $\left(\frac{7}{p}\right) = (-1)^{(p-1)/2}\left(\frac{p}{7}\right)$, $\left(\frac{7}{p}\right) = 1$ if and only if one of the following cases occurs:

(6-19) $$\frac{p-1}{2} = 2k \quad \text{and} \quad \left(\frac{p}{7}\right) = 1$$

or

(6-20) $$\frac{p-1}{2} = 2l + 1 \quad \text{and} \quad \left(\frac{p}{7}\right) = -1.$$

Case (6-19) arises when and only when $p \equiv 1 \pmod 4$ and $p \equiv 1, 2, 4 \pmod 7$. Using the Chinese Remainder Theorem, we see this can occur when and only when p is of one of the forms $p = 28M + 1, 28M + 9$, or $28M + 25$.

Case (6-20) occurs when and only when $p \equiv 3 \pmod 4$ and $p \equiv 3, 5, 6 \pmod 7$; that is, when and only when p is of one of the forms $p = 28M + 3$, $28M + 19, 28M + 27$. Hence, 7 is a quadratic residue of an odd prime p when and only when it is of one of the forms $28M \pm 1$, $28M \pm 3$, $28M \pm 9$.

Example 4 If p and q are primes such that $p = 4q + 1$, show that a quadratic nonresidue of p either belongs to exponent 4, modulo p, or is a primitive root of p.

Proof Let a be a quadratic nonresidue of p; thus $a^{(p-1)/2} \equiv a^{2q} \equiv -1 \pmod p$. Let a belong to exponent e, modulo p. Since e divides $p - 1 = 4q$, the only possible values for e are $1, 2, 4, q, 2q, 4q$. Since $1, 2, q, 2q$ all divide $2q$, e can have only the values 4 and $4q$.

6-7 Problems

1. Determine whether the following congruences are solvable:
    (a) $x^2 \equiv 15 \pmod{19}$;
    (b) $x^2 \equiv 6 \pmod{43}$;
    (c) $3x^2 \equiv 35 \pmod{73}$.

2. Determine which of the following congruences have solutions:
    (a) $x^2 + 2 \equiv 0 \pmod{33}$;
    (b) $2x^2 \equiv 37 \pmod{89}$;
    (c) $x^2 \equiv 67 \pmod{101}$;
    (d) $x^2 \equiv 57 \pmod{2993}$;
    (e) $x^2 - 3 \equiv 0 \pmod{253}$.

3. Determine the solvability of the congruences:
    (a) $x^2 + x + 2 \equiv 0 \pmod{67}$;
    (b) $x^2 + x + 3 \equiv 0 \pmod{343}$;
    (c) $x^2 + x + 9 \equiv 0 \pmod{221}$.

The Quadratic Reciprocity Law

4. Determine the solvability of the congruences:
   (a) $2x^2 - 3x + 1 \equiv 0 \pmod{89}$;
   (b) $x^2 + 5x \equiv 10 \pmod{101}$;
   (c) $x^2 - 891x + 24 \equiv 0 \pmod{943}$.

5. Determine the solvability of:
   (a) $x^2 \equiv 663 \pmod{4235}$;
   (b) $x^2 + 3 \equiv 0 \pmod{53}$;
   (c) $x^2 + 4x - 28 \equiv 0 \pmod{289}$;
   (d) $x^2 - x + 7 \equiv 0 \pmod{4819}$, where $4819 = 61 \cdot 79$.

6. Determine the solvability of the congruences:
   (a) $x^2 - 13x + 40 \equiv 0 \pmod{527}$, where $527 = 17 \cdot 31$;
   (b) $4x^2 + 3x - 5 \equiv 0 \pmod{97}$;
   (c) $3x^2 + 2x + 1 \equiv 0 \pmod{371}$.

7. Find the form of all primes p for which:
   (a) 5 is a quadratic residue;
   (b) 10 is a quadratic residue.

8. Find the form of all primes p for which:
   (a) 6 is a quadratic residue;
   (b) 15 is a quadratic residue.

9. (a) Prove that if c is a quadratic residue of an odd prime p, then c is not a primitive root of p.
   (b) If p is a prime of the form $2^s + 1$, prove that every quadratic nonresidue of p is a primitive root of p.

10. Let p and q be distinct odd primes. Then, if exactly one of the two congruences $x^2 \equiv p \pmod{q}$, $y^2 \equiv q \pmod{p}$ is solvable, prove that both p and q must be of the form $4k + 3$.

11. Find the form of all primes for which each of the following holds:
    (a) 9 is a quadratic residue and 11 is a quadratic nonresidue;
    (b) 8 is a quadratic residue and 13 is a quadratic nonresidue;
    (c) 11 and 13 are both quadratic residues;
    (d) 11 and 13 are both quadratic nonresidues.

12. (a) Prove that -3 is a quadratic residue exclusively of those odd primes of the form $6M + 1$.
    (b) Prove that there is an infinitude of primes of the form $6M + 1$.

13. (a) Show that the integer $n^2 + 1$ cannot have a prime factor of the form $4k + 3$.
    (b) Show that an odd integer of the form $n^2 + 2$ must have a prime factor of the form $8M + 1$ or $8M + 3$, but cannot have a prime factor of the form $8M + 5$ or $8M + 7$.

14. (a) Prove that 2 is a primitive root of $p = 4q + 1$ if p and q are primes.
 (b) Prove that 2 is a primitive root of $p = 8n + 3$ if p and $q = 4n + 1$ are primes.

15. Prove that 3 is a primitive root of a prime $p = 2^{2n} + 1$, $n \geq 1$.

16. Let p and $q = 2^a p + 1$ be odd primes. Prove that a quadratic nonresidue c of q is a primitive root of q if $c^{2^{a-1}} \not\equiv -1 \pmod{q}$.

17. Let p and q be primes such that $p = 8q + 1$ and $q = 3k + 2$, where $k > 1$. Prove that 3 is a primitive root of p.

18. Let p and q be primes such that $p = 8q + 1$ and $q = 5k + 2$ or $q = 5k + 4$. Prove that 5 is a primitive root of p.

19. Let p and q be primes such that $p = 4q + 1$ and $q = 13n + 1$. Prove that 13 is a primitive root of p.

20. Prove that the odd prime factors of $n^4 + 5$ which are different from 5 are of the form $20M + 1$, $20M + 3$, $20M + 7$, or $20M + 9$.

21. If p and q are primes such that $p = 8q + 1$ and $q = 13k + 4$, prove that 13 and 13^{13} are distinct primitive roots of p, but are not primitive roots of q.

22. If $p = 4k + 3$ is a prime, prove that a necessary and sufficient condition that $q = 2p + 1$ be a prime is that $2^p \equiv 1 \pmod{q}$. (That is, q divides the Mersenne number $M_p = 2^p - 1$.)

6-8 Another Proof of the Quadratic Reciprocity Law[3]

We shall now consider Zeller's interesting proof of the Quadratic Reciprocity Law, as modified by Frobenius.

Let p and q be distinct odd primes. Then, by Gauss's Lemma, $\left(\dfrac{p}{q}\right) = (-1)^\lambda$ and $\left(\dfrac{q}{p}\right) = (-1)^\mu$, where μ denotes the number of the least positive residues, modulo p, of

(6-21) $$1 \cdot q, 2 \cdot q, \ldots, \frac{p-1}{2} q$$

which exceed $p/2$; that is, the number of the numerically least residues modulo p of (6-21) which are negative. Moreover, λ denotes the number of numerically least residues, modulo q, of

(6-22) $$1 \cdot p, 2 \cdot p, \ldots, \frac{q-1}{2} p$$

Another Proof of the Quadratic Reciprocity Law

which are negative. Since $\left(\dfrac{p}{q}\right)\left(\dfrac{q}{p}\right) = (-1)^{\lambda+\mu}$, the Quadratic Reciprocity Law will be established if we can show that

$$\lambda + \mu \equiv \frac{p-1}{2} \frac{q-1}{2} \pmod{2}.$$

Let x range over the set

(6-23) $\qquad\qquad 1, 2, 3, \ldots, \dfrac{p-1}{2}.$

Then, for a specific value x of the set (6-23), y and r are *uniquely* determined such that $xq = yp + r$, where $-p/2 < r < p/2$ (See Theorem 2-2, Corollary). Here r is the numerically least residue obtained when xq is divided by p. Since p and q are distinct, let it be assumed that $p < q$.

Now, $p < q \leq xq < (pq)/2$; moreover, $-p/2 < -r < p/2$. On adding these inequalities, we obtain

$$\frac{p}{2} < xq - r < \frac{p(q+1)}{2}.$$

Noting that $xq - r = yp$, we obtain, on dividing these last inequalities through by p,

$$\frac{1}{2} < y < \frac{q+1}{2};$$

that is,

$$1 \leq y \leq \frac{q-1}{2}.$$

Therefore, y is a member of the set

(6-24) $\qquad\qquad 1, 2, 3, \ldots, \dfrac{q-1}{2}.$

The number μ denotes, then, the number of pairs x and y chosen from the sets (6-23) and (6-24), respectively, for which

(6-25) $\qquad\qquad -\dfrac{p}{2} < xq - yp < 0.$

Similarly, we can show that λ is the number of combinations of x and y from the sets (6-23) and (6-24), respectively, for which

(6-26) $\qquad\qquad -\dfrac{q}{2} < yp - xq < 0.$

For, if y denotes a number of the set (6-24), there are unique values X and R such that $yp = Xq + R$, where $-q/2 < R < q/2$ (cf. Theorem 2-2, Corollary). Then

$$p \leq yp < \frac{q}{2} p$$

and

$$-\frac{q}{2} < -R < \frac{q}{2}.$$

On adding, we have

$$p - \frac{q}{2} < yp - R < \frac{q(p+1)}{2};$$

that is, since p is positive,

$$-\frac{q}{2} < Xq < \frac{q(p+1)}{2}.$$

On dividing through by q, we get

$$-\frac{1}{2} < X < \frac{p+1}{2}.$$

Hence X is one of the numbers of the set $0, 1, 2, \ldots, (p-1)/2$. Moreover, we note that if $X = 0$, $R > 0$.

Hence, λ being the number of numerically least residues of (6-22) which are negative, we can say that λ is the number of combinations of x and y from the sets (6-23) and (6-24), respectively, for which $yp - xq$ is negative but greater than $-q/2$; that is, for which (6-26) is true.

Consequently, on combining (6-25) and (6-26), we see that since the combination $xq - yp = 0$ is impossible, $\mu + \lambda$ is the number of combinations of x and y from the respective sets (6-23) and (6-24) for which

(6-27)
$$-\frac{p}{2} < xq - yp < \frac{q}{2}.$$

Now, set

$$x' = \frac{p+1}{2} - x \quad \text{and} \quad y' = \frac{q+1}{2} - y.$$

As x and y take on the respective sequence of values (6-23) and (6-24), we see that x' and y' range, respectively, over the same sets of values, but in reverse order. Substituting in inequalities (6-27) x and y in terms of x' and y', we find that the inequalities (6-27) become the like inequalities

(6-28)
$$-\frac{p}{2} < x'q - y'p < \frac{q}{2}.$$

Hence, to every pair of values x and y satisfying (6-27) there corresponds another pair x' and y' satisfying the same inequalities. If each pair x', y'

The Jacobi Symbol 147

satisfying the inequalities (6-28) is different from the corresponding pair x, y satisfying the same inequalities, $\mu + \lambda$ would be even. Consider, then, the exceptional solitary case where we might have $y' = y$, $x' = x$, and, consequently, $x = x' = (p+1)/4$ and $y = y' = (q+1)/4$. Since both $(p+1)/4$ and $(q+1)/4$ are to be integers, the primes p and q must both be of the form $4M + 3$. Conversely, if p and q are distinct primes of the form $4M + 3$, $(p+1)/4$ and $(q+1)/4$ are integers. Moreover, if x and y are taken to be $(p+1)/4$ and $(q+1)/4$, respectively, then $x' = x = (p+1)/4$ and $y' = y = (q+1)/4$. Since, in this case, $\mu + \lambda$ is odd and

$$\frac{p-1}{2} \frac{q-1}{2} \equiv 1 \pmod{2},$$

we have

$$\mu + \lambda \equiv \frac{p-1}{2} \frac{q-1}{2} \pmod{2}.$$

Therefore, if merely one of the primes p and q is of the form $4M + 1$, this exceptional case cannot occur, and $\mu + \lambda$ would be even; likewise, $\{(p-1)/2\} \cdot \{(q-1)/2\}$ would be even. Hence, in every case

$$\lambda + \mu \equiv \frac{p-1}{2} \frac{q-1}{2} \pmod{2}.$$

Since the final statement of the Quadratic Reciprocity Law is symmetric in p and q, we would have obtained a like result if we had assumed $q < p$.

6-9 The Jacobi Symbol

A more general symbol, expressed in terms of Legendre's symbol, has been introduced by Jacobi. Let P be a positive odd number. Then either $P = 1$ or $P = p_1, \ldots, p_s$, where $p_1, p_2, \ldots, p_s$ are odd primes, not necessarily distinct. If m is relatively prime to P, the Jacobi symbol $\left(\frac{m}{P}\right)$ is defined as follows: $\left(\frac{m}{1}\right) = 1$, and, when $P > 1$,

$$\left(\frac{m}{P}\right) = \left(\frac{m}{p_1}\right)\left(\frac{m}{p_2}\right) \cdots \left(\frac{m}{p_s}\right).$$

Obviously, the value of the Jacobi symbol $\left(\frac{m}{P}\right)$ is either $+1$ or -1. If it is -1, then m is a quadratic nonresidue of P; but $\left(\frac{m}{P}\right)$ can be $+1$ when m is a quadratic nonresidue of P.

For example,

$$\left(\frac{3}{35}\right) = \left(\frac{3}{5}\right)\left(\frac{3}{7}\right)$$
$$= \left(\frac{2}{3}\right)(-1)\left(\frac{7}{3}\right) = (-1)(-1)\cdot 1 = 1.$$

But, since $x^2 \equiv 3 \pmod 5$ has no solution, 3 is a quadratic nonresidue of 35. On the other hand,

$$\left(\frac{3}{143}\right) = \left(\frac{3}{11}\right)\left(\frac{3}{13}\right) = 1\cdot 1 = 1;$$

and since both $x^2 \equiv 3 \pmod{11}$ and $x^2 \equiv 3 \pmod{13}$ are solvable, 3 is a quadratic residue of 143. However, if, when $P = p_1 \cdots p_s$, $\left(\frac{m}{P}\right) = -1$, it means that at least one Legendre symbol $\left(\frac{m}{p_i}\right) = -1$. Hence m is a quadratic nonresidue of P.

We shall now prove formulae for Jacobi's symbol similar to those for Legendre's symbol. We shall also prove an extended Quadratic Reciprocity Law in terms of Jacobi's symbol.

- **Theorem 6-10** Let P be positive and odd, $(m, P) = 1$, and $m \equiv n \pmod P$. Then

$$\left(\frac{m}{P}\right) = \left(\frac{n}{P}\right).$$

Proof If $P = 1$, the result is obvious. Let $P = p_1 p_2 \cdots p_s$, where the p_i are odd primes. Then

$$\left(\frac{m}{P}\right) = \left(\frac{m}{p_1}\right)\cdots\left(\frac{m}{p_s}\right)$$
$$= \left(\frac{n}{p_1}\right)\cdots\left(\frac{n}{p_s}\right) = \left(\frac{n}{P}\right).$$

- **Theorem 6-11** Let P and Q be positive odd integers and let $(m, P) = 1$ and $(n, PQ) = 1$. Then we have:

(i) $\left(\frac{m}{P}\right)\left(\frac{n}{P}\right) = \left(\frac{mn}{P}\right);$

(ii) $\left(\frac{n}{P}\right)\left(\frac{n}{Q}\right) = \left(\frac{n}{PQ}\right);$

(iii) $\left(\frac{-1}{P}\right) = (-1)^{\frac{P-1}{2}};$

(iv) $\left(\frac{2}{P}\right) = (-1)^{\frac{P^2-1}{8}}.$

Proof If $P = 1$, the parts of this theorem are obviously true. Moreover, if $Q = 1$, part (ii) is evidently true. Let, then, $P = p_1 p_2 \cdots p_s$, $Q = q_1 q_2 \cdots, q_t$, where the p_i and the q_j are odd primes.

Then
$$\left(\frac{m}{P}\right) = \left(\frac{m}{p_1}\right) \cdots \left(\frac{m}{p_s}\right),$$
$$\left(\frac{n}{P}\right) = \left(\frac{n}{p_1}\right) \cdots \left(\frac{n}{p_s}\right),$$

and
$$\left(\frac{n}{Q}\right) = \left(\frac{n}{q_1}\right) \cdots \left(\frac{n}{q_t}\right).$$

Then
$$\left(\frac{m}{P}\right)\left(\frac{n}{P}\right) = \left(\frac{m}{p_1}\right) \cdots \left(\frac{m}{p_s}\right)\left(\frac{n}{p_1}\right) \cdots \left(\frac{n}{p_s}\right)$$
$$= \left(\frac{m}{p_1}\right)\left(\frac{n}{p_1}\right)\left(\frac{m}{p_2}\right)\left(\frac{n}{p_2}\right) \cdots \left(\frac{m}{p_s}\right)\left(\frac{n}{p_s}\right)$$
$$= \left(\frac{mn}{p_1}\right)\left(\frac{mn}{p_2}\right) \cdots \left(\frac{mn}{p_s}\right)$$
$$= \left(\frac{mn}{P}\right).$$

Hence (i) is true.
Next,
$$\left(\frac{n}{PQ}\right) = \left(\frac{n}{p_1}\right)\left(\frac{n}{p_2}\right) \cdots \left(\frac{n}{p_s}\right)\left(\frac{n}{q_1}\right)\left(\frac{n}{q_2}\right) \cdots \left(\frac{n}{q_t}\right)$$
$$= \left(\frac{n}{P}\right)\left(\frac{n}{Q}\right);$$

thus (ii) is established.
Now,
$$P \equiv \{1 + (p_1 - 1)\}\{1 + (p_2 - 1)\} \cdots \{1 + (p_s - 1)\}$$
$$\equiv 1 + (p_1 - 1) + (p_2 - 1) + \ldots + (p_s - 1) \pmod{4};$$

from which it follows that

(6-29) $$\frac{P-1}{2} \equiv \frac{p_1 - 1}{2} + \frac{p_2 - 1}{2} + \ldots + \frac{p_s - 1}{2} \pmod{2}.$$

Hence
$$\left(\frac{-1}{P}\right) = \left(\frac{-1}{p_1}\right)\left(\frac{-1}{p_2}\right) \cdots \left(\frac{-1}{p_s}\right) = (-1)^{(p_1-1)/2 + (p_2-1)/2 + \ldots + (p_s-1)/2}$$
$$= (-1)^{(P-1)/2}.$$

Thus (iii) is proved.
Since $p_i^2 - 1$ is divisible by 8,

$$P^2 \equiv \{1 + (p_1^2 - 1)\}\{1 + (p_2^2 - 1)\}\ldots\{1 + (p_s^2 - 1)\}$$
$$\equiv 1 + (p_1^2 - 1) + (p_2^2 - 1) + \ldots + (p_s^2 - 1) \pmod{64}.$$

Hence,

$$\frac{P^2 - 1}{8} \equiv \frac{p_1^2 - 1}{8} + \frac{p_2^2 - 1}{8} + \ldots + \frac{p_s^2 - 1}{8} \pmod 8.$$

Therefore,

$$\left(\frac{2}{P}\right) = \left(\frac{2}{p_1}\right)\left(\frac{2}{p_2}\right)\ldots\left(\frac{2}{p_s}\right)$$
$$= (-1)^{(p_1^2-1)/8 + (p_2^2-1)/8 + \ldots + (p_s^2-1)/8}$$
$$= (-1)^{(P^2-1)/8}.$$

This completes the proof of (iv).

6-10 Generalized Quadratic Reciprocity Law

We shall state and prove a Generalized Quadratic Reciprocity Law in terms of Jacobi's Symbol.[4]

- **Theorem 6-12** If P and Q are positive, odd, relatively prime integers,

$$\left(\frac{P}{Q}\right)\left(\frac{Q}{P}\right) = (-1)^{\frac{P-1}{2} \cdot \frac{Q-1}{2}}$$

Proof If either $P = 1$ or $Q = 1$, the statement is obviously true. Let, then, $P = p_1 p_2 \ldots p_s$, and $Q = q_1 q_2 \ldots q_t$, where the p_i and q_j are odd primes and no p is a q.

Now,

$$\left(\frac{P}{Q}\right) = \left(\frac{P}{q_1}\right)\ldots\left(\frac{P}{q_t}\right) = \left(\frac{p_1}{q_1}\right)\ldots\left(\frac{p_s}{q_1}\right)\left(\frac{p_1}{q_2}\right)\ldots\left(\frac{p_s}{q_2}\right)$$
$$\ldots \left(\frac{p_1}{q_t}\right)\left(\frac{p_2}{q_t}\right)\ldots\left(\frac{p_s}{q_t}\right);$$

$$\left(\frac{Q}{P}\right) = \left(\frac{Q}{p_1}\right)\ldots\left(\frac{Q}{p_s}\right) = \left(\frac{q_1}{p_1}\right)\ldots\left(\frac{q_t}{p_1}\right)\left(\frac{q_1}{p_2}\right)\ldots\left(\frac{q_t}{p_2}\right)$$
$$\ldots \left(\frac{q_1}{p_s}\right)\ldots\left(\frac{q_t}{p_s}\right).$$

Hence

$$\left(\frac{P}{Q}\right)\left(\frac{Q}{P}\right) = \left(\frac{p_1}{q_1}\right)\left(\frac{q_1}{p_1}\right) \cdot \left(\frac{p_2}{q_1}\right)\left(\frac{q_1}{p_2}\right)\ldots\left(\frac{p_s}{q_1}\right)\left(\frac{q_1}{p_s}\right)$$
$$\times \left(\frac{p_1}{q_2}\right)\left(\frac{q_2}{p_1}\right)\ldots\left(\frac{p_s}{q_2}\right)\left(\frac{q_2}{p_s}\right)$$
$$\times \ldots \left(\frac{p_1}{q_t}\right)\left(\frac{q_t}{p_1}\right)\ldots\left(\frac{p_s}{q_t}\right)\left(\frac{q_t}{p_s}\right).$$
$$= (-1)^u,$$

The Jacobi Symbol

where

$$u = \frac{p_1-1}{2}\frac{q_1-1}{2} + \ldots + \frac{p_s-1}{2}\frac{q_1-1}{2} + \frac{p_1-1}{2}\frac{q_2-1}{2}$$
$$+ \ldots + \frac{p_s-1}{2}\frac{q_2-1}{2} + \ldots + \frac{p_1-1}{2}\frac{q_t-1}{2}$$
$$+ \ldots + \frac{p_s-1}{2}\frac{q_t-1}{2}$$
$$= \left(\frac{p_1-1}{2} + \frac{p_2-1}{2} + \ldots + \frac{p_s-1}{2}\right)$$
$$\cdot \left(\frac{q_1-1}{2} + \frac{q_2-1}{2} + \ldots + \frac{q_t-1}{2}\right).$$

Therefore, by (6-29) $u \equiv \{(P-1)/2\}\{(Q-1)/2\}$ (mod 2). Hence the theorem is established.

Example Using the Generalized Quadratic Reciprocity Law, determine whether the congruence $x^2 \equiv -22$ (mod 73) is solvable.
Now,

$$\left(\frac{-22}{73}\right) = \left(\frac{51}{73}\right) = \left(\frac{73}{51}\right) = \left(\frac{22}{51}\right) = \left(\frac{2}{51}\right)\left(\frac{11}{51}\right)$$
$$= (-1)\cdot(-1)\left(\frac{51}{11}\right) = \left(\frac{7}{11}\right) = -\left(\frac{11}{7}\right)$$
$$= -\left(\frac{4}{7}\right) = -1.$$

Hence the given congruence is not solvable.

6-11 Problems

1. Using the Generalized Quadratic Reciprocity Law, determine whether the following congruences are solvable:
    (a) $x^2 + 3 \equiv 0$ (mod 53);
    (b) $x^2 + 4x - 28 \equiv 0$ (mod 289);
    (c) $x^2 + 2x - 118 \equiv 0$ (mod 165).

2. Evaluate:
    (a) $\left(\frac{1326}{385}\right)$;
    (b) $\left(\frac{702}{535}\right)$;
    (c) $\left(\frac{86}{707}\right)$;
    (d) $\left(\frac{5900}{7007}\right)$.

3. Determine whether the following congruences are solvable:
   (a) $x^2 \equiv 663 \pmod{385}$;
   (b) $x^2 \equiv -172 \pmod{707}$.
4. Let p be a prime of the form
$$p = 2^{a_0} q_1^{a_1} \cdots q_t^{a_t} + 1,$$
where $a_0, a_1, \ldots, a_t$ are positive integers and the q's are distinct odd primes. Prove that when $(r, p) = 1$, the integer r is a primitive root of p if and only if no one of the following $t + 1$ congruences is solvable:
$$x^2 \equiv r \pmod{p},$$
$$x^{q_i} \equiv r \pmod{p} \text{ for } i = 1, 2, \ldots, t.$$
5. Prove that the prime $p = 4n + 1$ divides $n^n - 1$.

Elementary Considerations on the Distribution of Primes and Composites 7

7-1 Introduction

In the analytical theory of numbers, results from the theory of functions are applied to arithmetical problems. The analytical method, which really began with the work of P. G. L. Dirichlet, has been found to be a powerful one and is applicable to a wide range of difficult problems. This method has been successfully employed in treating the distribution of primes and in establishing the Prime Number Theorem, proved independently in 1896 by J. Hadamard and Ch. J. de la Vallée Poussin. Although the analytical approach is beyond the scope of this text, mention will be made of some of the results, concepts, and notations that have been used.

When we examine tables of some well-known arithmetical functions, we cannot help noticing their erratic behavior. For example, the function $\tau(n)$, the number of positive divisors of n, is especially erratic. Whenever n is a prime, its value is 2. Consider this function for a few consecutive values of n. We have:

$\tau(70) = 8$, $\tau(73) = 2$, $\tau(76) = 6$, $\tau(79) = 2$,
$\tau(71) = 2$, $\tau(74) = 4$, $\tau(77) = 4$, $\tau(80) = 10$,
$\tau(72) = 12$, $\tau(75) = 6$, $\tau(78) = 8$, $\tau(81) = 5$.

Where $\pi(x)$ denotes the number of primes not exceeding a real value x, we see that the number of positive integers n not exceeding x for which $\tau(n) = 2$ is the same as the value of $\pi(x)$.

7-2 The O-notation

A convenient notation for considering the size of a function $f(x)$ in terms of the size of x was introduced by Paul Bachmann[1] and was popularized by E. Landau and others.

Let $f(x)$ be a function assuming real or complex values for large real values of x; also let the function $g(x)$ be real and positive for large real values of x. If

$$\frac{|f(x)|}{g(x)}$$

is bounded for all large values of x, we may write $f(x) = O(g(x))$. More explicitly, we say that $f(x) = O(g(x))$ if there exist positive (real) constants α and β such that $|f(x)| < \beta g(x)$ whenever $x > \alpha$.

For example, taking $f(x) = x \sin x$ and $g(x) = x$, we have $x \sin x = O(x)$. Likewise,

$$\sqrt{x} = O(x), \; x^{-3} = O(x^{-5/2}), \; \phi(n) = O(n), \; \log_e x = O(\sqrt{x}),$$
$$\frac{\sin x}{x} = O(1), \; \frac{1+i}{x} = O(1), \; x = O(e^x), \; e^{ix} = O(1).$$

Let $f(x)$ and $g(x)$ be defined for large values of x and let $g(x)$ be positive for such large values of x. Then, if $\lim_{x \to \infty} \{|f(x)|/g(x)\} = 0$, we may write $f(x) = o(g(x))$. In other words, for every given $\epsilon > 0$ there exists a positive constant γ such that $|f(x)| < \epsilon g(x)$ whenever $x > \gamma$. It is to be noted that in the above statements, the constants α and β depend on the functions $f(x)$ and $g(x)$, and the constant γ depends on ϵ, $f(x)$ and $g(x)$.

For example,

$$x = o(x^{3/2}), \; \frac{x}{(\log x)^2} = o\left(\frac{x}{\log x}\right), \; x^4 = o(x^6 + 2), \; \sin x = o(\sqrt{x});$$

but $3x \neq o(x)$ and $\sqrt{x^4 + x + 1} \neq o(x^2)$.

By $O(g(x)) = O(h(x))$, we shall mean that a function $f(x)$ such that

$f(x) = O(g(x))$ is also $O(h(x))$. For example, $O(x) = O(x^2)$ means that a function $f(x)$, such as $2x + 3$, which is $O(x)$ is also $O(x^2)$. We note that, in general, we cannot say that $O(x^2) = O(x)$. For, $x^{4/3} = O(x^2)$, but $x^{4/3} \neq O(x)$. In other words, equations involving the O-symbol are to be read from left to right. Obviously, then, " $=$ " is not, in this context, an equivalence relation. Similar statements apply to the o-symbol. The O-symbol and the o-symbol may be used also where $f(x)$ and $g(x)$ are defined for only a subset of large real numbers, say for large positive integers.

We now note some obvious results. If $f(x) = o(g(x))$, then $f(x) = O(g(x))$; moreover, if $f_1(x) = O(g_1(x))$ and $f_2(x) = o(g_2(x))$, then $f_1(x) \cdot f_2(x) = o(g_1(x) \cdot g_2(x))$. Also, $O(g(x)) + o(g(x)) = O(g(x))$. Moreover, we note that if a is a positive constant and $f(x) = O(ag(x))$, then $f(x) = O(g(x))$.

We say that $f(x)$ is *of the order of* $g(x)$ when $f(x) = O(g(x))$, and $f(x)$ is *of lower order than* $g(x)$ when $f(x) = o(g(x))$. In the former case, $f(x)$ does not grow any faster than $g(x)$; in the latter case, it does not grow as fast as $g(x)$.

When $\lim_{x \to \infty} \{f(x)/g(x)\} = 1$, we say that $f(x)$ is *asymptotic* to $g(x)$ or that $f(x)$ is *asymptotically equal to* $g(x)$; symbolically, we write $f(x) \sim g(x)$. This is obviously an equivalence relation:

$$f_1(x) \sim f_1(x);$$

if $f_1(x) \sim f_2(x),$
then $f_2(x) \sim f_1(x);$
if $f_1(x) \sim f_2(x)$ and $f_2(x) \sim f_3(x),$
then $f_1(x) \sim f_3(x).$

Two functions $f_1(x)$ and $f_2(x)$ which are positive for sufficiently large x are said to have the *same order of magnitude* if there exist positive quantities α, β, and γ such that for $x > \alpha$

$$\beta < \frac{f_1(x)}{f_2(x)} < \gamma.$$

In order for $f_1(x)$ and $f_2(x)$ to be of the same order of magnitude, it is clearly necessary and sufficient that

$$f_1(x) = O(f_2(x))$$

and

$$f_2(x) = O(f_1(x)).$$

For example,

$$f_1(x) = 3x - 100 \quad \text{and} \quad f_2(x) = 2x + \sqrt{x}$$

are of the same order of magnitude: for $x \geq 52$,

$$\frac{1}{2} < \frac{f_1(x)}{f_2(x)} < \frac{3}{2}.$$

The usefulness of these symbols is seen when a complicated expression can be expressed as the sum of its dominant or principal part and an estimate of the error term. When several expressions containing an estimate of the error term are to be combined, we wish to express the result as the sum of the dominant part and an estimate of the error term. Consequently, we want to know how expressions involving these symbols may be combined. The following theorem gives some of the properties needed for this purpose.

- **Theorem 7-1** Let $f_1(x) = O(g_1(x))$, $f_2(x) = O(g_2(x))$, $g_1(x) = O(h_1(x))$, and $g_2(x) = o(h_2(x))$. Then we have the following results:
  - (i) $f_1(x) + f_2(x) = O(g_1(x) + g_2(x))$;
  - (ii) $f_1(x)f_2(x) = O(g_1(x)g_2(x))$;
  - (iii) $f_1(x) = O(h_1(x))$;
  - (iv) For a positive integer k, $\{f_1(x)\}^k = O(\{g_1(x)\}^k)$;
  - (v) $f_2(x) = o(h_2(x))$.

Proofs Now, for sufficiently large real x and for suitable positive quantities β_1 and β_2, $|f_1(x)| < \beta_1 g_1(x)$ and $|f_2(x)| < \beta_2 g_2(x)$. Moreover, where β_3 is the greater of β_1 and β_2, $|f_1(x) + f_2(x)| \leq |f_1(x)| + |f_2(x)| < \beta_3\{g_1(x) + g_2(x)\}$. Hence (i) is established. Now, for sufficiently large x, $|f_1(x)f_2(x)| = |f_1(x)| \cdot |f_2(x)| < \beta_1\beta_2 g_1(x)g_2(x)$. Thus (ii) is proved.

For $x > \alpha_1 > 0$, let $|f_1(x)| < \beta_1 g_1(x)$, and for $x > \alpha_2 > 0$, let $0 < g_1(x) < \gamma_1 h_1(x)$ for a suitable positive quantity γ_1. Then, where α_3 is the greater of α_1 and α_2, we see that for $x > \alpha_3$,

$$|f_1(x)| < \beta_1\gamma_1 h_1(x).$$

Consequently, (iii) is valid.

Taking $f_2(x)$ in (ii) to be $f_1(x)$, we obtain

$$\{f_1(x)\}^2 = O(\{g_1(x)\}^2).$$

The result in (iv) is true for $k = 1$ and 2. Assuming it to be true for $k = r$, we readily see it must be true for $k = r + 1$. Hence (iv) is established by mathematical induction on k.

For $x > \alpha_4 > 0$, let $|f_2(x)| < \beta_2 g_2(x)$. For arbitrarily given $\epsilon > 0$, let $0 < g_2(x) < \epsilon h_2(x)$ when $x > \alpha_5$. If α_6 is the larger of α_4 and α_5, we have $|f_2(x)| < \beta_2\epsilon h_2(x)$ for all $x > \alpha_6$. Since $\beta_2\epsilon$ can be arbitrarily small, we see that $f_2(x) = o(h_2(x))$. Thus part (v) is established.

7-3 Problems

1. Prove that $o(o(h(x))) = o(h(x))$.

2. Show that $f(x) \sim g(x)$, $f(x) = g(x)(1 + o(1))$, and $f(x) = g(x) + o(g(x))$ are equivalent assertions whenever $g(x) > 0$ for large x.

3. Show that $1 + 2 + 3 + \ldots + n = n^2/2 + O(n)$.

4. If $\lim_{x \to \infty} f(x)$ is infinite, show that $\lim_{x \to \infty} (\{\log f(x)\}/\log g(x)) = 1$ when $f(x) \sim g(x)$; then if $e^{f(x)} \sim e^{g(x)}$, show that $f(x) = g(x) + o(1)$.

5. Prove that
$$e^{-\sqrt{\log x}} = o(\{\log x\}^{-20}).$$

6. Prove that
$$e^{-x}(\log x)^{\log x} = O(x).$$

7. Show that
$$(x^{1/x} - 1)^x = o(1).$$

8. Show that for a positive quantity c,
$$e^{-c\sqrt{\log x \cdot \log \log x}} = O(\{\log x\}^{-4}).$$

7-4 Bertrand's Postulate

- **Theorem 7-2** (Bertrand's Postulate)[2] If the real quantity $\alpha \geq 1$, there is at least one prime p such that
$$\alpha < p \leq 2\alpha.$$

- **Corollary** If p_i is the ith prime, $p_{i+1} < 2p_i$ for $i \geq 1$.

We shall prove that for a positive integer n, there is at least one prime p satisfying the condition $n < p \leq 2n$. From this last statement, the above theorem immediately follows. For, if there exists a prime p satisfying $[\alpha] < p \leq 2[\alpha]$, then obviously $\alpha < p \leq 2\alpha$.

We shall consider the binomial coefficient
$$N = \binom{2n}{n} = \frac{(2n)!}{n! \, n!},$$

and, on the assumption that for some integer n greater than a certain specified positive integer n_0 there is *no* prime p satisfying $n < p \leq 2n$, we shall obtain one expression less than N and another greater than N. This resulting inequality, however, we shall see is impossible when $n > n_0$. Actually, we use $n_0 = 511$. Consequently, the theorem must be true for $n > n_0$. An inspection of a table of primes will verify the truth of the theorem for $1 \leq n \leq n_0$.

- **Lemma 7-3** (i) If p is an arbitrary prime $\leq 2n$, then the exponent of the highest power of p dividing $N = \binom{2n}{n}$ is given by
$$h_p = \sum_{k=1}^{\infty} \left(\left[\frac{2n}{p^k} \right] - 2 \left[\frac{n}{p^k} \right] \right).$$

(ii) Moreover,
$$\left[\frac{2n}{p^k}\right] - 2\left[\frac{n}{p^k}\right]$$
is either 1 or 0: it is 1 if $[(2n)/p^k]$ is odd, and 0 if $[(2n)/p^k]$ is even.

(iii) Also,
$$h_p \leq \left[\frac{\log 2n}{\log p}\right].$$

Proof Where
$$N = \binom{2n}{n} = \frac{(2n)!}{n!\, n!},$$
let p be an arbitrary prime $\leq 2n$. By Theorem 4-5, the exponents of the highest power of p dividing $(2n)!$ and $n!$ are, respectively,
$$\sum_{k=1}^{\infty} \left[\frac{2n}{p^k}\right] \quad \text{and} \quad \sum_{k=1}^{\infty} \left[\frac{n}{p^k}\right].$$
Hence, the exponent of the highest power of p dividing N is h_p as defined above, and part (i) is thereby proved.

Moreover,
$$0 \leq \left[\frac{2n}{p^k}\right] - 2\left[\frac{n}{p^k}\right] < \frac{2n}{p^k} - 2\left(\frac{n}{p^k} - 1\right) = 2.$$
If $[(2n)/p^k] = 2l + 1$, then $(2n)/p^k = 2l + 1 + \theta$, where $0 \leq \theta < 1$, and $n/p^k = l + (1 + \theta)/2$.

Hence,
$$\left[\frac{2n}{p^k}\right] - 2\left[\frac{n}{p^k}\right] = 2l + 1 - 2\left(l + \left[\frac{1+\theta}{2}\right]\right).$$
$$= 1.$$
On the other hand, if $[(2n)/p^k] = 2l$, then $(2n)/p^k = 2l + \theta_1$, where $0 \leq \theta_1 < 1$, and $n/p^k = l + \theta_1/2$. Therefore,
$$\left[\frac{2n}{p^k}\right] - 2\left[\frac{n}{p^k}\right] = 2l - 2\left(l + \left[\frac{\theta_1}{2}\right]\right) = 0.$$
Thus (ii) is proved.

In order to prove part (iii), let p^{r_p} be the highest power of p not exceeding $2n$. We then have $p^{r_p} \leq 2n < p^{r_p+1}$; where $r_p \log p \leq \log 2n < (r_p + 1) \log p$ and $r_p \leq (\log 2n)/\log p < r_p + 1$. Hence, since $[(2n)/p^k] - 2[n/p^k]$ is 1 or 0, and since it is obviously zero for $k > r_p$, we have

(7-1) $$h_p \leq r_p = \left[\frac{\log 2n}{\log p}\right].$$

- **Lemma 7-4** For $n \geq 3$, no prime p subject to the condition $(2/3)n < p \leq n$ can be a divisor of the binomial coefficient $N = \binom{2n}{n}$.

Proof For, if there were such a prime p, then $3p > 2n$; and p and $2p$ are the only terms of the product $1 \cdot 2 \cdot 3 \cdot \ldots \cdot (2n - 1) \cdot 2n$ in the numerator of $(2n)!/(n!\,n!)$ which are divisible by p (and, since $p > 2$, they are each divisible by the first power of p but not by p^2). Since $p \leq n$, the denominator is divisible by p^2. Consequently, when $n \geq 3$, the binomial coefficient $\binom{2n}{n}$ is *not* divisible by such a prime p subject to the restriction $(2/3)n < p \leq n$.

- **Lemma 7-5** If n is a positive integer, the product of all primes p not exceeding n is less then 2^{2n}; that is, $\prod_{p \leq n} p < 2^{2n}$.

Proof Since the result is true for $n = 1$ and $n = 2$, we shall prove the theorem by complete mathematical induction on n. Assume the result to be true for $n = 1, 2, \ldots, k$, where $k \geq 2$. If $k + 1$ is even, then

$$\prod_{p \leq k+1} p = \prod_{p \leq k} p;$$

and, by hypothesis,

$$\prod_{p \leq k} p < 2^{2k} < 2^{2k+2}.$$

Hence, the result is true for $n = k + 1$ when $k + 1$ is even. Next, let $n = k + 1$ be odd: $k + 1 = 2l + 1$. We note that obviously the integer

$$M = \binom{2l+1}{l} = \binom{2l+1}{l+1}.$$

This integer M occurs twice in the binomial expansion of $(1 + 1)^{2l+1}$; namely, as the $(l + 1)^{\text{st}}$ and $(l + 2)^{\text{nd}}$ terms of this expansion. Then, $2M < (1 + 1)^{2l+1}$ and $M < 2^{2l}$. Consequently, since the product of all the distinct primes p such that $l + 2 \leq p \leq 2l + 1$ is a divisor of M, we have, where $\prod_{a \leq p \leq b} p$ denotes the product of all primes p in the interval $a \leq p \leq b$, (Its value is 1 if there are no primes in this interval.)

(7-2) $$\prod_{l+2 \leq p \leq 2l+1} p \leq M < 2^{2l}.$$

But, by the hypothesis of the induction, since $l + 1 \leq 2l = k$,

$$\prod_{p \leq l+1} p < 2^{2l+2}.$$

Hence, using (7-2), we have

$$\prod_{p \leq 2l+1} p < 2^{2l+2} 2^{2l} = 2^{2(2l+1)}.$$

Hence, the lemma is true for $n = k + 1 = 2l + 1$. Thus, the induction is complete.

- **Lemma 7-6** For an integer $n > 1$,
$$2^{2n} < 2n\binom{2n}{n}.$$

Proof This statement is true for $n = 2$. Assuming the result to be true for $n = k$, we desire to prove it true for $n = k + 1$. Assume
$$2^{2k} < 2k\frac{(2k)!}{k!\,k!}.$$

Multiplying through by 2^2, we get
$$2^{2k+2} < 2^2 \cdot 2k\frac{(2k)!}{k!\,k!}$$
$$< 2^2 \cdot 2k\frac{(2k)!}{k!\,k!}\frac{(k+1)(2k+2)(2k+1)}{2^2 \cdot k(k+1)(k+1)}$$
$$= 2(k+1)\frac{(2k+2)!}{(k+1)!\,(k+1)!},$$

since
$$\frac{(k+1)(2k+2)(2k+1)}{2^2 k(k+1)(k+1)} = \frac{2k+1}{2k} > 1.$$

Hence the result is true for $n = k + 1$, and the theorem follows by induction on n.

Now,
$$N = \binom{2n}{n} = \frac{(2n)!}{n!\,n!} = \prod_{p \leq 2n} p^{h_p},$$

where the product is extended to every prime $p \leq 2n$. (If p does not divide N, then, of course, $h_p = 0$.)

We note that the totality of the real values η such that $1 < \eta \leq 2n$ is, for $n > 9/2$, the same as the totality of real values η given by the four non-overlapping intervals: $1 < \eta \leq \sqrt{2n}$, $\sqrt{2n} < \eta \leq (2/3)n$, $(2/3)n < \eta \leq n$, $n < \eta \leq 2n$.

We have already observed that, since each term $([(2n)/p^k] - 2[n/p^k])$ in the sum h_p of Lemma 7-3 is either 1 or 0, the exponent given by h_p cannot be greater than the largest integer r_p such that $p^{r_p} \leq 2n$. That is, the highest power of the prime p which occurs as a factor of the integer N is $\leq 2n$. From this it also follows that the prime numbers $p > \sqrt{2n}$ occur at most to the first power in the integer N (since, in this case, the r_p in the inequality $p^{r_p} \leq 2n$ is 1 if $p \leq 2n$, otherwise it is zero).

Moreover, let us now *make the assumption* that for some $n \geq 10$, there is no prime p subject to the condition $n < p \leq 2n$. In view of

Bertrand's Postulate

Lemma 7-4 and this assumption, we see that for this $n \geq 10$ there are no primes p dividing N in either of the two intervals $(2/3)n < p \leq n$ and $n < p \leq 2n$. Hence all primes $p \leq 2n$ dividing N will lie in the two intervals $1 < p \leq \sqrt{2n}$ and $\sqrt{2n} < p \leq (2/3)n$.

Hence,
$$N \leq \left(\prod_{p \leq \sqrt{2n}}(2n)\right)\left(\prod_{\sqrt{2n}<p\leq(2/3)n}p\right),$$

where in the first product we have the factor $2n$ for each prime $p \leq \sqrt{2n}$. Since there are at least two positive integers $\leq \sqrt{2n}$ (namely, 1 and 4) which are not primes, the number of primes $p \leq \sqrt{2n}$ is not greater than $\sqrt{2n} - 2$.

Hence,
$$\prod_{p \leq \sqrt{2n}}(2n) < (2n)^{\sqrt{2n}-1};$$

and by Lemma 7-5,
$$\prod_{\sqrt{2n}<p\leq(2/3)n} p \leq \prod_{p \leq (2/3)n} p$$
$$= \prod_{p \leq [(2/3)n]} p < 2^{2[(2/3)n]} \leq 2^{(4/3)n}.$$

For $n \geq 10$, we now have, employing Lemma 7-6,
$$\frac{2^{2n}}{2n} < N < (2n)^{\sqrt{2n}-1} 2^{(4/3)n}.$$

Hence,
$$2^{(2/3)n} < (2n)^{\sqrt{2n}}.$$

Taking natural logarithms, we obtain
$$\frac{2n}{3} \log 2 < \sqrt{2n} \log (2n),$$

$$\sqrt{2n} \frac{\log 2}{3} < \log (2n) = 2 \log \sqrt{2n},$$

and
$$\sqrt{2n} \log 2 < 6 \log \sqrt{2n}.$$

Thus,

(7-3) $$\sqrt{2n} \log 2 - 6 \log \sqrt{2n} < 0.$$

But, for $n = 512 = 2^9$, the left member of the inequality (7-3) becomes positive:
$$2^5 \log 2 - 6 \cdot 5 \log 2 = 2 \log 2.$$

Also, for a real variable x,

$$\frac{d}{dx}\{\sqrt{2x}\log 2 - 6\log\sqrt{2x}\} = \frac{\sqrt{2x}\log 2 - 6}{2x}$$

is positive whenever

$$x > \frac{18}{(\log 2)^2} = 37.4646.$$

Consequently, for all $n \geq 512$, the left member of the inequality (7-3) is positive. This contradicts inequality (7-3) when $n > 511$. Hence our assumption that there is no prime p satisfying the condition $n < p \leq 2n$ is untenable, and Bertrand's Postulate is true when $n \geq 512$.

In order to arrive at the theorem that, for $n \geq 1$, there exists at least one prime p satisfying the condition $n < p \leq 2n$, it is not necessary to make individual tests for $n = 1, 2, \ldots, 511$. It is sufficient to notice that of the primes

(7-4)   2, 3, 5, 7, 13, 23, 43, 83, 163, 317, 521,

each is greater than its predecessor but less than double its predecessor, and that the last exceeds 511. Now, if $1 \leq n < 512$ and if p is the first number of the sequence (7-4) greater than n and if p' denotes the immediate predecessor of p in (7-4) (and in the case $p = 2$, $p' = 1$), then $n < p \leq 2p' \leq 2n$. This completes the proof of Bertrand's Postulate (Theorem 7-2).

7-5 Problems

1. Prove that $(2n)!/(n!)^2$ is divisible by a prime greater than n, but is not divisible by the square of any such prime.

2. Show that, when $n > 1$, $n!$ cannot be an integer of the form m^s, $s \geq 2$.

3. Prove that when n is composite,

$$\frac{\pi(n-1)}{n-1} > \frac{\pi(n)}{n};$$

and when n is a prime,

$$\frac{\pi(n-1)}{n-1} < \frac{\pi(n)}{n}.$$

4. Show that when the sequence of odd numbers $n, n+2, n+4, \ldots, n+2k$, for a positive integer k, includes a prime number, there is always a number in the sequence relatively prime to all other numbers of this sequence.

7-6 Bounds³ for $\pi(x)$

- **Theorem 7-7** For $\eta \geq 2$ there are positive constants, α and β, such that
$$\alpha \frac{\eta}{\log \eta} < \pi(\eta) < \beta \frac{\eta}{\log \eta}.$$

Theorem 7-7 will follow as a consequence of Lemma 7-8.

- **Lemma 7-8**

(7-5) $$2^{(5/4)n} < \binom{2n}{n} < 2^{2n}$$

for $n \geq 2$.

Proof First,
$$2^{(5/4)n} < \binom{2n}{n}$$
is true for $n = 2$.

Let us assume that, for $k \geq 2$,
$$2^{(5/4)k} < \binom{2k}{k}.$$

Multiply through by $2^{5/4}$. Then
$$2^{5/4} 2^{(5/4)k} = 2^{(5/4)(k+1)} < 2^{5/4} \binom{2k}{k}.$$

Now
$$\binom{2k+2}{k+1} = \frac{(2k+2)(2k+1)2k \ldots (k+2)}{(k+1)k \ldots 2 \cdot 1} \cdot \frac{k+1}{k+1}$$
$$= \frac{(2k+2)(2k+1)}{(k+1)(k+1)} \cdot \frac{2k \ldots (k+1)}{k \ldots 2 \cdot 1}$$
$$= \frac{(2k+2)(2k+1)}{(k+1)(k+1)} \cdot \binom{2k}{k}.$$

We have $2^{5/4} = 2 \cdot 2^{1/4} = \frac{2k+2}{k+1} 2^{1/4} < \frac{(2k+2)(2k+1)}{(k+1)(k+1)}$

since
$$2^{1/4} < \frac{2k+1}{k+1}$$

when $k \geq 1$.

That is,
$$2^{(5/4)(k+1)} < \frac{(2k+2)(2k+1)}{(k+1)(k+1)} \binom{2k}{k} = \binom{2k+2}{k+1}.$$

Therefore, by induction on n, we have
$$2^{(5/4)n} < \binom{2n}{n}.$$

Now,
$$\binom{2n}{n} < 2^{2n}$$
is true for $n = 1$. Assume it to be true for $n = k \geq 1$; that is, assume
$$\binom{2k}{k} < 2^{2k}.$$

Then
$$\binom{2k+2}{k+1} = \frac{(2k+2)(2k+1)}{(k+1)(k+1)}\binom{2k}{k} < \frac{(2k+2)(2k+1)}{(k+1)(k+1)} 2^{2k}$$
$$< 2^2 2^{2k} = 2^{2(k+1)}.$$

Hence the result is true for $n = k + 1$. Thus, Lemma 7-8 is proved.

We shall now establish Theorem 7-7.

Since every prime p greater than n but not greater than $2n$ will divide the numerator but not the denominator of $(2n)!/(n!\,n!)$, and since the highest power of any prime dividing $\binom{2n}{n}$ is $\leq 2n$, for $n \geq 1$, we obviously have
$$\prod_{n<p\leq 2n} p \leq \binom{2n}{n} \leq \prod_{p\leq 2n} 2n,$$
where in the last product, $2n$ is taken as many times as there are primes $p \leq 2n$. That is,
$$n^{\pi(2n)-\pi(n)} \leq \binom{2n}{n} \leq (2n)^{\pi(2n)}.$$

Using natural logarithms, we get, for $n \geq 2$,

(7-6) $\qquad \{\pi(2n) - \pi(n)\} \log n \leq \log \binom{2n}{n} \leq \pi(2n) \cdot \log (2n).$

Thus,

(7-7) $\qquad \dfrac{\log \binom{2n}{n}}{\log (2n)} \leq \pi(2n),$

and
$$\{\pi(2n) - \pi(n)\} \log n \leq \log \binom{2n}{n} < 2n \log 2$$
(by (7-5)); consequently,

(7-8) $\qquad \pi(2n) - \pi(n) < 2 \log 2 \dfrac{n}{\log n} = \alpha_1 \dfrac{n}{\log n},$

where $\alpha_1 = 2 \log 2$.

Bounds for $\pi(x)$

Now, taking natural logarithms of (7-5), we have

$$\frac{5}{4}n \log 2 < \log \binom{2n}{n} < 2n \log 2.$$

Using (7-7), we have

$$\pi(2n) \geq \frac{\log \binom{2n}{n}}{\log 2n} > \frac{\frac{5}{4}n \log 2}{\log 2n}$$

$$= \left(\frac{5}{8} \log 2\right) \cdot \frac{2n}{\log 2n}$$

$$= \alpha_2 \frac{2n}{\log 2n},$$

where

$$\alpha_2 = \frac{5}{8} \log 2.$$

Let $\eta \geq 4$ and $\eta/2 = [\eta/2] + \theta$, where $0 \leq \theta < 1$.
Then

$$\pi(\eta) = \pi\left(\frac{2\eta}{2}\right) \geq \pi\left(2\left[\frac{\eta}{2}\right]\right) > \alpha_2 \frac{2[\eta/2]}{\log 2[\eta/2]}.$$

Now,

$$\frac{[\eta/2]}{\log 2[\eta/2]} = \frac{\eta/2 - \theta}{\log (\eta - 2\theta)} > \frac{\eta/2 - 1}{\log \eta} = \frac{\eta - 2}{2 \log \eta} \geq \frac{1}{4} \frac{\eta}{\log \eta}.$$

Hence, for $\eta \geq 4$,

$$\pi(\eta) > 2\alpha_2 \frac{1}{4} \frac{\eta}{\log \eta} = \alpha_3 \frac{\eta}{\log \eta},$$

where $\alpha_3 = \alpha_2/2$.
Thus for $\eta \geq 4$,

$$\pi(\eta) > \alpha_3 \frac{\eta}{\log \eta}.$$

We note that $\eta/\log \eta$ decreases in the interval $2 \leq \eta < e$, increases in the interval $e < \eta < 4$, and has a minimum value at $\eta = e$. Consequently, for $2 \leq \eta < 4$, $\pi(\eta) \geq 1$ and $\eta/\log \eta \leq 2/\log 2 = 4/\log 4$; this gives us

$$\pi(\eta) > \alpha_4 \frac{\eta}{\log \eta} \quad \text{when} \quad 0 < \alpha_4 < \frac{\log 2}{2}.$$

Hence, for $\eta \geq 2$, $\pi(\eta) > \alpha_3 (\eta/\log \eta)$, where $\alpha_3 = (5/16) \log 2$.

Next, we shall show that $\pi(\eta) < \beta\,(\eta/\log \eta)$ for a suitable positive quantity β. Let $\eta \geq 4$ and let m be the least integer not less than $\eta/2$. Thus, where the least possible value of m is 2, $m - 1 < \eta/2 \leq m$ and $2m < \eta + 2 \leq 2m + 2$. Then $\pi(\eta) = \pi(2m)$ if $2m - 1 \leq \eta \leq 2m$ and $\pi(\eta) \leq \pi(2m)$ if $2m - 2 < \eta < 2m - 1$. Moreover, $\pi(m) \leq \pi(\eta/2) + 1$. Then

$$\pi(\eta) - \pi\left(\frac{\eta}{2}\right) \leq \pi(2m) - \pi(m) + 1$$

$$< (2 \log 2)\frac{m}{\log m} + 1 \quad \text{(by (7-8))}$$

$$\leq \frac{(\log 2)2m}{\log (\eta/2)} + 1$$

$$< \frac{(\log 2)(\eta + 2)}{\log (\eta/2)} + 1$$

$$= \frac{(\log 2)\eta + (\log 2)2}{\log (\eta/2)} + 1$$

$$\leq \frac{(\log 2)\eta + (\log 2)2}{(1/2) \log \eta} + 1$$

$$= \frac{(\log 2)2\eta + 4(\log 2)}{\log \eta} + 1$$

$$\leq \frac{(3 \log 2)\eta + \log \eta}{\log \eta} = \lambda, \text{ say}.$$

Since $\eta/\log \eta$ is an increasing function of η when $\eta > e$ and since, for $\eta \geq 4$,

$$\frac{\eta}{\log \eta} \geq \frac{4}{\log 4} = \frac{2}{\log 2},$$

we have

$$(\tfrac{1}{2} \log 2)\eta \geq \log \eta.$$

Hence,

$$\lambda \leq \frac{(3 \log 2)\eta + (\tfrac{1}{2} \log 2)\eta}{\log \eta}$$

$$= \frac{7}{2} \log 2 \frac{\eta}{\log \eta}.$$

Thus, for $\eta \geq 4$,

(7-9) $$\pi(\eta) - \pi\left(\frac{\eta}{2}\right) < \frac{7}{2} \log 2 \frac{\eta}{\log \eta}.$$

Now, noting that $\pi(\eta/2) < \eta/2$ when $\eta > 0$, we have

$$\pi(\eta) \log \eta - \pi\left(\frac{\eta}{2}\right) \log \left(\frac{\eta}{2}\right)$$

$$= \pi(\eta) \log \eta - \pi\left(\frac{\eta}{2}\right)\{\log \eta - \log 2\}$$

$$= \left\{\pi(\eta) - \pi\left(\frac{\eta}{2}\right)\right\} \log \eta + \pi\left(\frac{\eta}{2}\right) \log 2$$

$$< \beta_1 \frac{\eta}{\log \eta} \log \eta + \pi\left(\frac{\eta}{2}\right) \log 2, \text{ where } \beta_1 = \frac{7}{2} \log 2, \text{ (by (7-9))}$$

$$= \beta_1 \eta + \pi\left(\frac{\eta}{2}\right) \log 2$$

$$< \beta_1 \eta + \frac{\eta}{2} \log 2$$

$$= \left(\beta_1 + \frac{\log 2}{2}\right)\eta = \beta_2 \eta, \text{ where } \beta_2 = 4 \log 2 .$$

Let $2^{s+1} \leq \eta < 2^{s+2}$, where $s \geq 1$.

Thus, since

$$\pi\left(\frac{\eta}{2^s}\right) \log \left(\frac{\eta}{2^s}\right) \leq 2 \cdot \log \left(\frac{\eta}{2^s}\right) < 2 \cdot 2 \log 2$$

$$= \beta_2 < \beta_2 \frac{\eta}{2^s} \text{ and } \pi\left(\frac{\eta}{2^{s+1}}\right) = 0,$$

$$\pi(\eta) \log \eta - \pi\left(\frac{\eta}{2}\right) \log \left(\frac{\eta}{2}\right) < \beta_2 \eta,$$

$$\pi\left(\frac{\eta}{2}\right) \log \left(\frac{\eta}{2}\right) - \pi\left(\frac{\eta}{2^2}\right) \log \left(\frac{\eta}{2^2}\right) < \beta_2 \frac{\eta}{2},$$

$$\pi\left(\frac{\eta}{2^2}\right) \log \left(\frac{\eta}{2^2}\right) - \pi\left(\frac{\eta}{2^3}\right) \log \left(\frac{\eta}{2^3}\right) < \beta_2 \frac{\eta}{2^2},$$

$$\vdots \qquad \vdots \qquad \vdots \qquad \vdots \qquad \vdots$$

$$\pi\left(\frac{\eta}{2^s}\right) \log \left(\frac{\eta}{2^s}\right) - \pi\left(\frac{\eta}{2^{s+1}}\right) \log \left(\frac{\eta}{2^{s+1}}\right) < \beta_2 \frac{\eta}{2^s} .$$

Adding these inequalities, we get:

$$\pi(\eta) \log \eta < \beta_2 \left(1 + \frac{1}{2} + \cdots + \frac{1}{2^s}\right)\eta$$

$$= \beta_2 \cdot 2\left(1 - \frac{1}{2^{s+1}}\right)\eta$$

$$< 8 \log 2 \cdot \eta .$$

Thus, for $\eta \geq 4$, $\pi(\eta) < (8 \log 2) \eta/\log \eta$. For $2 \leq \eta < 4$, $\pi(\eta) \leq 2$ and $2 < e \leq \eta/\log \eta$. Therefore, the result $\pi(\eta) < (8 \log 2) \eta/\log \eta$ is true for all $\eta \geq 2$.

Hence, we have proved the result (Theorem 7-7) that, for $\eta \geq 2$,

$$\frac{5}{16} \log 2 \frac{\eta}{\log \eta} < \pi(\eta) < 8 \log 2 \frac{\eta}{\log \eta}.$$

Theorem 7-7 tells us, then, that not only is $\pi(\eta)$ of the same order of magnitude as $\eta/\log \eta$, but that the proportion of primes in the sequence of positive integers $\leq \eta$ becomes less as η increases. In fact, since for a suitable positive constant β, $\pi(\eta)/\eta < \beta/\log \eta$, the proportion of primes tends to zero as η increases: $\pi(\eta) = o(\eta)$. We often express this fact by saying that almost all positive integers are composite.

7-7 Remarks on the Prime Number Theorem

The function $\pi(x)$, the number of primes not exceeding x, grows very irregularly with increasing x. That a simple expression can be found for $\pi(x)$ is hardly to be expected.[4] However, as early as Legendre's time, a suitable function $f(x)$ was sought to which $\pi(x)$ would be asymptotically equal; in other words, a function $f(x)$ was desired such that

$$\lim_{x \to \infty} \frac{\pi(x)}{f(x)} = 1.$$

In 1798, Legendre conjectured that $x/\log_e x$ would serve as such a function $f(x)$. More precisely, he ventured the conjecture that for large values of x, the number $\pi(x)$ is given approximately by

$$\frac{x}{\log x - 1.08366}.$$

In this connection, there are two questions involved: first, does

$$\frac{\pi(x)}{x/\log x}$$

approach a limit as x increases without limit; secondly, is this limit equal to unity? Gauss, as well as others, believed that

$$\lim_{x \to \infty} \frac{\pi(x)}{x/\log x}$$

existed and had the value 1.

P. L. Tchebychef was the first mathematician to obtain significant results concerning the behavior of $\pi(x)$ as x increases without limit. Among other results, he established that for x sufficiently large,

$$0.921 \frac{x}{\log x} < \pi(x) < 1.106 \frac{x}{\log x},$$

and that if
$$\lim_{x\to\infty}\frac{\pi(x)}{x/\log x}$$
exists, the limit is unity.

Employing the theory of functions of a complex variable, J. Hadamard and Ch. de la Vallée Poussin independently proved[5] the *Prime Number Theorem:*

$$\lim_{x\to\infty}\frac{\pi(x)}{x/\log x}=1.$$

De la Vallée Poussin proved that

$$Li(x) = \lim_{\delta\to 0}\left\{\int_0^{1-\delta}\frac{du}{\log u} + \int_{1+\delta}^x\frac{du}{\log u}\right\},$$

where $x > 1$, represents $\pi(x)$ more accurately than does $x/\log x$.

For many years, analytical methods were considered essential to a proof of this Prime Number Theorem. In 1948, however, A. Selberg and P. Erdös,[6] using elementary methods, developed an "elementary" proof of the theorem without employing the theory of analytic functions. Although technically referred to as elementary, these proofs are actually quite complicated.

It is to be noted that although the Prime Number Theorem states that $\pi(x) \sim x/\log x$, it is not to be inferred that the difference $\pi(x) - x/\log x$ becomes small, but rather that this difference is small with regard to $x/\log x$:

$$\pi(x) - \frac{x}{\log x} = o\left(\frac{x}{\log x}\right).$$

An examination of existing tables of primes reveals that, on the whole, the gap between consecutive primes tends to increase. There are also instances of very wide gaps between consecutive primes; on the other hand, instances of pairs of primes differing by 2 (such as 5 and 7, 11 and 13) occur as far as the tables extend. We see that in the sequence of consecutive positive integers there are intervals of arbitrary length without a single prime. For example, the following sequence of n consecutive integers consists entirely of composite numbers:

$$(n+1)! + 2, (n+1)! + 3, \ldots, (n+1)! + n, (n+1)! + n + 1.$$

7-8 Primes in Arithmetical Progressions

The first recorded proof of an infinitude of primes is that of Euclid (see Theorem 2-13). The question arises whether there is an infinite number of primes in an infinite arithmetical progression $l, l+k, l+2k, \ldots, l+xk,$ ... when $(k, l) = 1$. L. Euler had stated that such an arithmetical progression with $l = 1$ does contain an infinitude of primes. However, P. G. L. Dirichlet,[7] using analytical methods, was the first to prove the classical result that the infinite progression $kx + l$, where $(k, l) = 1$, contains an infinite number of primes.

Let k and l be relatively prime and positive. Let $\pi(x; k, l)$ denote the number of primes not exceeding x in the infinite arithmetical progression $ku + l$, where $u = 0, 1, 2, \ldots$. Then, where $h = \phi(k)$, the following generalization of the Prime Number Theorem has been proved:[8]

$$\pi(x; k, l) \sim \frac{1}{h} \frac{x}{\log x}.$$

Where $l_1, l_2, \ldots, l_h$ form a reduced set of (positive) residues, modulo k, then we have, as a consequence of this last statement,

$$\pi(x; k, l_i) \sim \pi(x; k, l_j).$$

7-9 Highly Composite Numbers

Of those integers exceeding one, the primes possess the least number (namely, two) of positive divisors. We shall now consider numbers which are the antithesis of primes: numbers which have many divisors. S. Ramanujan has considered[9] at considerable length numbers which he called highly composite. A *highly composite number* is an integer exceeding one which has more divisors than any preceding positive integer. The first six highly composite numbers are 2, 4, 6, 12, 24, and 36.

- **Theorem 7-9** A highly composite number is of the form

$$2^{a_2} 3^{a_3} 5^{a_5} \cdots p_m^{a_{p_m}},$$

where $a_2 \geq a_3 \geq a_5 \geq \cdots \geq a_{p_m} \geq 1$.

Proof If $h = 2^{a_2} 3^{a_3} \cdots p_m^{a_{p_m}}$ is highly composite, suppose that $a_i < a_j$ when $i < j$. Then

$$i^{a_j} j^{a_i} = i^{a_j - a_i}(ij)^{a_i} < j^{a_j - a_i}(ij)^{a_i} = i^{a_i} j^{a_j}.$$

Hence, if we interchange the exponents a_i and a_j in h, we obtain the number h' having the same number of divisors as h, yet h' is less than h. Consequently, we must have $a_2 \geq a_3 \geq \cdots \geq a_{p_m}$.

- **Theorem 7-10** For all highly composite numbers $h = 2^{a_2} 3^{a_3} \cdots p_m^{a_{p_m}}$, except 4 and 36, $a_{p_m} = 1$. For 4 and 36, $a_{p_m} = 2$.

Proof If h is highly composite, then $\tau(h') < \tau(h)$ when $h' < h$. Let $h = 2^{a_2} 3^{a_3} 5^{a_5} \cdots p_m^{a_{p_m}}$. If a prime p does not divide h, assume that $a_p = 0$. If h is a multiple of 3,

$$\frac{5}{6}h = 2^{a_2-1}3^{a_3-1}5^{a_5+1}\cdots p_m{}^{a_{p_m}}.$$

Since $\tau(\frac{5}{6}h) < \tau(h)$, $a_2 a_3(a_5+2) < (a_2+1)(a_3+1)(a_5+1)$.

Dividing the last inequality through by $a_2 a_3(a_5+1)$, we have

$$\left(1 + \frac{1}{1+a_5}\right) < \left(1 + \frac{1}{a_2}\right)\left(1 + \frac{1}{a_3}\right).$$

Now, we shall prove that $a_{p_m} \leq 2$ for all values of p_m. For, if p_{m+1} be the next prime following p_m, we see by Bertrand's Postulate that $p_{m+1} < 2p_m \leq p_m^2$. If $a_{p_m} > 2$, set

$$h_1 = \frac{hp_{m+1}}{p_m^2} < h.$$

Hence, $2(a_{p_m} - 1) < (1 + a_{p_m})$; therefore $a_{p_m} < 3$. This contradicts our hypothesis that $a_{p_m} > 2$. Hence, for all values of p_m, $a_{p_m} \leq 2$.

If $p_m \geq 5$, let $p_{m-2}, p_{m-1}, p_m, p_{m+1}, p_{m+2}$ be five consecutive primes in ascending order. Then we see that $a_{p_{m-2}} \leq 4$. For, if $a_{p_{m-2}} > 4$, consider

$$h_2 = \frac{hp_{m+1}}{p_{m-2}^3}.$$

Since[10] $p_{m+1} < p_{m-2}^3$, $h_2 < h$; and, since h is highly composite, $2(a_{p_{m-2}} - 2) < (a_{p_{m-2}} + 1)$, and $a_{p_{m-2}} \leq 4$. This contradicts the assumption that $a_{p_{m-2}} > 4$. Consequently, if $p_m \geq 5$, then $a_{p_{m-2}} \leq 4$. Let $h_3 = (hp_{m-2}p_{m+1})/(p_{m-1}p_m)$. If $5 \leq p_m \leq 19$, we can readily verify that $p_{m-1}p_m > p_{m-2}p_{m+1}$. Hence, $h_3 < h$ and $\tau(h_3) < \tau(h)$, and $2(2 + a_{p_{m-2}})a_{p_{m-1}}a_{p_m} < (1 + a_{p_{m-2}}) \cdot (1 + a_{p_{m-1}})(1 + a_{p_m})$. Dividing this inequality through by $(1 + a_{p_{m-2}})a_{p_{m-1}} \cdot a_{p_m}$, we obtain

$$2\left(1 + \frac{1}{1 + a_{p_{m-2}}}\right) < \left(1 + \frac{1}{a_{p_{m-1}}}\right)\left(1 + \frac{1}{a_{p_m}}\right).$$

Since $1 + a_{p_{m-2}} \leq 5$, from the last inequality we have

$$\frac{12}{5} < \left(1 + \frac{1}{a_{p_{m-1}}}\right)\left(1 + \frac{1}{a_{p_m}}\right).$$

Consequently, $a_{p_m} < 2$. Hence, if $5 \leq p_m \leq 19$, then $a_{p_m} = 1$.

Next, let

$$h_4 = \frac{hp_{m+1}p_{m+2}}{p_m p_{m-1}p_{m-2}}.$$

Then, when $p_m \geq 11$, we see that

(7-10) $$p_{m+1}p_{m+2} < p_m p_{m-1}p_{m-2}.$$

For, we verify its truth for $11 \leq p_m \leq 19$, and using Bertrand's Postulate, we have $p_{m+1}p_{m+2} < 2p_m 2p_{m+1} < 4p_m 2p_m < 16 p_m p_{m-1}$. But $16 p_m p_{m-1} < p_m p_{m-1} p_{m-2}$ when $16 < p_{m-2}$; that is, when $p_m \geq 23$. Hence (7-10) is true for every $p_m \geq 11$. Then, since $h_4 < h$, $\tau(h_4) < \tau(h)$; and $4 a_{p_{m-2}} a_{p_{m-1}} a_{p_m} < (1 + a_{p_{m-2}})(1 + a_{p_{m-1}})(1 + a_{p_m})$. Dividing through by $a_{p_{m-2}} a_{p_{m-1}} a_{p_m}$, we obtain

$$4 < \left(1 + \frac{1}{a_{p_{m-2}}}\right)\left(1 + \frac{1}{a_{p_{m-1}}}\right)\left(1 + \frac{1}{a_{p_m}}\right).$$

Here a_{p_m} cannot be as large as 2. Hence, again $a_{p_m} = 1$. Thus, when $p_m \geq 5$, $a_{p_m} = 1$. If $p_m = 2$ or 3, then $a_{p_m} = 1$ or 2. Now, $a_{p_m} = 2$ for the highly composite number $h = 4 = 2^2$. On the other hand, $2^a 3^2$ cannot be highly composite when $a \geq 3$. For, $2^{a-1} \cdot 3 \cdot 5 < 2^a 3^2$; and $\tau(2^{a-1} \cdot 3 \cdot 5) < \tau(2^a 3^2)$. That is, $4a < (a + 1)3$; whence $a < 3$. This contradicts our assumption. Consequently, there remains only the possibility $2^2 3^2$ which is the highly composite number 36. Every highly composite number exceeding 36 has the form $p_1^{a_1} p_2^{a_2} \cdots p_r^{a_r}$, where p_j is the jth prime and $a_1 \geq a_2 \geq \cdots \geq a_r = 1$.

Ramanujan showed that two consecutive highly composite numbers are asymptotically equal; moreover, when $\eta \geq 1$, there is always a highly composite number h satisfying the condition $\eta < h \leq 2\eta$. However, the apparent analogy with the theory of primes does not always hold. Although the infinite series formed by the reciprocals of the prime numbers is divergent, that formed by the reciprocals of the highly composite numbers is convergent.[11] It is also known that no arithmetical progression $ax + b$, where $1 \leq b \leq a - 1$ and $x = 0, 1, 2, \ldots$, contains more than a finite number of highly composite numbers.

7-10 Relatively Highly Composite Numbers

A class of numbers including the highly composite numbers of Ramanujan as a sub-class has been studied by Hardy and Ramanujan.[12]

Let $\{a_i\}$ be a sequence containing an infinitude of primes, in ascending order of magnitude. Then an integer n greater than unity is said to be (*relatively*) *highly composite* relative to this sequence if all the prime factors of n are contained in $\{a_i\}$ and if the number of divisors of n is greater than that of every smaller positive integer all of whose prime factors are contained in the given sequence. If this sequence contains all rational primes, then the corresponding relatively highly composite numbers are the highly composite numbers of Ramanujan.

Relatively Highly Composite Numbers **173**

Relative to an arbitrary sequence containing an infinitude of primes, the series of reciprocals of the relatively highly composite numbers is convergent.[13] On the other hand, unlike the case of Ramanujan's highly composite numbers, there are arithmetical progressions with a common difference greater than unity which contain an infinitude of certain relatively highly composite numbers. It has been shown that, when p is the least of the infinitude of primes contained in the given sequence and if η is a real quantity not less than unity, there exists, relative to this sequence, at least one relatively highly composite number n such that $\eta < n \leq p\eta$. All sufficiently large relatively highly composite numbers contain as a factor an arbitrary power of an arbitrarily chosen prime of the basic sequence. With respect to the basic arithmetic progression $kw + 1$, $k = 1$ or 2, $w = 0, 1, 2, \ldots$, consecutive relatively highly composite numbers are asymptotically equal.

For algebraic number fields *Highly Composite Ideals* have been defined and properties determined.[14]

7-11 Problems

1. Prove that every prime is a divisor of some highly composite number.

2. Prove that if $\{c_i\}$ is a sequence containing an infinitude of primes, in ascending order of magnitude, and if p_j denotes the jth prime of this sequence, then relative to $\{c_i\}$ a relatively highly composite number is of the form

$$n = \prod_{j=1}^{r} p_j^{a_j}, \ a_1 \geq a_2 \geq \cdots \geq a_r \geq 1.$$

3. If $p_1, p_2, \ldots, p_j \ldots$ is the infinite sequence of successive primes contained in the sequence $\{c_i\}$, prove that there exists a relatively highly composite number n satisfying the condition $\eta < n \leq p_1\eta$, where $\eta \geq 1$.

Continued Fractions 8

8-1 Introduction

We have already briefly considered, in Section 3-9, a few elementary but basic properties of simple continued fractions. The concept of a continued fraction plays an important and useful role in certain branches of the theory of numbers. We shall, in this chapter, further develop the theory and show how continued fractions can be employed in the problem of determining solutions of Pell's Equation.

Where the letters a_i and b_j ($i = 1, 2, \ldots, n$; $j = 1, 2, \ldots, n-1$) may denote complex numbers such that $a_i \neq 0$ when $i \geq 2$, an expression of the form

$$a_1 + \cfrac{b_1}{a_2 + \cfrac{b_2}{a_3 + \cfrac{b_3}{a_4 + \cfrac{\ddots}{ + \cfrac{b_{n-1}}{a_n}}}}},$$

is called a *finite continued fraction*. We shall restrict ourselves to the case where all the b's are unity. The finite continued fraction

(8-1)
$$a_1 + \cfrac{1}{a_2 + \cfrac{1}{a_3 + \cfrac{1}{a_4 + \cfrac{\ddots}{ + \cfrac{1}{a_n}}}}},$$

as well as its value, will be denoted[1] by the symbol $[a_1, a_2, \ldots, a_n]$. When a_1 is an integer ($\geqq 0$) and the $a_i (i \geq 2)$ are positive integers, the fraction is said to be a *finite simple continued fraction*. In what follows, we shall be considering simple continued fractions, often referred to simply as *continued fractions*.

For example,

$$[-1, 2, 1, 3] = -1 + \cfrac{1}{2 + \cfrac{1}{1 + \cfrac{1}{3}}} = -\frac{7}{11};$$

$$[2, 3, 5, 2] = 2 + \cfrac{1}{3 + \cfrac{1}{5 + \cfrac{1}{2}}} = \frac{81}{35};$$

$$[0, 3, 1, 4] = 0 + \cfrac{1}{3 + \cfrac{1}{1 + \cfrac{1}{4}}} = \frac{5}{19}.$$

We recall from Section 3-9 that the value obtained by stopping with the partial quotient a_i is called the ith convergent, C_i, to the simple continued fraction; also, from Theorem 3-13,

(8-2) $$C_i = \frac{p_i}{q_i},$$

where $p_1 = a_1$, $p_2 = a_2 a_1 + 1$, $p_i = a_i p_{i-1} + p_{i-2}$ ($i \geq 3$), $q_1 = 1$, $q_2 = a_2$, $q_i = a_i q_{i-1} + q_{i-2}$ ($i \geq 3$). We recall also, from Theorem 3-14, that

Finite Continued Fractions

(8-3) $\qquad p_n q_{n-1} - p_{n-1} q_n = (-1)^n \quad \text{when } n \geq 2.$

8-2 Finite Continued Fractions

It is obvious that the value of every finite simple continued fraction is a unique rational number. Conversely, we readily see that every rational number can be written as a finite simple continued fraction.

Consider the rational number $\alpha_1 = a/c, c > 0$. Using Euclid's algorithm for finding the greatest common divisor of a and c, we have, for $r \geq 1$:

(8-4)
$$\begin{aligned} a &= a_1 c + c_1, \quad 0 < c_1 < c\,; \\ c &= a_2 c_1 + c_2, \quad 0 < c_2 < c_1\,; \\ c_1 &= a_3 c_2 + c_3, \quad 0 < c_3 < c_2\,; \\ &\vdots \\ c_{r-3} &= a_{r-1} c_{r-2} + c_{r-1}, \quad 0 < c_{r-1} < c_{r-2}\,; \\ c_{r-2} &= a_r c_{r-1} \end{aligned}$$

(where $c_r = 0$, $c_{-1} = a$, and $c_0 = c$). Since the c's are positive, the a_i, $i \geq 2$, are all positive. Hence, dividing the r equations (8-4) by $c, c_1, \ldots, c_{r-2}, c_{r-1}$, respectively, we have

$$\frac{a}{c} = a_1 + \frac{1}{c/c_1}, \quad \frac{c}{c_1} = a_2 + \frac{1}{c_1/c_2}, \quad \frac{c_1}{c_2} = a_3 + \frac{1}{c_2/c_3}, \ldots,$$

$$\frac{c_{r-3}}{c_{r-2}} = a_{r-1} + \frac{1}{c_{r-2}/c_{r-1}}, \quad \frac{c_{r-2}}{c_{r-1}} = a_r.$$

Therefore,

$$\alpha_1 = \frac{a}{c} = [a_1, a_2, \ldots, a_r].$$

- **Theorem 8-1** A rational number can be represented as a finite simple continued fraction in two ways only: one with an even number of partial quotients, the other with an odd number of partial quotients.

Proof If the rational number a/c is an integer, we have, using the greatest integer function, the two representations

$$\frac{a}{c} = \left[\frac{a}{c}\right] = \left(\left[\frac{a}{c}\right] - 1\right) + \frac{1}{1}.$$

Assume that a/c is not an integer, and that, as determined above,

$$\alpha_1 = \frac{a}{c} = [a_1, a_2, \ldots, a_r], \text{ where } r \geq 2 \text{ and } a_r \geq 2.$$

We define for $1 \leq i \leq r$ the *i*th *complete quotient* (See Section 3-9)

(8-5) $$\alpha_i = [a_i, a_{i+1}, \ldots, a_r],$$

where $\alpha_r = a_r$. Then, where $i \geq 2$,

(8-6) $$\alpha_{i-1} = a_{i-1} + \frac{1}{\alpha_i},$$

which we may write

$$\alpha_{i-1} = [a_{i-1}, \alpha_i].$$

Hence, for $r \geq 2$, we have the value

(8-7) $$[a_1, a_2, \ldots, a_r] = [a_1, a_2, \ldots, a_{r-2}, \alpha_{r-1}]$$

and the value

(8-8) $$[a_1, a_2, \ldots, a_r] = a_1 + \frac{1}{\alpha_2} = a_1 + \frac{1}{[a_2, a_3, \ldots a_r]}.$$

More generally, where $2 \leq s \leq r$, we have

(8-9) $$[a_1, a_2, \ldots, a_r] = [a_1, a_2, \ldots, a_{s-1}, [a_s, \ldots, a_r]]$$
$$= [a_1, a_2, \ldots, a_{s-1}, \alpha_s].$$

Obviously, when $r > 1$, $a/c > a_1$. We also immediately see that $\alpha_i > 1$ when $i \geq 2$. For, when $i = r$, $\alpha_r = a_r \geq 2$; when $2 \leq i < r$, $\alpha_i > a_i \geq 1$. Now, a_1 is the uniquely determined greatest integer in $\alpha_1 = a/c$: $a_1 = [a/c]$. Next, we see from (8-6) that a_{i-1} is determined uniquely as the greatest integer in α_{i-1}, $i \geq 2$. For $i = 2$, the equation (8-6) uniquely determines the quantity α_2; and a_2 in turn is uniquely determined as the greatest integer in α_2. Continuing in this manner, we see that when $a_r \geq 2$, the a's are uniquely determined.

Now,

$$a_r = (a_r - 1) + \frac{1}{1} = [a_r - 1, 1]$$

and

$$\alpha_1 = \frac{a}{c} = [a_1, a_2, \ldots, a_r] = [a_1, a_2, \ldots, a_{r-1}, a_r - 1, 1].$$

The expression $[a_1, a_2, \ldots, a_{r-1}, a_r - 1, 1]$ contains one more partial quotient than does $[a_1, a_2, \ldots, a_r]$. One expression represents α_1 in terms of an even number of partial quotients and the other in terms of an odd number. These, moreover, are the only representations of α_1 as a simple continued fraction with $a_i > 0$ for $2 \leq i \leq r$.

For, if $\alpha_1 = a/c$ has the representation

$$\alpha_1 = [b_1, b_2, , \ldots b_s, 1],$$

then it has also the representation

Convergents and Their Limits **179**

$$\alpha_1 = [b_1, b_2, \ldots, b_{s-1}, (b_s + 1)],$$

where $b_s + 1 \geq 2$. However, as we have just seen, α_1 has the unique representation

$$\alpha_1 = [a_1, a_2, \ldots, a_r]$$

when $a_r \geq 2$. Hence, $s = r$, $b_i = a_i$ for $1 \leq i \leq s - 1$, and $b_s + 1 = a_r$. There is, then, only one representation in which the last partial quotient is unity.

For example,

$$\frac{10}{37} = [0, 3, 1, 2, 3] = [0, 3, 1, 2, 2, 1];$$

$$-\frac{17}{15} = [-2, 1, 6, 2] = [-2, 1, 6, 1, 1];$$

$$\frac{29}{23} = [1, 3, 1, 5] = [1, 3, 1, 4, 1].$$

8-3 Convergents and Their Limits[2]

Where a_1 is real and $a_2, a_3, \ldots, a_n, \ldots$ are positive real values, the expression

(8-10)
$$a_1 + \cfrac{1}{a_2 + \cfrac{1}{a_3 + \cdots}},$$

denoted by $[a_1, a_2, a_3, \ldots]$, is called an *infinite continued fraction*. We shall now assume that a_1 is an integer (≥ 0) and that $a_2, a_3, \ldots, a_n, \ldots$ are positive integers. In this case, the expression (8-10) is called an *infinite simple continued fraction*. If $C_n = [a_1, a_2, \ldots, a_n]$, we shall see that C_n approaches a finite limit as n increases without limit. Consequently, we define the value of this infinite simple continued fraction as $\lim_{n \to \infty} C_n$. We shall, then, employ $[a_1, a_2, a_3, \ldots]$ to denote not only the infinite simple continued fraction, but also its value.

The following result is valid whether the continued fraction is finite or infinite.

• **Theorem 8-2** The convergents of an odd order of a simple continued fraction form a strictly increasing sequence and those of an even order form a strictly decreasing sequence; moreover, every odd convergent is less than every even convergent.

Proof Now, since $p_n q_{n-1} - p_{n-1} q_n = (-1)^n$ (See Theorem 3-14),

(8-11) $$C_n - C_{n-1} = \frac{(-1)^n}{q_n q_{n-1}}$$

for $n \geq 2$; moreover, since

$$p_n q_{n-2} - p_{n-2} q_n = (a_n p_{n-1} + p_{n-2}) q_{n-2} - p_{n-2}(a_n q_{n-1} + q_{n-2})$$
$$= a_n(-1)^{n-1},$$

(8-12) $$C_n - C_{n-2} = \frac{(-1)^{n-1} a_n}{q_n q_{n-2}}$$

for $n \geq 3$. We see, therefore, that $C_n - C_{n-1}$ and $C_n - C_{n-2}$ have opposite signs. Consequently, from (8-11) and (8-12),

$$C_{2m-1} < C_{2m} < C_{2m-2}$$

(8-13) and

$$C_{2m-1} < C_{2m+1} < C_{2m}.$$

Hence, when $n \geq 3$, every convergent C_n lies between its two preceding convergents.

From (8-11), $C_1 < C_2$. Then, since from (8-13), $C_1 < C_3 < C_2$, $C_3 < C_4 < C_2$, $C_3 < C_5 < C_4$, $C_5 < C_6 < C_4$, $C_5 < C_7 < C_6, \ldots$, we see that, on continuing these inequalities, we obtain $C_1 < C_3 < C_5 < \ldots$, and $\ldots < C_6 < C_4 < C_2$. Also, from the definition of a simple continued fraction, every convergent C_i, $i \geq 3$, must lie between C_1 and C_2.

We shall now prove that every odd convergent is less than every even convergent; that is, we shall show that

(8-14) $$C_{2k-1} < C_{2l}$$

when $k \geq 1$, $l \geq 1$. Since $C_{2l} < C_{2l-2} < \ldots < C_4 < C_2$, it suffices to prove (8-14) when $2l > 2k - 1$ (that is, when $l \geq k$). Now, since $C_1 < C_3 < \ldots < C_{2k-1}$, we see from (8-13) that $C_{2k-1} \leq C_{2l-1} < C_{2l+1} < C_{2l}$; that is, $C_{2k-1} < C_{2l}$. Consequently, every odd convergent is less than every even convergent.

• **Theorem 8-3** Where a_1 is an integer (≥ 0) and $a_i (i \geq 2)$ is a positive integer, consider the infinite sequence $a_1, a_2, \ldots, a_n, \ldots$. If $C_n = p_n/q_n$ is the nth convergent of the simple continued fraction $[a_1, a_2, \ldots, a_n]$, as n increases without limit, C_n approaches a finite limit which is greater than every odd convergent C_{2k-1} and less than every even convergent C_{2l}.

Proof Since the odd convergents $C_{2k-1}(k \geq 1)$ form an increasing sequence each term of which is less than C_2, we see that $\lim_{k \to \infty} C_{2k-1} = L$ exists. Similarly, the decreasing sequence of the even convergents C_{2k}

Convergents and Their Limits

($k \geq 1$) has a limit L'. Since $q_n = a_n q_{n-1} + q_{n-2}$, $q_n > q_{n-1} \geq n - 2$ when $n \geq 3$, and q_i increases without limit as i increases without limit. Therefore, since $p_{2k} q_{2k-1} - p_{2k-1} q_{2k} = (-1)^{2k}$,

$$\lim_{k \to \infty} C_{2k} - \lim_{k \to \infty} C_{2k-1} = \lim_{k \to \infty} (C_{2k} - C_{2k-1})$$

$$= \lim_{k \to \infty} \frac{1}{q_{2k} q_{2k-1}} = 0.$$

Hence, the sequence of odd convergents and the sequence of even convergents converge to the same limit L; and, consequently, when $k \geq 1$ and $l \geq 1$,

$$C_{2k-1} < L < C_{2l}.$$

The following result we have already established in the case where β is a positive integer. However, the present more general statement will be found useful.

• **Theorem 8-4** Let a_1 be an arbitrary integer (≥ 0) and let $a_2, a_3, \ldots, a_n, \ldots$ be a sequence of positive integers. Then, for an arbitrary positive quantity β, we have

$$[a_1, a_2, \ldots, a_{n-1}, \beta] = \frac{\beta p_{n-1} + p_{n-2}}{\beta q_{n-1} + q_{n-2}}$$

when $n \geq 3$.

Proof In the proof, we shall employ induction on n. We have $p_1 = a_1$, $p_2 = a_2 a_1 + 1$, $q_1 = 1$, and $q_2 = a_2$. For $n = 3$,

$$[a_1, a_2, \beta] = \frac{\beta(a_2 a_1 + 1) + a_1}{\beta a_2 + 1} = \frac{\beta p_2 + p_1}{\beta q_2 + q_1}.$$

For every positive real quantity β, assume the result to be true for $n = k \geq 3$; namely, assume that

$$[a_1, a_2, \ldots, a_{k-1}, \beta] = \frac{\beta p_{k-1} + p_{k-2}}{\beta q_{k-1} + q_{k-2}}.$$

Now,

$$[a_1, a_2, \ldots, a_k, \beta] = \left[a_1, a_2, \ldots, a_{k-1}, a_k + \frac{1}{\beta}\right].$$

By hypothesis,

$$\left[a_1, a_2, \ldots, a_{k-1}, a_k + \frac{1}{\beta}\right] = \frac{(a_k + 1/\beta) p_{k-1} + p_{k-2}}{(a_k + 1/\beta) q_{k-1} + q_{k-2}}$$

$$= \frac{\beta(a_k p_{k-1} + p_{k-2}) + p_{k-1}}{\beta(a_k q_{k-1} + q_{k-2}) + q_{k-1}}$$

$$= \frac{\beta p_k + p_{k-1}}{\beta q_k + q_{k-1}}.$$

Hence, the result is now true for $n = k + 1$, and the theorem follows by mathematical induction.

- **Corollary 1** Where $\alpha_m = \alpha_m(n) = [a_m, a_{m+1}, \ldots, a_n]$, $m < n$, and $\lim_{n \to \infty} \alpha_m(n) = \beta_m$, we have, for $m \geq 3$,

$$\xi = [a_1, a_2, \ldots, a_n, \ldots] = [a_1, a_2, \ldots, a_{m-1}, \beta_m]$$
$$= \frac{\beta_m p_{m-1} + p_{m-2}}{\beta_m q_{m-1} + q_{m-2}}.$$

Proof Let m be a fixed integer ≥ 2. Then by definition,

$$\xi = \lim_{n \to \infty} [a_1, a_2, \ldots, a_{m-1}, [a_m, a_{m+1}, \ldots, a_n]]$$
$$= \lim_{n \to \infty} [a_1, a_2, \ldots, a_{m-1}, \alpha_m].$$

Since $[a_1, a_2, \ldots, a_{m-1}, \eta]$ is a continuous function of the variable η,

$$\xi = \lim_{n \to \infty} [a_1, a_2, \ldots, a_{m-1}, \alpha_m(n)]$$
$$= [a_1, a_2, \ldots, a_{m-1}, \lim_{n \to \infty} \alpha_m(n)]$$
$$= [a_1, a_2, \ldots, a_{m-1}, \beta_m].$$

Hence,

$$\xi = [a_1, a_2, \ldots, a_{m-1}, \beta_m]$$

when $m \geq 2$. But, since by Theorem 8-4 (when $m \geq 3$)

$$[a_1, a_2, \ldots, a_{m-1}, \beta_m] = \frac{\beta_m p_{m-1} + p_{m-2}}{\beta_m q_{m-1} + q_{m-2}},$$

the corollary is established.

We shall refer to β_m as the mth *complete quotient* of the infinite simple continued fraction $[a_1, a_2, \ldots, a_n, \ldots]$.

- **Corollary 2**

$$\xi = [a_1, a_2, \ldots, a_n, \ldots]$$
$$= [a_1, \beta_2]$$
$$= a_1 + \frac{1}{[a_2, a_3, \ldots]}.$$

Where $a_1 > 0$, the simple continued fractions $[a_1, a_2, \ldots, a_n]$ and $[a_n, a_{n-1}, \ldots, a_2, a_1]$ are called *inverses* of each other. If a simple continued fraction is identically the same as its inverse, it is called *symmetric*.

For example, $11/4$ and $11/3$ give rise to simple continued fractions which are inverses of each other; namely,

Convergents and Their Limits

$$\frac{11}{4} = [2, 1, 3],$$

$$\frac{11}{3} = [3, 1, 2].$$

Also, 224/97 may be represented by the symmetric continued fraction [2, 3, 4, 3, 2].

Example 1 Consider the simple continued fraction $[1, 2, 3, \ldots, n, \ldots]$. We find that the first four odd convergents are: $C_1 = 1.0$, $C_3 = 10/7 = 1.\overline{428571}$, $C_5 = 225/157 = 1.4331210191\ldots$, $C_7 = 9976/6961 = 1.4331274242\ldots$. The first four even convergents are: $C_2 = 3/2 = 1.50$, $C_4 = 43/30 = 1.4\overline{3}$, $C_6 = 1393/972 = 1.4331275720\ldots$, $C_8 = 81201/56660 = 1.433127426756\ldots$. The value of this infinite simple continued fraction is greater than 1.4331274242 by less than $3 \cdot 10^{-9}$.

Example 2 Consider the simple continued fraction $[1, 1, 1, \ldots]$. The successive convergents are:

$$C_1 = \frac{1}{1}, \quad C_2 = \frac{2}{1}, \quad C_3 = \frac{3}{2}, \quad C_4 = \frac{5}{3}, \quad C_5 = \frac{8}{5},$$

$$C_6 = \frac{13}{8}, \quad C_7 = \frac{21}{13}, \ldots.$$

Since the numerator p_i of C_i is $p_{i-1} + p_{i-2}$ and the denominator q_i is $q_{i-1} + q_{i-2}$, where $i \geq 3$, we note that

$$C_i = \frac{u_{i+1}}{u_i},$$

where u_i is the ith Fibonacci number. (See Section 4-11.) Let the value of this infinite simple continued fraction be $\xi = \lim_{i \to \infty} C_i$. Then

$$\xi = [1, 1, 1, \ldots] = 1 + \frac{1}{[1, 1, 1, \ldots]}$$

$$= 1 + \frac{1}{\xi}.$$

Hence $\xi^2 - \xi - 1 = 0$ and, since $\xi > 1$,

$$\xi = \frac{1 + \sqrt{5}}{2}.$$

Since $1/\xi = \xi - 1$, $1/\xi = (\sqrt{5} - 1)/2 = 0.6180339887\ldots$. Hence the limit of the ratio of consecutive Fibonacci numbers is

$$\lim_{i \to \infty} \frac{u_{i-1}}{u_i} = \frac{1}{\xi} = 0.6180339887\ldots.$$

8-4 Problems

1. Write as a simple continued fraction:

   (a) $\frac{73}{53}$; (b) $\frac{8}{15}$; (c) $-\frac{23}{29}$.

2. Find the value of each of the following continued fractions:
   (a) $[-1, 2, 1, 3, 4, 1, 2]$;
   (b) $[0, 1, 1, 3, 4, 5, 1]$;
   (c) $[0, 1, 1, 3, 4, 6]$;
   (d) $[2, 3, 1, 4, 5, 2, 2]$.

3. Find, to the nearest fifth decimal place, the value of the infinite continued fraction $[1, 2, 1, 2, 1, 2, \ldots]$.

4. Prove that the value of the infinite simple continued fraction $[1, 1, 2, 1, 2, 1, 2, \ldots]$ is $\sqrt{3}$.

5. Where u_n denotes the nth Fibonacci number (Section 4-11), prove that, for an arbitrary infinite simple continued fraction $[a_1, a_2, a_3, \ldots]$ with sth convergent $C_s = p_s/q_s$, $q_n \geq u_n$ for positive integers n.

6. Prove that the value of the infinite continued fraction $[3, 2, 2, 2, \ldots]$ is $2 + \sqrt{2}$.

7. If p_s/q_s is the sth convergent of $p/q = [a_1, a_2, \ldots, a_n]$, where $a_1 > 0$, prove that

   (a) $\dfrac{p_n}{p_{n-1}} = [a_n, a_{n-1}, \ldots a_2, a_1]$;

   (b) $\dfrac{q_n}{q_{n-1}} = [a_n, a_{n-1}, \ldots a_3, a_2]$.

8. Show that the following rational numbers can be represented as symmetric simple continued fractions:

   (a) $\frac{25}{18}$; (b) $\frac{238}{69}$; (c) $\frac{10105}{7051}$.

8-5 Representation of Irrational Numbers

We have seen that the value of a finite simple continued fraction is a rational number and that every rational number can be represented as a finite simple continued fraction. Let us now consider irrational values.

• **Theorem 8-5** The value ξ of every infinite simple continued fraction is an irrational quantity.

Representation of Irrational Numbers

Proof Since
$$\xi = [a_1, a_2, \ldots, a_n, \ldots]$$
$$= \lim_{n \to \infty} [a_1, a_2, \ldots, a_n]$$
$$= \lim_{n \to \infty} \left(a_1 + \frac{1}{[a_2, \ldots, a_n]} \right)$$
$$= a_1 + \frac{1}{\lim_{n \to \infty} [a_2, a_3, \ldots, a_n]}$$
$$= a_1 + \frac{1}{[a_2, a_3, \ldots a_n, \ldots]}$$

and since $1 \leq a_2 < [a_2, a_3, \ldots, a_n, \ldots] < a_2 + 1$, we have $a_1 < [a_1, a_2, \ldots, a_n, \ldots] < a_1 + 1$. Hence ξ cannot be an integer.

Next, assume that ξ is a rational number but not an integer. Therefore, ξ can be represented as a simple continued fraction:

$$\xi = [b_1, b_2, \ldots, b_s], \text{ where } s > 1 \text{ and } b_s > 1.$$

Consequently,

(8-15) $$\xi = [a_1, a_2, \ldots, a_s, \ldots] = [b_1, b_2, \ldots, b_s].$$

The greatest integer not exceeding ξ in (8-15) is a_1 as well as b_1. Thus, $a_1 = b_1$ and $\xi_1 = [a_2, \ldots, a_s, \ldots] = [b_2, \ldots, b_s]$. The greatest integer in ξ_1 is a_2 and is also b_2. Hence $a_2 = b_2$ and $\xi_2 = [a_3, \ldots, a_s, \ldots] = [b_3, \ldots, b_s]$. Continuing in this manner, we see that $a_i = b_i$ ($i = 1, 2, \ldots, s-1$) and

$$\frac{1}{[a_s, a_{s+1}, \ldots]} = \frac{1}{b_s}.$$

Therefore, $2 \leq b_s = [a_s, a_{s+1}, \ldots]$. However, since $a_s < [a_s, a_{s+1}, \ldots] < a_s + 1$, $[a_s, a_{s+1}, \ldots]$ cannot be an integer b_s. Consequently, the value of the infinite simple continued fraction is not rational.

- **Theorem 8-6** If two infinite simple continued fractions $[a_1, a_2, \ldots]$ and $[b_1, b_2, \ldots]$ have the same value, then $b_s = a_s$ for all $s \geq 1$.

In other words, there is *at most one* infinite simple continued fraction representing a given irrational number.

Proof If the continued fractions $[a_k, a_{k+1}, \ldots, a_n, \ldots]$ and $[b_k, b_{k+1}, \ldots, b_n, \ldots]$ are equal, it follows that

(8-16) $$\lim_{n \to \infty} [a_k, a_{k+1}, \ldots, a_n] = \lim_{n \to \infty} [b_k, b_{k+1}, \ldots, b_n];$$

consequently,

$$\lim_{n \to \infty} \left(a_k + \frac{1}{[a_{k+1}, \ldots, a_n]} \right) = \lim_{n \to \infty} \left(b_k + \frac{1}{[b_{k+1}, \ldots, b_n]} \right)$$

and

$$a_k + \frac{1}{\lim_{n\to\infty} [a_{k+1}, \ldots a_n]} = b_k + \frac{1}{\lim_{n\to\infty} [b_{k+1}, \ldots, b_n]}.$$

Since $\lim_{n\to\infty} [a_{k+1}, \ldots, a_n] > 1$ and $\lim_{n\to\infty} [b_{k+1}, \ldots, b_n] > 1$, the (unique) greatest integer not exceeding the common value of the two continued fractions $[a_k, a_{k+1}, \ldots a_n, \ldots]$ and $[b_k, b_{k+1}, \ldots, b_n, \ldots]$ is both a_k and b_k; therefore $a_k = b_k$ and $\lim_{n\to\infty} [a_{k+1}, \ldots, a_n] = \lim_{n\to\infty} [b_{k+1}, \ldots, b_n]$. Hence, whenever (8-16) is true for k (giving us $a_k = b_k$), (8-16) must also be true for $k+1$ (giving us $a_{k+1} = b_{k+1}$).

But, by hypothesis, (8-16) is true for $k = 1$ (then $a_1 = b_1$). Hence it follows by induction on k that (8-16) is true for all positive integral values of k, and $a_s = b_s$ for all $s \geq 1$.

We have seen (Theorem 8-5) that the value of every infinite simple continued fraction is irrational, and that, if an irrational quantity can be represented by a simple continued fraction, that fraction must be a unique (Theorem 8-6) infinite simple continued fraction. We shall now seek an infinite simple continued fraction whose value is a given irrational value ξ.

Let ξ be a given irrational number. Let $a_1 = [\xi]$, where $a_1 < \xi < a_1 + 1$. Then

$$\xi = a_1 + \frac{1}{1/(\xi - a_1)}.$$

Since $0 < \xi - a_1 < 1$, write the irrational number $\xi_1 = 1/(\xi - a_1) > 1$. Let $a_2 = [\xi_1]$ and write

$$\xi_1 = a_2 + \frac{1}{1/(\xi_1 - a_2)} \quad \text{and} \quad \xi_2 = \frac{1}{\xi_1 - a_2} > 1,$$

where ξ_2 is irrational. Continuing in this manner, by inductive definition we obtain irrational values $\xi_k > 1$, $k = 1, 2, \ldots$; consequently we have $a_1 = [\xi]$ and positive integers $a_k = [\xi_{k-1}]$, $k = 2, 3, \ldots$, such that $\xi_i = a_{i+1} + 1/\xi_{i+1}$, $i = 1, 2, 3, \ldots$. Thus, by successive substitution, we obtain $\xi = a_1 + 1/\xi_1 = [a_1, \xi_1] = [a_1, a_2, \xi_2] = \ldots = [a_1, a_2, \ldots, a_n, \xi_n]$ for an arbitrarily large positive integer n.

- **Theorem 8-7** If ξ is an irrational number, $a_1 = [\xi]$, and $a_k = [\xi_{k-1}]$, $k = 2, 3, \ldots$, where $\xi = a_1 + 1/\xi_1$ and $\xi_i = a_{i+1} + 1/\xi_{i+1}$ ($i = 1, 2, 3, \ldots$), then $\xi = [a_1, a_2, \ldots, a_n, \ldots]$.

Proof The first n convergents $C_k = p_k/q_k$, $1 \leq k \leq n$, of

(8-17) $\qquad [a_1, a_2, \ldots, a_n, \ldots]$

are obviously the same as the first n convergents of

(8-18) $$[a_1, a_2, \ldots, a_n, \xi_n].$$

Thus the $(n+1)^{\text{st}}$ convergent of (8-18) (by Theorem 8-4) is

$$\xi = \frac{\xi_n p_n + p_{n-1}}{\xi_n q_n + q_{n-1}}.$$

However,

$$\xi - C_n = \frac{\xi_n p_n + p_{n-1}}{\xi_n q_n + q_{n-1}} - \frac{p_n}{q_n} = \frac{(-1)(p_n q_{n-1} - p_{n-1} q_n)}{(\xi_n q_n + q_{n-1})q_n}$$

$$= \frac{(-1)^{n+1}}{(\xi_n q_n + q_{n-1})q_n}.$$

Since (for $n > 1$) $n - 1 \leq (n-1)^2 \leq q_n^2 < (\xi_n q_n + q_{n-1})q_n$, we see that the denominator $(\xi_n q_n + q_{n-1})q_n$ becomes infinite as n increases without limit. Consequently,

$$\xi - \lim_{n \to \infty} C_n = \lim_{n \to \infty} (\xi - C_n) = 0.$$

Hence the irrational number ξ has the unique (see Theorem 8-6) representation (8-17) as an infinite simple continued fraction.

The following theorem expresses the relation between the simple continued fraction representations of a positive irrational number ξ and its reciprocal.

- **Theorem 8-8** For an irrational number $\xi > 1$, the $(n+1)^{\text{st}}$ convergent of $1/\xi$ is the reciprocal of the nth convergent of ξ.

Proof Let $\xi = [a_1, a_2, \ldots]$, where the a_i $(i = 1, 2, 3, \ldots)$ are positive integers. Then

$$\frac{1}{\xi} = 0 + \frac{1}{[a_1, a_2, \ldots]} = \lim_{n \to \infty} \left(0 + \frac{1}{[a_1, a_2, \ldots, a_n]}\right)$$

$$= \lim_{n \to \infty} [0, a_1, a_2, \ldots, a_n]$$

$$= [0, a_1, a_2, \ldots, a_n, \ldots].$$

The $(n+1)^{\text{st}}$ convergent of $1/\xi$ is $[0, a_1, \ldots, a_n] = 0 + 1/[a_1, \ldots, a_n]$, and the nth convergent of ξ is $[a_1, a_2, \ldots, a_n]$; from this the result follows.

- **Corollary** For an irrational number ξ for which $0 < \xi < 1$, the $(n+1)^{\text{st}}$ convergent of ξ is the reciprocal of the nth convergent of $1/\xi$.

Example Expand $\sqrt{3}$ and $1/\sqrt{3}$ as infinite continued fractions.

Solution First, take $a_1 = [\sqrt{3}] = 1$; then
$$\sqrt{3} = 1 + \frac{1}{\xi_1} \quad \text{where} \quad \xi_1 = \frac{1}{\sqrt{3}-1} = \frac{\sqrt{3}+1}{2} > 1.$$
Then, since
$$a_2 = \left[\frac{\sqrt{3}+1}{2}\right] = 1, \quad \frac{\sqrt{3}+1}{2} = 1 + \frac{1}{\xi_2},$$
where
$$\xi_2 = \frac{2}{\sqrt{3}-1} = \sqrt{3}+1.$$
Next, $a_3 = [\sqrt{3}+1] = 2$, and
$$\xi_2 = \sqrt{3} + 1 = 2 + \frac{1}{\xi_3},$$
where
$$\xi_3 = \frac{1}{\sqrt{3}-1} = \xi_1.$$

Since the complete quotients ξ_3 and ξ_1 are equal, we now obviously have repetition. Hence $\sqrt{3} = [1, 1, 2, 1, 2, \ldots]$ with the pair 1, 2 repeating indefinitely. We say that the expansion of $\sqrt{3}$ as a simple continued fraction is *periodic*, and to indicate the repetition of the block 1, 2, we write this in the more compact form $\sqrt{3} = [1, \overline{1, 2}]$. The first seven convergents of $\sqrt{3}$ are
$$\frac{1}{1}, \frac{2}{1}, \frac{5}{3}, \frac{7}{4}, \frac{19}{11}, \frac{26}{15}, \frac{71}{41}.$$

Next, take
$$b_1 = \left[\frac{1}{\sqrt{3}}\right] = \left[\frac{\sqrt{3}}{3}\right] = 0.$$
Then
$$\frac{1}{\sqrt{3}} = 0 + \frac{1}{\eta_1},$$
where $\eta_1 = \sqrt{3}$. Using the representation found for $\sqrt{3}$, we see that
$$\frac{1}{\sqrt{3}} = 0 + \frac{1}{[1, \overline{1, 2}]} = [0, 1, \overline{1, 2}].$$
The first seven convergents of $1/\sqrt{3}$ are
$$\frac{0}{1}, \frac{1}{1}, \frac{1}{2}, \frac{3}{5}, \frac{4}{7}, \frac{11}{19}, \frac{15}{26}.$$

8-6 Approximation by Rational Numbers

We shall now consider a number of results on approximation by rational numbers.

• **Theorem 8-9** If p_n/q_n is the nth convergent of the simple continued fraction (finite or infinite) representing ξ and if the continued fraction has at least two convergents beyond the nth, then

(8-19) $$|\xi q_{s-1} - p_{s-1}| > |\xi q_s - p_s|$$

when $2 \leq s \leq n$.

Proof Let $\xi = [a_1, a_2, \ldots, a_{s-1}, \alpha_s]$, where α_s is the sth complete quotient; that is, $\alpha_s = [a_s, \ldots, a_m]$, $m \geq n + 2 \geq s + 2 \geq 4$, when ξ is rational, and $\alpha_s = \lim_{m \to \infty} [a_s, \ldots, a_m]$ when ξ is irrational. Then $0 < a_s < \alpha_s < a_s + 1$. Therefore, when $s \geq 3$, we have:

$$\alpha_s q_{s-1} + q_{s-2} < (a_s + 1)q_{s-1} + q_{s-2}$$
$$= q_s + q_{s-1}$$
$$\leq a_{s+1} q_s + q_{s-1} = q_{s+1}.$$

Hence,

(8-20) $$\alpha_s q_{s-1} + q_{s-2} < q_{s+1}.$$

Now,

$$\xi - \frac{p_{s-1}}{q_{s-1}} = \frac{\alpha_s p_{s-1} + p_{s-2}}{\alpha_s q_{s-1} + q_{s-2}} - \frac{p_{s-1}}{q_{s-1}}$$
$$= \frac{(-1)^s}{(\alpha_s q_{s-1} + q_{s-2})q_{s-1}}.$$

Consequently, by (8-20)

(8-21) $$\left|\xi - \frac{p_{s-1}}{q_{s-1}}\right| > \frac{1}{q_{s-1} q_{s+1}}.$$

Multiplying (8-21) through by q_{s-1}, we have

$$|\xi q_{s-1} - p_{s-1}| > \frac{1}{q_{s+1}}.$$

But, since

$$\left|\xi - \frac{p_s}{q_s}\right| < \left|\frac{p_{s+1}}{q_{s+1}} - \frac{p_s}{q_s}\right| < \frac{1}{q_s q_{s+1}},$$

then

$$|\xi q_{s-1} - p_{s-1}| > |\xi q_s - p_s|.$$

Hence the theorem is established for $s \geq 3$.

Next, since
$$\xi = [a_1, a_2, \ldots, a_{s-1}, \alpha_s] = a_1 + \frac{1}{\alpha_2},$$
where $0 < a_2 < \alpha_2 < a_2 + 1$,
and since $p_1 = a_1$, $p_2 = a_2 a_1 + 1$, $q_1 = 1$, $q_2 = a_2$, we have
$$\xi q_1 - p_1 = \left(a_1 + \frac{1}{\alpha_2}\right) - a_1 = \frac{1}{\alpha_2};$$
also
$$\xi q_2 - p_2 = \left(a_1 + \frac{1}{\alpha_2}\right) a_2 - (a_2 a_1 + 1) = -\frac{\alpha_2 - a_2}{\alpha_2}.$$
Consequently, since $0 < \alpha_2 - a_2 < 1$,
$$|\xi q_1 - p_1| = \frac{1}{\alpha_2} > |\xi q_2 - p_2|.$$
This establishes the case $s = 2$.

- **Corollary 1**

(8-22) $$|\xi q_1 - p_1| > |\xi q_2 - p_2| > |\xi q_3 - p_3| > \cdots$$
$$> |\xi q_{n-1} - p_{n-1}| > |\xi q_n - p_n|.$$

Proof Let t be a fixed value $1 \leq t \leq n - 2$. Consider
(8-23) $$|\xi q_1 - p_1| > |\xi q_2 - p_2| > \cdots > |\xi q_{1+t} - p_{1+t}|.$$
Since (8-23) is true for $t = 1$ and since (8-23) is true for $t + 1$ whenever it is true for t, (8-22) follows by induction on t.

- **Corollary 2** Each convergent p_s/q_s of a simple continued fraction is closer to the value of the continued fraction than is its immediately preceding convergent.

Proof Since $|\xi q_s - p_s| < |\xi q_{s-1} - p_{s-1}|$ and $q_{s-1} \leq q_s$ $(s \geq 2)$, we have
$$\left|\xi - \frac{p_s}{q_s}\right| < \frac{q_{s-1}}{q_s}\left|\xi - \frac{p_{s-1}}{q_{s-1}}\right| \leq \left|\xi - \frac{p_{s-1}}{q_{s-1}}\right|.$$

- **Corollary 3** For $s \geq 1$,
$$\frac{1}{2 q_s q_{s+1}} < \left|\xi - \frac{p_s}{q_s}\right| < \frac{1}{q_s q_{s+1}}.$$

Proof Since, by Theorem 3-14,
$$\frac{1}{q_s q_{s+1}} = \left|\frac{p_{s+1}}{q_{s+1}} - \frac{p_s}{q_s}\right| = \left|\left(\frac{p_{s+1}}{q_{s+1}} - \xi\right) + \left(\xi - \frac{p_s}{q_s}\right)\right|$$
$$= \left|\frac{p_{s+1}}{q_{s+1}} - \xi\right| + \left|\xi - \frac{p_s}{q_s}\right| < 2\left|\xi - \frac{p_s}{q_s}\right|,$$

Approximation by Rational Numbers

we have

$$\frac{1}{2q_s q_{s+1}} < \left|\xi - \frac{p_s}{q_s}\right|.$$

Moreover,

$$\left|\xi - \frac{p_s}{q_s}\right| < \left|\frac{p_{s+1}}{q_{s+1}} - \frac{p_s}{q_s}\right| = \frac{1}{q_s q_{s+1}}.$$

From these inequalities the statement follows. We note that, since $q_s \leq q_{s+1}$, the following weaker relationship is true:

$$\frac{1}{2q_s q_{s+1}} < \left|\xi - \frac{p_s}{q_s}\right| < \frac{1}{q_s^2}.$$

We shall now prove a result[3] useful in showing that the "best" approximations to an irrational number are the convergents of its continued fraction.

- **Theorem 8-10** If $n > 2$ and if $0 < q \leq q_n$, then, when $(p, q) = 1$ and p/q is distinct from the nth convergent of the continued fraction for the quantity ξ, it follows that

(8-24) $$|p_n - q_n \xi| < |p - q\xi|.$$

Proof Let $3 \leq k \leq n$. We desire to prove that

$$|p_n - q_n \xi| < |p - q\xi|$$

when $q_{k-1} < q \leq q_k$; and then also to show that (8-24) is valid when $1 = q_1 \leq q \leq q_2$. Since q must occur in one of the non-overlapping intervals $q_1 \leq x \leq q_2, q_2 < x \leq q_3, \ldots, q_{n-1} < x \leq q_n$, the result (8-24) would then be valid for all positive integers q not exceeding q_n.

Consider, therefore, $q_{k-1} < q \leq q_k$, $2 \leq k \leq n$. First of all, if $k < n$, $q = q_k$, and $p/q = p_k/q_k$, then obviously $p = p_k$ and (by Theorem 8-9, Corollary 1)

$$|p_n - q_n \xi| < |p_k - q_k \xi| = |p - q\xi|.$$

Second, if $2 \leq k \leq n$, $q = q_k$, and $p/q \neq p_k/q_k$, then, since (by hypothesis) $p \neq p_k$,

$$\left|\frac{p_k}{q_k} - \frac{p}{q_k}\right| \geq \frac{1}{q_k}.$$

However, we have, by Theorem 8-9, Corollary 3 (when $k \geq 2$)

$$\left|\frac{p_k}{q_k} - \xi\right| < \frac{1}{q_{k+1} q_k} \leq \frac{1}{2q_k}.$$

We now have

$$\frac{1}{q_k} \leq \left|\frac{p_k}{q_k} - \frac{p}{q_k}\right|$$
$$= \left|\left(\frac{p_k}{q_k} - \xi\right) + \left(\xi - \frac{p}{q_k}\right)\right|$$
$$\leq \left|\frac{p_k}{q_k} - \xi\right| + \left|\xi - \frac{p}{q_k}\right|$$
$$< \frac{1}{2q_k} + \left|\xi - \frac{p}{q_k}\right|.$$

Therefore,

$$\left|\frac{p_k}{q_k} - \xi\right| < \frac{1}{2q_k} < \left|\xi - \frac{p}{q_k}\right|.$$

Hence, on multiplying through by q_k, we have (since $q = q_k$)

$$|p_k - q_k\xi| < |p - q_k\xi| = |p - q\xi|;$$

consequently,

$$|p_n - q_n\xi| < |p - q\xi|.$$

Third, let $q_{k-1} < q < q_k$ (where $2 \leq k \leq n$). Since q does not divide q_{k-1}, we have $p/q \neq p_{k-1}/q_{k-1}$; similarly, since q_k does not divide q, $p/q \neq p_k/q_k$.

Consider the two simultaneous equations in α and β:

(8-25) $$\alpha p_k + \beta p_{k-1} = p,$$
(8-26) $$\alpha q_k + \beta q_{k-1} = q.$$

Since $p_k q_{k-1} - p_{k-1} q_k = (-1)^k$, we find on solving that

$$\alpha = (-1)^k(pq_{k-1} - qp_{k-1}) \quad \text{and} \quad \beta = (-1)^k(qp_k - pq_k).$$

Hence α and β are both non-zero integers. Since $0 < q = \alpha q_k + \beta q_{k-1} < q_k$, α and β must have opposite signs. Moreover, since ξ lies between p_{k-1}/q_{k-1} and p_k/q_k, $p_k - q_k\xi$ and $p_{k-1} - q_{k-1}\xi$ have opposite signs. Hence,

$$\alpha(p_k - q_k\xi) \quad \text{and} \quad \beta(p_{k-1} - q_{k-1}\xi)$$

have the same sign and the numerical value of their sum is the sum of their numerical values. Consequently,

$$|p - q\xi| = |\alpha p_k + \beta p_{k-1} - (\alpha q_k + \beta q_{k-1})\xi|$$
$$= |\alpha(p_k - q_k\xi) + \beta(p_{k-1} - q_{k-1}\xi)|$$
$$= |\alpha| \cdot |p_k - q_k\xi| + |\beta| \cdot |p_{k-1} - q_{k-1}\xi|$$
$$\geq |p_k - q_k\xi| + |p_{k-1} - q_{k-1}\xi|$$
$$> |p_k - q_k\xi| \geq |p_n - q_n\xi|.$$

Approximation by Rational Numbers **193**

Finally, let $q_1 \leq q \leq q_2$. We have already seen from the cases above (when $k = 2$) that (8-24) follows when $q = q_2$ and when $q_1 < q < q_2$. Also, (8-24) follows when $q = q_1$ and $p/q = p_1/q_1$. For, $q = q_1 = 1$ and $p = p_1$; from which it follows that

$$|p_n - q_n\xi| < |p_1 - q_1\xi| = |p - q\xi|.$$

For the case where $q = q_1$ and $p/q \neq p_1/q_1$, we have

$$\left|\frac{p}{q} - \frac{p_1}{q_1}\right| \geq \frac{1}{q_1} = 1\ ;$$

that is,

$$|p - p_1| \geq 1.$$

Writing

$$\xi = [a_1, \alpha_2] = p_1 + \frac{1}{\alpha_2},$$

we see that

(8-27)
$$\begin{aligned}|p_2 - q_2\xi| &= \left|a_2 p_1 + 1 - a_2\left(p_1 + \frac{1}{\alpha_2}\right)\right| \\ &= 1 - \frac{a_2}{\alpha_2},\end{aligned}$$

and, since

$$|p - \xi| + |\xi - p_1| \geq |(p - \xi) + (\xi - p_1)| = |p - p_1| \geq 1,$$

we have

(8-28) $\quad |p - q\xi| = |p - \xi| \geq 1 - |\xi - p_1| = 1 - \frac{1}{\alpha_2}.$

Hence, if $n > 2$, $\alpha_2 > a_2$ and $0 < 1 - a_2/\alpha_2 \leq 1 - 1/\alpha_2$. Consequently, since $|p_3 - q_3\xi| < |p_2 - q_2\xi|$, we have, by (8-27) and (8-28),

$$|p_n - q_n\xi| < |p - q\xi|.$$

- **Corollary 1** If $n > 2$ and $0 < q \leq q_n$ and if $p/q \neq p_n/q_n$, then

$$\left|\frac{p_n}{q_n} - \xi\right| < \left|\frac{p}{q} - \xi\right|.$$

- **Corollary 2** If, for $n > 2$ and for some positive integer q, we have

$$\left|\frac{p}{q} - \xi\right| < \left|\frac{p_n}{q_n} - \xi\right|,$$

then $q > q_n$.

These corollaries show that there is no rational number with positive denominator $\leq q_n$ that gives a better approximation to the value of the continued fraction than the convergent p_n/q_n.

For example, the first seven convergents for $\sqrt{3} = [1, \overline{1, 2}]$ are

$$\frac{1}{1}, \frac{2}{1}, \frac{5}{3}, \frac{7}{4}, \frac{19}{11}, \frac{26}{15}, \frac{71}{41}.$$

Since $71/41 = 1.\overline{73170}$, the seventh convergent differs from $\sqrt{3}$ by $0.00034\ldots$. However, $72/41 = 1.\overline{75609}$ (which is not a convergent) differs from $\sqrt{3}$ by 0.02405. Although $433/250$ differs from the value of $\sqrt{3}$ by $0.00005\ldots$, we note that the denominator of this fraction exceeds $q_7 = 41$.

- **Theorem 8-11** If there are at least four convergents to the continued fraction representing the quantity ξ, of any two consecutive convergents to ξ at least one, p/q, satisfies the inequality

(8-29)
$$\left| \frac{p}{q} - \xi \right| < \frac{1}{2q^2}.$$

Proof Since

$$\left| \frac{p_{k+1}}{q_{k+1}} - \frac{p_k}{q_k} \right| = \left| \frac{p_{k+1}q_k - p_k q_{k+1}}{q_k q_{k+1}} \right| = \frac{1}{q_k q_{k+1}}$$

and since ξ lies between any two consecutive convergents,

$$\frac{1}{q_k q_{k+1}} = \left| \frac{p_{k+1}}{q_{k+1}} - \frac{p_k}{q_k} \right| = \left| \left(\frac{p_{k+1}}{q_{k+1}} - \xi \right) + \left(\xi - \frac{p_k}{q_k} \right) \right|$$

$$= \left| \frac{p_{k+1}}{q_{k+1}} - \xi \right| + \left| \frac{p_k}{q_k} - \xi \right|.$$

If the theorem were false for the two consecutive convergents p_k/q_k and p_{k+1}/q_{k+1}, then

$$\left| \frac{p_{k+1}}{q_{k+1}} - \xi \right| \geq \frac{1}{2q_{k+1}^2} \quad \text{and} \quad \left| \frac{p_k}{q_k} - \xi \right| \geq \frac{1}{2q_k^2};$$

this yields

$$\frac{1}{q_k q_{k+1}} \geq \frac{1}{2q_{k+1}^2} + \frac{1}{2q_k^2}.$$

Consequently

$$0 \geq \frac{1}{2} \left(\frac{q_k - q_{k+1}}{q_k q_{k+1}} \right)^2.$$

That is, $q_{k+1} = q_k$. If we have $k > 1$, $q_{k+1} = a_{k+1} q_k + q_{k-1}$; as a result, $a_{k+1} q_k + q_{k-1} = q_k$ and $q_{k-1} = q_k(1 - a_{k+1})$. Since $q_i \geq 1$ for all $i > 0$, this last equation is impossible. Consequently, we must have $k = 1$

Approximation by Rational Numbers **195**

and $a_2 = q_2 = q_1 = 1$. Moreover, $\xi = [a_1, 1, \alpha_3] = \{a_1 + \alpha_3(a_1 + 1)\}/\{\alpha_3 + 1\}$, where $\alpha_3 > a_3 \geq 1$. Consider the consecutive convergents $p_1/q_1 = a_1/1$ and $p_2/q_2 = (a_1 + 1)/1$, neither of which, by hypothesis, satisfies (8-29). However,

$$\left|\frac{p_2}{q_2} - \xi\right| = \left|\frac{a_1 + 1}{1} - \frac{a_1 + \alpha_3(a_1 + 1)}{\alpha_3 + 1}\right| = \frac{1}{\alpha_3 + 1} < \frac{1}{2} = \frac{1}{2 \cdot q_2^2}.$$

Thus, p_2/q_2 *does* satisfy (8-29). Consequently $k \neq 1$. The assumption that the theorem is not true leads to a contradiction. This establishes the theorem.

As a result of this theorem, we see that every irrational number (having an infinite continued fraction representation) has an infinite number of convergents which give closer and closer rational approximations to this irrational value.[4] We shall now prove that this inequality is a characteristic property only of convergents to ξ.

- **Theorem 8-12** If the rational number p/q, $q > 0$, $(p, q) = 1$, satisfies the condition

$$\left|\frac{p}{q} - \xi\right| < \frac{1}{2q^2},$$

then p/q is a convergent to the irrational number ξ.

Proof The rational number p/q can be expressed as a finite simple continued fraction

$$\frac{p}{q} = [a_1, a_2, \ldots, a_n],$$

with convergents p_k/q_k, $1 \leq k \leq n$; and since $p/q = p_n/q_n$ (thus $p = p_n$ and $q = q_n$),

$$\left|\xi - \frac{p}{q}\right| < \frac{1}{2q_n^2}.$$

Write $p/q - \xi = \pm \theta/q^2$, where $0 < \theta < \frac{1}{2}$. Since n may be considered odd or even, as we desire (Theorem 8-1), let n be odd when $\xi > p/q$ and even when $\xi < p/q$. Hence, we may write

$$\frac{p_n}{q_n} - \xi = \frac{p}{q} - \xi = \frac{(-1)^n \theta}{q^2}.$$

If, after making the choice of the value of n, we have $n = 1$, we see that this occurs only when $\xi > p/q$ and $q = 1$; that is, by the hypothesis of the theorem, $0 < \xi - p/q < \frac{1}{2}$; thus, $[\xi] = p/q$, and p/q is the first convergent in the infinite continued fraction representation of ξ. Consequently, the theorem is true in the case where $n = 1$.

Let us now assume that $n > 1$. Define a real (irrational) value β by the equation

$$\xi = \frac{\beta p_n + p_{n-1}}{\beta q_n + q_{n-1}};$$

from which it follows (provided $\beta > 0$) that (Theorem 8-4)

$$\xi = [a_1, a_2, \ldots, a_n, \beta] \quad \text{and} \quad \beta = \frac{p_{n-1} - \xi q_{n-1}}{\xi q_n - p_n}.$$

Hence,

$$\frac{p}{q} - \xi = \frac{(-1)^n \theta}{q^2} = \frac{(-1)^n \theta}{q_n^2} = \frac{p_n}{q_n} - \xi$$

$$= \frac{p_n}{q_n} - \frac{\beta p_n + p_{n-1}}{\beta q_n + q_{n-1}} = \frac{(-1)^n}{q_n(\beta q_n + q_{n-1})};$$

thus, since $1/\theta > 2$ and $q_{n-1}/q_n \leq 1$, we obtain on solving for β

$$\beta = \frac{1}{\theta} - \frac{q_{n-1}}{q_n} > 1.$$

Writing the irrational value β as an infinite simple continued fraction, we have

$$\beta = [b_1, b_2, \ldots, b_m, \ldots],$$

where $b_1 = [\beta] \geq 1$. Consider, then, the infinite simple continued fraction

$$\eta = [a_1, a_2, \ldots, a_n, b_1, \ldots, b_m, \ldots]$$
$$= [a_1, a_2, \ldots, a_n, [b_1, \ldots, b_m, \ldots]]$$
$$= [a_1, a_2, \ldots, a_n, \beta]$$
$$= \xi.$$

Since the expansion of ξ is unique, its representation as a simple continued fraction is

$$\xi = [a_1, a_2, \ldots, a_n, b_1, b_2, \ldots],$$

with nth convergent $[a_1, a_2, \ldots, a_n] = p/q$.

8-7 Problems

1. Express $\sqrt{6}$ as a simple continued fraction.

2. Find the representation of $\sqrt{21}$ as a simple continued fraction.

3. For the infinite continued fraction $[1, 2, 1, 3, 1, 4, 1, 5, \ldots]$ find the best rational approximation p/q:
    (a) with denominator $q < 15$;
    (b) with denominator $q < 10$;
    (c) with denominator $q < 475$.

Quadratic Irrational Numbers

4. For the infinite continued fraction $[1, 2, 2, 2, \ldots]$, find the best rational approximation p/q with denominator $q < 44$.

5. (a) Show that each convergent p/q to $\sqrt{2}$ with $1 < q < 400$ differs from $\sqrt{2}$ by less than $2/(5q^2)$ (and, consequently, by less than $1/(\sqrt{5}q^2)$).
   (b) Find the first convergent that differs from $\sqrt{2}$ by less than $1/3200$.

6. Consider the infinite sequence: $p_1/q_1 = 0/1$, $p_2/q_2 = 1/1$, $p_3/q_3 = 1/2, \ldots, p_k/q_k, \ldots$, where $p_k = p_{k-1} + p_{k-2}$ and $q_k = q_{k-1} + q_{k-2}$ when $k \geq 3$. Show that each of the rational numbers p_k/q_k is in its lowest terms and less than 1 when $k \geq 3$; also, show that this sequence converges to a limit. Then determine the value of this limit, and evaluate it to five decimal places.

8-8 Quadratic Irrational Numbers

This section briefly considers a concept that will be useful in treating periodic continued fractions.

Let d be a positive integer which is not a perfect square, and let r and s be rational numbers. If $\xi = r + s\sqrt{d}$, the number $\xi' = r - s\sqrt{d}$ is called the *conjugate* of ξ.

We note that the conjugate of $\xi' = r - s\sqrt{d}$ is ξ; that is, $(\xi')' = \xi$. Moreover, $\xi' = \xi$ when and only when $s = 0$.

We define a *quadratic irrational number* as one which may be expressed in the form $(e + \sqrt{d})/f$, where e is an integer, f a non-zero integer, and d a positive integer which is not a perfect square.

If $\xi = (e + \sqrt{d})/f$ is a quadratic irrational number, then so is

$$\xi' = \frac{e - \sqrt{d}}{f} = \frac{-e + \sqrt{d}}{-f}.$$

• **Theorem 8-13** If $\xi = r_1 + s_1\sqrt{d}$ and $\eta = r_2 + s_2\sqrt{d}$, where d is a positive integer not a perfect square and r_1, s_1, r_2, and s_2 are rational numbers, we have the following results:

(i) $(\xi + \eta)' = \xi' + \eta'$;
(ii) $(\xi\eta)' = \xi'\eta'$;
(iii) $\left(\dfrac{\xi}{\eta}\right)' = \dfrac{\xi'}{\eta'}$ when $\eta \neq 0$;
(iv) If $\theta = \dfrac{r + s\xi}{t + u\xi}$, where r, s, t, u are rational numbers, then

$$\theta' = \frac{r + s\xi'}{t + u\xi'}.$$

Proof Since $\xi + \eta = r_1 + r_2 + (s_1 + s_2)\sqrt{d}$,
$$(\xi + \eta)' = r_1 + r_2 - (s_1 + s_2)\sqrt{d}$$
$$= (r_1 - s_1\sqrt{d}) + (r_2 - s_2\sqrt{d})$$
$$= \xi' + \eta'.$$

Hence part (i) is true.

Next, since
$$(\xi\eta)' = (r_1 r_2 + d s_1 s_2) - (r_1 s_2 + r_2 s_1)\sqrt{d}$$
$$= (r_1 - s_1\sqrt{d})(r_2 - s_2\sqrt{d})$$
$$= \xi'\eta'.$$

Thus part (ii) is established.

Moreover, since
$$\frac{\xi}{\eta} = \frac{r_1 + s_1\sqrt{d}}{r_2 + s_2\sqrt{d}} \cdot \frac{r_2 - s_2\sqrt{d}}{r_2 - s_2\sqrt{d}}$$
$$= \frac{r_1 r_2 - d s_1 s_2}{r_2^2 - d s_2^2} + \frac{-r_1 s_2 + r_2 s_1}{r_2^2 - d s_2^2}\sqrt{d},$$
$$\left(\frac{\xi}{\eta}\right)' = \frac{r_1 r_2 - d s_1 s_2}{r_2^2 - d s_2^2} + \frac{r_1 s_2 - r_2 s_1}{r_2^2 - d s_2^2}\sqrt{d}.$$

Also,
$$\frac{\xi'}{\eta'} = \frac{r_1 - s_1\sqrt{d}}{r_2 - s_2\sqrt{d}} \cdot \frac{r_2 + s_2\sqrt{d}}{r_2 + s_2\sqrt{d}}$$
$$= \frac{r_1 r_2 - d s_1 s_2}{r_2^2 - d s_2^2} + \frac{r_1 s_2 - r_2 s_1}{r_2^2 - d s_2^2}\sqrt{d}.$$

Hence, when $\eta \neq 0$, part (iii) is true.

Next,
$$\theta' = \frac{(r + s\xi)'}{(t + u\xi)'} = \frac{r + s\xi'}{t + u\xi'}.$$

Hence part (iv) is true.

We shall now prove a few simple results which will be useful in proving subsequent theorems.

• **Theorem 8-14** (i) If ξ is a quadratic irrational number, it is a solution of a quadratic equation with integral coefficients, the other solution being ξ'.

(ii) Every quadratic irrational number can be written in the form
$$\xi = \frac{e + \sqrt{d}}{f},$$

Quadratic Irrational Numbers

where e and f ($f \neq 0$) are integers, d a positive integer which is not a perfect square, and $f|(d - e^2)$.

Proof We note that $\xi = (e + \sqrt{d})/f$ and $\xi' = (e - \sqrt{d})/f$ are the solutions of the equation
$$f^2 x^2 - 2efx + e^2 - d = 0.$$
Now, if ξ is a solution of the quadratic equation
$$ax^2 + bx + c = 0,$$
where a, b, c are integers, then
$$a\xi^2 + b\xi + c = 0.$$
Taking conjugates of each side, we obtain
$$(a\xi^2)' + (b\xi)' + c = 0;$$
this gives us
$$a(\xi')^2 + b\xi' + c = 0.$$
Hence $\xi' \neq \xi$ is also a solution of the quadratic equation. Consequently part (i) is proved.

From our definition, we know that a quadratic irrational number is of the form
$$\frac{e_0 + \sqrt{d_0}}{f_0},$$
where e_0 and f_0 ($f_0 \neq 0$) are integers and d_0 is a positive integer not a perfect square.

Now,
$$\frac{e_0 + \sqrt{d_0}}{f_0} = \frac{e_0|f_0| + |f_0|\sqrt{d_0}}{f_0|f_0|}$$
$$= \frac{e + \sqrt{d}}{f},$$
where $e = e_0|f_0|$, $d = f_0^2 d_0$, $f = f_0|f_0|$. Since $d - e^2 = f_0^2(d_0 - e_0^2)$, f divides $d - e^2$ and d is a positive integer which is not a perfect square.

- **Theorem 8-15** If $\xi = (e + \sqrt{d})/f$ is a quadratic irrational number such that $f|(d - e^2)$, and if ξ has the infinite continued fraction representation given in Theorem 8-7, then, where e_j and f_j are defined recursively by the relations

$$e_1 = a_1 f - e, \qquad f_1 = \frac{d - e_1^2}{f},$$

(8-30) $$e_j = a_j f_{j-1} - e_{j-1} \quad (j = 2, 3, \ldots),$$

(8-31) $$f_j = \frac{d - e_j^2}{f_{j-1}} \quad (j = 2, 3, \ldots),$$

the following results follow for positive integral values of j:
(i) e_j and f_j are integers with $f_j \neq 0$;
(ii) $f_j | (d - e_j^2)$;
(iii) $\xi_j = \dfrac{e_j + \sqrt{d}}{f_j}$.

Proof Parts (i) and (ii) are obviously true when $j = 1$. Now (from Theorem 8-7),

$$\xi_1 = \frac{1}{\xi - a_1} = \frac{1}{(e + \sqrt{d})/f - a_1} = \frac{f}{\sqrt{d} - e_1}.$$

Moreover, since $ff_1 = d - e_1^2$,

$$\frac{f}{\sqrt{d} - e_1} = \frac{\sqrt{d} + e_1}{f_1}.$$

Hence (iii) is valid when $j = 1$.

We shall prove the theorem by mathematical induction on j. Assuming parts (i), (ii), and (iii) to be true for a fixed positive integer $j = m$, we desire to establish the truth of the results for $j = m + 1$.

Since e_m and f_m are assumed to be integers, we see from (8-30) that e_{m+1} is an integer. Since d is not a perfect square, $f_{m+1} \neq 0$. Also, from (8-30),

$$e_{m+1}^2 - e_m^2 = f_m(a_{m+1}^2 f_m - 2a_{m+1} e_m);$$

and from (8-31),

$$f_{m+1} = \frac{d - e_{m+1}^2}{f_m}$$
$$= \frac{d - e_m^2}{f_m} - \frac{(e_{m+1}^2 - e_m^2)}{f_m}.$$

Therefore, since by the hypothesis of the induction $f_m | (d - e_m^2)$, f_{m+1} is an integer dividing $d - e_{m+1}^2$. Hence parts (i) and (ii) are true by induction on j.

Next, we know that

$$\xi_{m+1} = \frac{1}{\xi_m - a_{m+1}} = \frac{1}{(e_m + \sqrt{d})/f_m - a_{m+1}} = \frac{f_m}{\sqrt{d} - e_{m+1}}$$
$$= \frac{f_m(e_{m+1} + \sqrt{d})}{d - e_{m+1}^2}.$$

Since $d - e_{m+1}^2 = f_{m+1} f_m$, we see that

$$\xi_{m+1} = \frac{e_{m+1} + \sqrt{d}}{f_{m+1}};$$

consequently, part (iii) follows by induction on j.

Periodic Continued Fractions **201**

• **Theorem 8-16** Let $\xi = [a_1, a_2, \ldots]$ and ξ_j $(j = 1, 2, \ldots)$ be the complete quotients of the quadratic irrational number ξ (as in Theorem 8-7), and let e_j, f_j be defined as in Theorem 8-15. Then, when n is an integer > 1 such that $\xi'_{n-1} < 0$, we have:

(i) $-1 < \xi'_n < 0$;
(ii) $0 < e_n < \sqrt{d}$;
(iii) $0 < f_n < 2\sqrt{d}$.

Proof Since
$$\xi_{n-1} = a_n + \frac{1}{\xi_n} \quad (n = 2, 3, \ldots),$$
on taking conjugates we obtain
$$\xi'_{n-1} = a_n + \frac{1}{\xi'_n}.$$
Hence, since $a_n > 0$ and $\xi'_{n-1} < 0$ (by hypothesis), we have $\xi'_n < 0$. Moreover, since
$$\frac{1}{\xi'_n} = \xi'_{n-1} - a_n < -a_n \leq -1,$$
$-1 < \xi'_n < 0$; thus part (i) is established.

Next, ξ_n being > 1, we have by (i), $\xi_n + \xi'_n > 0$ and $\xi_n - \xi'_n > 0$. Since
$$\xi_n = \frac{e_n + \sqrt{d}}{f_n} \quad \text{and} \quad \xi'_n = \frac{e_n - \sqrt{d}}{f_n},$$
we have
$$\xi_n - \xi'_n = \frac{2\sqrt{d}}{f_n} > 0;$$
consequently, $f_n > 0$.
Also, since
$$\xi_n + \xi'_n = \frac{2e_n}{f_n} > 0, \quad e_n > 0.$$
Since $\xi'_n < 0$, $e_n < \sqrt{d}$; moreover, since $\xi_n > 1$, we have
$$2\sqrt{d} > e_n + \sqrt{d} > f_n.$$
Hence parts (ii) and (iii) are established.

8-9 Periodic Continued Fractions

In considering the representation of $\sqrt{3} = [1, 1, 2, 1, 2, \ldots]$, we found that the pair 1, 2 repeated indefinitely in the infinite simple continued fraction representing $\sqrt{3}$. We say that this continued fraction has the period 1, 2 and that the period begins after the first partial quotient.

If an infinite continued fraction contains a constantly repeating block of positive partial quotients $b_1, b_2, \ldots, b_m$, we call it a *periodic* continued fraction. If the period begins with the first partial quotient (assumed positive), the continued fraction is called a *purely periodic* continued fraction. We write it $[\overline{b_1, b_2, \ldots, b_m}]$. If there are n partial quotients preceding the repeating block, we write the continued fraction $[a_1, a_2, \ldots, a_n, \overline{b_1, b_2, \ldots, b_m}]$. If $b_1, \ldots, b_m$ is the smallest block of integers that repeat, we say that the *period* of the continued fraction is $b_1, \ldots, b_m$ and the *length of the period* is m.

Example 1 Expanding $(\sqrt{13} + 1)/2$, we have

$$\frac{\sqrt{13}+1}{2} = 2 + \frac{1}{2/(\sqrt{13}-3)} = 2 + \frac{1}{(\sqrt{13}+3)/2}$$

$$= 2 + \frac{1}{3 + \dfrac{1}{(\sqrt{13}+3)/2}} = [2, 3, 3, \ldots] = [2, \overline{3}].$$

Example 2 Considering $\sqrt{5} + 2$, we see that the period begins with the first partial quotient. Hence, we obtain a purely periodic continued fraction:

$$\sqrt{5} + 2 = 4 + \frac{1}{\sqrt{5}+2} = [4, 4, 4, \ldots] = [\overline{4}].$$

Example 3 Find the value of the periodic continued fraction $[2, \overline{3, 1}]$

Solution Let $\xi = [2, \overline{3, 1}] = 2 + 1/\eta$, where

$$\eta = 3 + \frac{1}{1 + 1/\eta}.$$

Hence, $\eta^2 - 3\eta - 3 = 0$; consequently,

$$\eta = \frac{3 + \sqrt{21}}{2} \quad \text{and} \quad \xi = \frac{9 + \sqrt{21}}{6}.$$

Example 4 Find the value of the purely periodic continued fraction $[\overline{2, 3}]$.

Solution Let

$$\xi = [\overline{2, 3}] = 2 + \frac{1}{3 + 1/\xi},$$

which yields $3\xi^2 - 6\xi - 2 = 0$ and $\xi = (3 + \sqrt{15})/3$.

Periodic Continued Fractions

- **Theorem 8-17** The value of every periodic continued fraction is a quadratic irrational number.

    Proof Let $\alpha = [a_1, \ldots, a_n, \overline{b_1, \ldots, b_m}]$, and let $\beta = [\overline{b_1, \ldots, b_m}]$. Then $\beta = [b_1, \ldots, b_m, [\overline{b_1, \ldots, b_m}]] = [b_1, \ldots, b_m, \beta]$. Let $\bar{p}_i/\bar{q}_i$ be the ith convergent of the continued fraction $[b_1, \ldots, b_m]$. If $m = 1$, then $\beta^2 - b_1\beta - 1 = 0$. If $m \geq 2$, then (by Theorem 8-4)

$$\beta = \frac{\beta \bar{p}_m + \bar{p}_{m-1}}{\beta \bar{q}_m + \bar{q}_{m-1}};$$

consequently,

$$\bar{q}_m \beta^2 + (\bar{q}_{m-1} - \bar{p}_m)\beta - \bar{p}_{m-1} = 0.$$

In both cases, β is a (unique) positive solution of the respective quadratic equation. Since β is the value of an infinite continued fraction, β must be irrational. Consequently, being of the form $(e + \sqrt{d})/f$, β is a quadratic irrational number.

If the continued fraction is purely periodic, then the a's are absent and $\alpha = \beta = [b_1, \ldots, b_m, \beta]$. We have just seen that, in this instance, β (and therefore α) is a quadratic irrational number.

Next, if the a's are present in the periodic continued fraction, $\alpha = [a_1, \ldots, a_n, \beta]$. Then, where p_i/q_i denotes the ith convergent of $[a_1, \ldots, a_n]$, we shall see that α is a quadratic irrational number.

If $n = 1$, $\alpha = a_1 + 1/\beta$, which is a quadratic irrational number when $1/\beta$ is. Now, when $\beta = (e + \sqrt{d})/f$,

(8-32)
$$\frac{1}{\beta} = \frac{f}{e + \sqrt{d}} \cdot \frac{e - \sqrt{d}}{e - \sqrt{d}}$$
$$= \frac{ef - f\sqrt{d}}{e^2 - d}.$$

If $n \geq 2$,

$$\alpha = \frac{\beta p_n + p_{n-1}}{\beta q_n + q_{n-1}}.$$

This last fraction is

(8-33)
$$\alpha = \frac{p_n((e + \sqrt{d})/f) + p_{n-1}}{q_n((e + \sqrt{d})/f) + q_{n-1}}$$
$$= \frac{(p_n e + p_{n-1}f + p_n\sqrt{d})(q_n e + q_{n-1}f - q_n\sqrt{d})}{(q_n e + q_{n-1}f)^2 - q_n^2 d}.$$

Now (8-32) is of the form $(A + B\sqrt{d})/C$ with $A = ef$, $B = -f$, $C = e^2 - d \neq 0$. Also, (8-33) is of this same form, $(A + B\sqrt{d})/C$, with

$A = p_n q_n (e^2 - d) + (p_n q_{n-1} + p_{n-1} q_n)ef + p_{n-1} q_{n-1} f^2,$
$B = (-1)^n f,$
$C = (q_n e + q_{n-1} f)^2 - q_n^2 d \neq 0.$

We see that $(A + B\sqrt{d})/C = (A + \sqrt{B^2 d})/C$ when $B > 0$, and $(A + B\sqrt{d})/C = (-A + \sqrt{B^2 d})/-C$ when $B < 0$. Now, $B^2 d$ is not a perfect square since d is not.

Consequently, both (8-32) and (8-33) are quadratic irrational numbers. Thus, every periodic continued fraction is a quadratic irrational number.

We shall now prove the converse of Theorem 8-17.

- **Theorem 8-18** The infinite continued fraction which represents a quadratic irrational number ξ is periodic.

Proof Let $\xi = [a_1, a_2, \ldots, a_n, \ldots]$ be a given quadratic irrational number, and let its convergents be $C_j = p_j/q_j$, $j = 1, 2, \ldots$. Then, by Theorem 8-4, Cor. 1, where the complete quotients of ξ are $\xi, \xi_1, \ldots, \xi_m, \ldots$, we have for $m \geq 2$,

$$\xi = \frac{\xi_m p_m + p_{m-1}}{\xi_m q_m + q_{m-1}}.$$

In view of Theorems 8-13 (iv) and 8-15 (iii), we obtain on taking conjugates

(8-34) $$\xi' = \frac{\xi'_m p_m + p_{m-1}}{\xi'_m q_m + q_{m-1}}.$$

Solving (8-34) for ξ'_m in terms of ξ', we have

$$\xi'_m = -\frac{q_{m-1}}{q_m} \cdot \frac{\xi' - C_{m-1}}{\xi' - C_m}.$$

Since ξ is irrational, $\xi' \neq \xi$; moreover,

$$\lim_{m \to \infty} \frac{\xi' - C_{m-1}}{\xi' - C_m} = \frac{\xi' - \xi}{\xi' - \xi} = 1.$$

Thus, ξ'_m remains negative for sufficiently large values of m. That is, there exists a positive integer N such that for all $m \geq N$, ξ'_m always remains negative. Consequently, by Theorem 8-16 (ii) and (iii), (in conjunction with Theorem 8-15), for sufficiently large values of m ($m \geq N$) the integers e_m and f_m satisfy the restrictions

$$0 < e_m < \sqrt{d} \quad \text{and} \quad 0 < f_m < 2\sqrt{d}.$$

Since d is independent of the value of m, and since only a finite number of distinct pairs of positive integers e_m, f_m can satisfy the above inequalities, there must exist positive integers r and s ($r < s$) such that $e_r = e_s$ and $f_r = f_s$. From this it follows, by Theorem 8-15 (iii), that $\xi_r = \xi_s$; consequently, $[a_{r+1}, a_{r+2}, \ldots] = [a_{s+1}, a_{s+2}, \ldots]$. That is, that part

Periodic Continued Fractions

of the continued fraction expansion of ξ beginning with a_{r+1} is the same as that part beginning with a_{s+1}; namely,

$$a_{r+1} = a_{s+1},$$
$$a_{r+2} = a_{s+2},$$
$$\vdots \quad \vdots \quad \vdots$$
$$a_{r+s-r} = a_{2s-r},$$
$$a_{r+s-r+1} = a_{r+1} = a_{2s-r+1}.$$

Hence, the expansion of the quadratic irrational number ξ is periodic, with the length of the period a divisor of $s - r$.

From Theorems 8-17 and 8-18, we see that those real numbers which are quadratic irrational numbers, and only such real numbers, can be the values of periodic continued fractions.

We shall now consider those periodic continued fractions which are purely periodic.

A quadratic irrational number ξ is said to be a *reduced* quadratic irrational number if $\xi > 1$ and if its conjugate ξ' satisfies the restriction $-1 < \xi' < 0$. We note that $\sqrt{5} + 2$, in Example 2 preceding Theorem 8-17, is a reduced quadratic irrational number. Let us consider the following interesting result regarding reduced quadratic irrational numbers.

- **Theorem 8-19** An infinite continued fraction representing a quadratic irrational number ξ is purely periodic when and only when ξ is a reduced quadratic irrational number.

Proof First, assume that $\xi = [a_1, a_2, \ldots]$ is a reduced quadratic irrational number; that is, we assume that $\xi > 1$ and $-1 < \xi' < 0$. Now, $\xi = a_1 + 1/\xi_1$ and $a_1 = [\xi] \geq 1$. Since $\xi' < 0$ and $a_1 \geq 1$, the equation

$$\xi' = a_1 + \frac{1}{\xi'_1}$$

yields the result that $-1 < \xi'_1 < 0$. An inductive argument based on Theorem 8-16 (i) shows that

$$-1 < \xi'_n < 0$$

for every $n > 0$.

From $\xi' = a_1 + 1/\xi'_1$, we have $a_1 - \xi' = -1/\xi'_1$, where $0 < -\xi' < 1$; also, from $\xi'_n = a_{n+1} + 1/\xi'_{n+1}$, we have $a_{n+1} - \xi'_n = -1/\xi'_{n+1}$, where $0 < -\xi'_n < 1$. Hence,

(8-35) $$a_n = \left[-\frac{1}{\xi'_n}\right]$$

for $n > 0$.

In the periodic continued fraction expansion of ξ, let us assume that ξ_n, $n > 0$, is the first complete quotient which recurs. Let m be the least positive integer such that $\xi_{n+m} = \xi_n$. Then $\xi'_{n+m} = \xi'_n$ and $[-1/\xi'_n] = [-1/\xi'_{n+m}]$. Hence $a_n = a_{n+m}$.

If $n = 1$, then ξ, and not ξ_n, would be the first complete quotient to be repeated. Hence $n > 1$; and, since $\xi'_{n-1} = a_n + 1/\xi'_n$, $\xi'_{n-1} = a_{n+m} + 1/\xi'_{n+m} = \xi'_{n+m-1}$. Therefore, $\xi_{n-1} = \xi_{n+m-1}$; and, consequently, ξ_n would not be the first complete quotient to be repeated. As a result, ξ is the first complete quotient to be repeated, and the continued fraction is purely periodic.

Conversely, let us now assume that the continued fraction representation of the quadratic irrational number ξ is purely periodic; that is, $\xi = [\overline{a_1, \ldots, a_m}]$, where $a_1 = a_{m+1} \geq 1$; consequently $\xi > 1$.

If $m = 1$, then $a_j = a_1 \geq 1$ for $j \geq 1$, and $1 < \xi = a_1 + 1/\xi_1 = a_1 + 1/\xi$; thus, $\xi^2 - a_1\xi - 1 = 0$ and $(\xi')^2 - a_1\xi' - 1 = 0$. Consequently, since the solutions ξ and ξ' of this quadratic equation satisfy $\xi > 1$ and $-1 < \xi' < 0$, ξ is a reduced quadratic irrational number.

Obviously, since $\xi_m = \xi$, when $m > 1$, we have (by Theorem 8-4)

$$\xi = \frac{\xi_m p_m + p_{m-1}}{\xi_m q_m + q_{m-1}};$$

thus it follows that

$$q_m \xi^2 + (q_{m-1} - p_m)\xi - p_{m-1} = 0.$$

Since $x = \xi$ is a solution of the quadratic equation

$$f(x) = q_m x^2 + (q_{m-1} - p_m)x - p_{m-1} = 0$$

with integral coefficients (Theorem 8-14 (i)), the other solution is the conjugate $x = \xi'$. We note that $p_1, p_2, \ldots$ is a strictly monotonic increasing sequence of positive integers, as is also $q_1, q_2, \ldots$ (except that we might have $q_1 = q_2 = 1$). Therefore,

$$f(-1) = (q_m - q_{m-1}) + (p_m - p_{m-1}) > 0.$$

Since $f(0) < 0$, the equation $f(x) = 0$ must have a real solution between -1 and 0. Since $\xi > 1$, the real solution ξ' must be such that $-1 < \xi' < 0$. Consequently, ξ is a reduced quadratic irrational number.

We shall now consider the infinite continued fraction expansion of $\sqrt{d}$, where d is a positive integer not a perfect square.

- **Theorem 8-20** Let d be a positive integer which is not a perfect square. The simple continued fraction expansion of $\sqrt{d}$ has a period beginning with the second partial quotient[5]; that is,

Periodic Continued Fractions

$$\sqrt{d} = [a_1, \overline{a_2, \ldots, a_{m+1}}]$$

for some positive integer m.

Proof We have $\xi = \sqrt{d} > 1$ and $a_1 = [\sqrt{d}] \geq 1$. Since the conjugate $\xi' = -\sqrt{d} < -1$, we note that ξ is not a reduced quadratic irrational number, and hence (Theorem 8-19) the expansion of the continued fraction is not purely periodic. On the other hand, since $0 < \sqrt{d} - a_1 < 1$, the equation $\xi = a_1 + 1/\xi_1$ yields

$$\xi_1 = \frac{1}{\sqrt{d} - a_1} \quad \text{and} \quad \xi_1' = \frac{1}{-\sqrt{d} - a_1};$$

therefore, $\xi_1 > 1$ and $-1 < \xi_1' < 0$. Hence ξ_1 is a reduced quadratic irrational number; consequently, since (Theorem 8-19) ξ_1 is purely periodic, $\xi = \sqrt{d}$ has a period beginning with the second partial quotient.

• **Theorem 8-21** Let d be a positive integer which is not a perfect square, and let the continued fraction expansion of $\sqrt{d}$ (in Theorem 8-20), with convergents p_j/q_j, have the length of the period m. Then, using the notation already employed in Theorems 8-15 and 8-16, we have the following results:

(i) $p_j^2 - dq_j^2 = (-1)^j f_j$ $(j = 1, 2, \ldots)$,
(ii) $f_j > 0$ $(j = 1, 2, \ldots)$,
(iii) $f_j = 1$ if and only if $m | j$.

Proof For $\sqrt{d} = [a_1, \overline{a_2, \ldots, a_{m+1}}]$, we have, when $j \geq 2$,

(8-36) $$\sqrt{d} = \frac{\xi_j p_j + p_{j-1}}{\xi_j q_j + q_{j-1}}.$$

Now, substituting

$$\xi_j = \frac{e_j + \sqrt{d}}{f_j}$$

in (8-36) and simplifying, we obtain

$$\sqrt{d}(e_j q_j + f_j q_{j-1} - p_j) + dq_j - e_j p_j - f_j p_{j-1} = 0.$$

Since $\sqrt{d}$ is irrational, this last equation implies that

(8-37) $$e_j q_j + f_j q_{j-1} = p_j,$$
(8-38) $$e_j p_j + f_j p_{j-1} = dq_j.$$

Multiplying (8-37) by p_j and (8-38) by $-q_j$ and adding, we obtain, by Theorem 3-14,

$$p_j^2 - dq_j^2 = (-1)^j f_j$$

for $j \geq 2$.

Noting that $\xi = \sqrt{d} = (e + \sqrt{d})/f$ (with $e = 0$ and $f = 1$), $e_1 = a_1 f - e$, and $f_1 = (d - e_1^2)/f$, we see that

$$p_1^2 - dq_1^2 = a_1^2 - d = -(d - a_1^2) = (-1)f_1.$$

Hence, (i) is established for $j \geq 1$.

Since $p_1 = a_1 = [\sqrt{d}] \geq 1$, we see that $p_j > 0$ and $q_j > 0$ for all $j \geq 1$. Moreover, since $p_j/q_j < \sqrt{d}$ when j is odd and $p_j/q_j > \sqrt{d}$ when j is even, we have

$$p_j^2 - dq_j^2 < 0 \quad \text{when } j \text{ is odd,}$$

and

$$p_j^2 - dq_j^2 > 0 \quad \text{when } j \text{ is even.}$$

These last two inequalities in conjunction with (i) show the validity of (ii).

Since $e = 0, f = 1, e_1 = a_1 f - e, f_1 = (d - e_1^2)/f, e_j = a_j f_{j-1} - e_{j-1}$ and $f_j = (d - e_j^2)/f_{j-1}$ when $j \geq 2$, we see that (by Theorem 8-20) in the continued fraction representation of $\sqrt{d}$, where $\xi_j = (e_j + \sqrt{d})/f_j$, complete quotients satisfy the following equations:

(8-39) $$\xi_1 = \xi_{m+1} = \xi_{2m+1} = \xi_{3m+1} = \ldots.$$

Since $\sqrt{d}$ is irrational, we immediately obtain (for a positive integer t) from (8-39)

$$\frac{e_1 + \sqrt{d}}{f_1} = \frac{e_{tm+1} + \sqrt{d}}{f_{tm+1}},$$

which yields

$$\sqrt{d}(f_{tm+1} - f_1) + e_1 f_{tm+1} - e_{tm+1} f_1 = 0;$$

that is,

$$e_{tm+1} = e_1 = a_1 \quad \text{and} \quad f_{tm+1} = f_1 = d - a_1^2.$$

Hence,

$$f_1 = d - a_1^2 = d - e_1^2 = d - e_{tm+1}^2 = f_{tm} f_{tm+1} = f_{tm} f_1;$$

thus $f_{tm} = 1$. This shows that part (iii) is true when $m | j$.

Next, assume that j is a positive integer such that $f_j = 1$. We desire to prove that j is a multiple of m.

Since $f_j = 1$, $\xi_j = e_j + \sqrt{d}$ and $[\xi_j] = [\sqrt{d}] + e_j = a_1 + e_j$. Hence, we have $\xi_j = [\xi_j] + 1/\xi_{j+1} = a_1 + e_j + 1/\xi_{j+1}$; this yields

$$\sqrt{d} = \xi_j - e_j = a_1 + \frac{1}{\xi_{j+1}}.$$

Therefore, since $\sqrt{d} = a_1 + 1/\xi_1$, we see that $\xi_{j+1} = \xi_1$. That is, the block $a_2, \ldots, a_{j+1}$ is continually repeated in the continued fraction rep-

Pell's Equation

resentation of $\sqrt{d}$. Consequently, the length of the period is a divisor of j.

8-10 Problems

1. Find the infinite continued fraction representation of the following irrational numbers:
    (a) $1 + \sqrt{8}$; (b) $2 + \sqrt{23}$; (c) $\sqrt{29}$.
2. Find the value of the following continued fractions:
    (a) $[3, \overline{2}]$; (b) $[\overline{3, 2}]$; (c) $[3, \overline{4, 5}]$.
3. Determine which of the following irrational numbers have purely periodic representations as infinite continued fractions:
    (a) $\sqrt{3}$; (b) $6 + \sqrt{10}$; (c) $3 + \sqrt{10}$; (d) $3 + \sqrt{8}$;
    (e) $(5 + \sqrt{43})/2$; (f) $4 + \sqrt{23}$; (g) $(2 + \sqrt{22})/3$.
4. Find the length of the period in the representation of each of the following as an infinite continued fraction:
    (a) $\sqrt{8}$; (b) $\sqrt{24}$; (c) $7 + \sqrt{51}$.

8-11 Pell's Equation

The Diophantine equation

$$x^2 - dy^2 = 1 \tag{8-40}$$

is commonly, but inaccurately, known as Pell's equation. L. Euler mistakenly attributed a method of solving this equation to John Pell, who apparently had no connection with the problem. Appropriately it should be known as Fermat's equation. However, the name 'Pell's equation' seems well established in the extensive literature on the subject. It was J. L. Lagrange who first proved that when d is a positive integer which is not a perfect square, (8-40) has integral solutions with $y \neq 0$.

Consider first the more general form[6]

$$x^2 - dy^2 = m. \tag{8-41}$$

Let $m = m_0 l^2$, where m_0 is not divisible by a perfect square greater than one. Integers x and y satisfying (8-41) constitute a *solution* of this equation. If we have a solution for which $(x, y) = 1$, the solution is said to be *primitive*. Since $(\pm x)^2 = x^2$ and $(\pm y)^2 = y^2$, we may, without loss of generality, restrict ourselves to *positive* solutions; that is, to solutions in which both x and y are positive. If (8-41) has a positive solution with $(x, y) = k > 1$ and $x = kx_0$, $y = ky_0$, then $k|l$ and $x_0^2 - dy_0^2 = m_0 l_0^2$, where $l = kl_0$ and

$(x_0, y_0) = 1$. Hence, solutions of (8-41) other than primitive solutions can be obtained from primitive solutions x_0, y_0 of equations of the form

(8-42) $$x_0^2 - dy_0^2 = m_0\left(\frac{l}{k}\right)^2,$$

where k ranges over the various divisors of l, by multiplying (8-42) through by k^2. Consequently, we shall restrict our consideration[7] to primitive solutions of (8-41).

If x, y is a primitive solution of (8-41), then $(y, m) = 1$ and there is, modulo m, a unique solution w of

(8-43) $$wy \equiv x \pmod{m}.$$

Primitive solutions of (8-41) employed in congruence (8-43) are said to *belong* to the solution w of (8-43). From (8-41) and (8-43) we have, since $(y, m) = 1$, $w^2 \equiv d \pmod{m}$; hence $z = w$ is a solution of the congruence

(8-44) $$z^2 \equiv d \pmod{m}.$$

If the congruence (8-44) has no solution, there is no primitive solution of (8-41). If (8-44) does have a solution, $z = w$, all primitive solutions of (8-41) (if there are any) are classified into sets, all those in the same set belonging to the same solution w of (8-43) (and, therefore, of (8-44)).

Let x, y and x_1, y_1 be two (distinct) primitive solutions of (8-41) which belong to the same solution of (8-44). Then, since $x \equiv wy \pmod{m}$ and $x_1 \equiv wy_1 \pmod{m}$, $xy_1 - x_1 y \equiv 0 \pmod{m}$ and, by (8-44), $xx_1 \equiv w^2 yy_1 \equiv dyy_1 \pmod{m}$; thus we have $xx_1 - dyy_1 = ms$ and $xy_1 - x_1 y = mt$.

Now,

$$\frac{x_1 + y_1\sqrt{d}}{x + y\sqrt{d}} = \frac{(x_1 + y_1\sqrt{d})(x - y\sqrt{d})}{(x + y\sqrt{d})(x - y\sqrt{d})}$$

(8-45) $$= \frac{xx_1 - dyy_1 + (xy_1 - x_1 y)\sqrt{d}}{x^2 - dy^2}.$$

But, since $x^2 - dy^2 = m$, we see that (8-45) reduces to

$$\frac{x_1 + y_1\sqrt{d}}{x + y\sqrt{d}} = s + t\sqrt{d};$$

consequently,

(8-46) $$x_1 + y_1\sqrt{d} = (s + t\sqrt{d})(x + y\sqrt{d}).$$

Similarly, taking conjugates of (8-46), we have

$$x_1 - y_1\sqrt{d} = (s - t\sqrt{d})(x - y\sqrt{d}).$$

On multiplying the last two equations, we obtain

(8-47) $$x_1^2 - dy_1^2 = (s^2 - dt^2)(x^2 - dy^2).$$

On cancelling $x^2 - dy^2 = x_1^2 - dy_1^2$, we have

(8-48) $$s^2 - dt^2 = 1.$$

Pell's Equation

Thus, two primitive solutions x, y and x_1, y_1 of (8-41) belonging to the same solution of (8-44) are bound by the relation (8-46), in which s and t are integral solutions of (8-48). Conversely, we shall see that for any solution s, t of (8-48) and a primitive solution x, y of (8-41), x_1, y_1 as determined by (8-46), give us a primitive solution of (8-41) belonging to the same solution w of (8-44) as that to which x, y belong. Since, from (8-47), $x_1^2 - dy_1^2 = m$, we need merely to show that x_1, y_1 are relatively prime and that $x_1 \equiv wy_1 \pmod{m}$.

From (8-46), we see that

$$x_1 = sx + dty, \qquad y_1 = tx + sy,$$

and also that

$$x = sx_1 - dty_1, \qquad y = -tx_1 + sy_1.$$

Hence, since $(x, y) = 1$, we see that $(x_1, y_1) = 1$. Moreover,

$$x_1 - wy_1 = sx + dty - twx - swy = s(x - wy) + t(-wx + dy).$$

But

$$-wx + dy \equiv -wx + w^2y \equiv -w(x - wy) \pmod{m}.$$

Consequently,

$$\begin{aligned} x_1 - wy_1 &\equiv s(x - wy) - wt(x - wy) \\ &\equiv (s - wt)(x - wy) \\ &\equiv 0 \pmod{m}, \end{aligned}$$

and x_1, y_1 is a primitive solution of (8-41) belonging to the same solution w of (8-44) to which the primitive solution x, y belongs.

Therefore, if we know *one* primitive solution x, y of (8-41) belonging to a solution w of (8-44) and if we can find all solutions of (8-48), we can then find all primitive solutions of (8-41) belonging to the solution w of (8-44).

We now return to the consideration of equation (8-40) in which d is a positive integer which is not a perfect square. We desire to obtain all its positive solutions. If it does have a positive solution, there must exist a positive solution x_1, y_1 such that $x_1 < x$ for every other positive solution x, y. Hence also, since $x_1^2 - dy_1^2 = x^2 - dy^2$, we have $x^2 - x_1^2 = d(y^2 - y_1^2)$ and $0 < y_1 < y$. Such a positive solution x_1, y_1 is called the *fundamental solution* of (8-40).

- **Theorem 8-22** If x_1, y_1 is the fundamental solution of Pell's equation (8-40), then (where $n \geq 1$) every pair of integers x_n, y_n defined by the equation

(8-49) $$x_n + y_n\sqrt{d} = (x_1 + y_1\sqrt{d})^n$$

is a positive solution of (8-40).

Proof Since $x_n + y_n\sqrt{d} = (x_1 + y_1\sqrt{d})^n$, we obtain, on taking conjugates,

(8-50) $$x_n - y_n\sqrt{d} = (x_1 - y_1\sqrt{d})^n.$$

Moreover, since x_1 and y_1 are positive, x_n and y_n are both positive. Hence, on multiplying (8-49) and (8-50), we obtain

$$x_n^2 - dy_n^2 = (x_1^2 - dy_1^2)^n = 1^n = 1.$$

Moreover, adding and subtracting (8-49) and (8-50), we obtain

(8-51) $$x_n = \tfrac{1}{2}\{(x_1 + y_1\sqrt{d})^n + (x_1 - y_1\sqrt{d})^n\}$$

and

(8-52) $$y_n = \frac{1}{2\sqrt{d}}\{(x_1 + y_1\sqrt{d})^n - (x_1 - y_1\sqrt{d})^n\}.$$

We shall now see that the x_n, y_n of Theorem 8-22 comprise all the positive solutions of (8-40).

• **Theorem 8-23** Where x_1, y_1 is the fundamental solution of (8-40), every positive solution is given, for a suitable positive integer n, by x_n, y_n defined by the equation

$$x_n + y_n\sqrt{d} = (x_1 + y_1\sqrt{d})^n.$$

Proof Let X, Y be an arbitrary positive solution of (8-40). We desire to show that there exists a positive integer n such that $X = x_n$ and $Y = y_n$, where x_n and y_n are determined from equation (8-49). Since x_1, y_1 is the fundamental solution and $\sqrt{d} > 1$, $1 < x_1 + y_1\sqrt{d}$. There is, therefore, a positive integer n such that

$$(x_1 + y_1\sqrt{d})^n \leq X + Y\sqrt{d} < (x_1 + y_1\sqrt{d})^{n+1}.$$

If $X + Y\sqrt{d} = (x_1 + y_1\sqrt{d})^n$, then $X = x_n$, $Y = y_n$, and the theorem is established in this case. Let us now assume that

(8-53) $$(x_1 + y_1\sqrt{d})^n < X + Y\sqrt{d} < (x_1 + y_1\sqrt{d})^{n+1}.$$

Since $(x_1 + y_1\sqrt{d})(x_1 - y_1\sqrt{d}) = 1$, $(x_1 - y_1\sqrt{d}) > 0$. Hence, multiplying (8-53) through by $(x_1 - y_1\sqrt{d})^n$, we obtain

$$1 < (X + Y\sqrt{d})(x_1 - y_1\sqrt{d})^n < 1^n \cdot (x_1 + y_1\sqrt{d})$$
$$= x_1 + y_1\sqrt{d}.$$

Where $X_1 = Xx_n - Ydy_n$, $Y_1 = -Xy_n + Yx_n$, these inequalities become (since $(X + Y\sqrt{d})(x_1 - y_1\sqrt{d})^n = (X + Y\sqrt{d})(x_n - y_n\sqrt{d})$)

(8-54) $$1 < X_1 + Y_1\sqrt{d} < x_1 + y_1\sqrt{d}.$$

Pell's Equation

On substituting for X_1 and Y_1, we see that
$$X_1^2 - dY_1^2 = (X^2 - dY^2)(x_n^2 - dy_n^2) = 1.$$

We see next that X_1, Y_1 is a *positive* solution of (8-40). Since

(8-55) $\qquad (X_1 + Y_1\sqrt{d})(X_1 - Y_1\sqrt{d}) = 1,$

and since from (8-54)

(8-56) $\qquad X_1 + Y_1\sqrt{d} > 1,$

it follows that

(8-57) $\qquad X_1 - Y_1\sqrt{d} > 0.$

On adding (8-56) and (8-57), we observe that X_1 is positive. Moreover, from (8-56) and (8-55) we obtain
$$X_1 + Y_1\sqrt{d} > 1$$
and
$$-X_1 + Y_1\sqrt{d} > -1.$$

On adding these last two inequalities, we see that Y_1 also is positive. Consequently, where x_1, y_1 is the fundamental solution of (8-40), we have found a positive solution X_1, Y_1 such that $X_1 + Y_1\sqrt{d} < x_1 + y_1\sqrt{d}$. This contradiction shows that there is no positive solution X, Y satisfying inequalities (8-53). Thus the theorem is established.

In view of the following theorem, we see that, when d is a positive integer not a perfect square, the equation (8-40) has an infinite number of positive solutions, obtainable from the convergents of a continued fraction.

- **Theorem 8-24** Let d be a positive integer which is not a perfect square, and let m be the length of the period of the continued fraction representation of $\sqrt{d}$ with convergents p_j/q_j. We have the following results:

  (i) When m is even, p_j, q_j is a solution of (8-40) if and only if j is a multiple of m.

  (ii) When m is odd, p_j, q_j is a solution of (8-40) if and only if j is a multiple of $2m$.

  Proof We see, from Theorem 8-21 (i), that p_j, q_j is a solution of (8-40) if and only if $(-1)^j f_j = 1$ (where $j \geq 1$); that is, if and only if $(-1)^j = 1$ and (since $f_j > 0$) $m|j$. In case m is even, these conditions are satisfied if and only if j is a multiple of m; in case m is odd, these conditions are satisfied if and only if j is an even multiple of m.

We shall see next that every positive solution p, q of (8-40) gives rise to a convergent p/q of $\sqrt{d}$.

- **Theorem 8-25** Let d be a positive integer which is not a perfect square and let M be an integer numerically less than $\sqrt{d}$. If p and q are relatively prime positive integers satisfying
$$p^2 - dq^2 = M,$$
then p/q is a convergent of the continued fraction representing $\sqrt{d}$.

Proof[8] First, consider $M > 0$. Since
$$(p - q\sqrt{d})(p + q\sqrt{d}) = M,$$
we have
$$p > q\sqrt{d} \quad \text{and} \quad \frac{p}{q} - \sqrt{d} = \frac{M}{q(p + q\sqrt{d})};$$
also, $p + q\sqrt{d} > 2q\sqrt{d}$. Therefore,
$$0 < \frac{p}{q} - \sqrt{d} < \frac{M}{2q^2\sqrt{d}} < \frac{\sqrt{d}}{2q^2\sqrt{d}} = \frac{1}{2q^2}.$$

By Theorem 8-12, p/q is a convergent to the irrational number $\sqrt{d}$. Next, let $M = -M_0 < 0$; that is, $M_0 < \sqrt{d}$, and $dq^2 - p^2 = M_0$. Hence,
$$(-p + q\sqrt{d})(p + q\sqrt{d}) = M_0 > 0$$
implies that $p < q\sqrt{d}$ and
$$-\left(\frac{1}{\sqrt{d}} - \frac{q}{p}\right) = \frac{M_0}{p(qd + p\sqrt{d})} < \frac{\sqrt{d}}{p(p\sqrt{d} + p\sqrt{d})} = \frac{1}{2p^2}.$$

Again, by Theorem 8-12, q/p is a convergent to $1/\sqrt{d}$; consequently, by Theorem 8-8, Corollary, p/q is a convergent to $\sqrt{d}$.

- **Corollary** If p and q are positive integers satisfying
$$p^2 - dq^2 = \pm 1,$$
then p/q is a convergent of the continued fraction representation of $\sqrt{d}$.

The following results are consequences of Theorem 8-24 and Theorem 8-25, Corollary.

- **Theorem 8-26** Where d, m, p_j, and q_j are defined as in Theorem 8-24, the fundamental solution of (8-40) is p_m, q_m when m is even, and p_{2m}, q_{2m} when m is odd.

- **Theorem 8-27** Where d, m, p_j and q_j are defined as in Theorem 8-24, the positive solutions of
$$x^2 - dy^2 = -1$$

Pell's Equation

are p_j, q_j when j is an odd multiple of the odd number m. If m is even, there is no integral solution.

Proof Since (from Theorem 8-21(i)) we desire that $(-1)^j f_j = -1$, we see that we must have either $(-1)^j = 1$ and $f_j = -1$ or $(-1)^j = -1$ and $f_j = 1$. By Theorem 8-21(ii) the former case is excluded. Consequently, j must be odd and (by Theorem 8-21(iii)) divisible by m.

As a consequence of Theorem 8-25, we obtain the following result.

• **Theorem 8-28** Let a_1 be an integer exceeding one, M an integer numerically greater than one and without a square factor greater than one. If $M^2 < d = a_1^2 + 1$, the equation

(8-58) $$x^2 - dy^2 = M$$

has no integral solution.

Proof If it has an integral solution, we see that it must be a primitive solution. For, if $(x, y) = k > 1$, then k^2 would be a factor of M, contrary to the hypothesis. Moreover, d is not a perfect square. By Theorem 8-25, any existing solution yields a convergent to $\sqrt{d}$. Employing Theorem 8-15 with $\xi = \sqrt{d}$, we see that (Since $e_1 = a_1 = [\xi]$, $\xi_1 = a_1 + \sqrt{d}$, $a_2 = [\xi_1] = 2a_1$, and $\xi_2 = \xi_1$.) the irrational number $\sqrt{d}$ has the representation $\sqrt{d} = [a_1, \overline{2a_1}]$. Since the period has length $m = 1$, by Theorem 8-21 we have $p_j^2 - dq_j^2 = (-1)^j$ for $j \geq 1$. Hence (8-58) has no solution in integers.

Example 1 Find the fundamental solution of the equation

(8-59) $$x^2 - 33y^2 = 1.$$

Solution We employ formulae (8-30) and (8-31) from Theorem 8-15 with $d = 33$ and $\xi = \sqrt{33}$. Since $\xi = (0 + \sqrt{33})/1$, we have $e = 0, f = 1$, $a_1 = [\xi] = [\sqrt{33}] = 5$. Consequently, employing $e_1 = a_1 f - e, f_1 = (d - e_1^2)/f$, $\xi_j = (e_j + \sqrt{d})/f_j, a_{j+1} = [\xi_j], e_{j+1} = a_{j+1} f_j - e_j, f_{j+1} = (d - e_{j+1}^2)/f_j$ (for $j > 0$), we may obtain the partial quotients in the continued fraction representation of $\sqrt{33}$ from the accompanying tabulation (Table 8-1).

Table 8-1

j	1	2	3	4	5	6
e_j	5	3	3	5	5	3
f_j	8	3	8	1	8	3
ξ_j	$\dfrac{5+\sqrt{33}}{8}$	$\dfrac{3+\sqrt{33}}{3}$	$\dfrac{3+\sqrt{33}}{8}$	$\dfrac{5+\sqrt{33}}{1}$	$\dfrac{5+\sqrt{33}}{8}$	$\dfrac{3+\sqrt{33}}{3}$
a_j	5	1	2	1	10	1

First, from $a_1 = [\sqrt{33}] = 5$ we obtain e_1, f_1, and ξ_1. Next, we get $a_2 = [\xi_1]$, and then e_2, f_2, and ξ_2, followed by $a_3 = [\xi_2]$. Continuing in this manner, we see that $\sqrt{33} = [5, \overline{1, 2, 1, 10}]$ and the length of the period is $m = 4$. The first four convergents to $\sqrt{33}$ are

$$\frac{5}{1}, \frac{6}{1}, \frac{17}{3}, \frac{23}{4}.$$

By Theorem 8-26, the fundamental solution of (8-59) is $x_1 = 23, y_1 = 4$.

All positive solutions are given by x_n, y_n as n takes on positive integral values in the equation $x_n + y_n\sqrt{33} = (23 + 4\sqrt{33})^n$. For example, $n = 2$ gives the positive solution $x_2 = 1057, y_2 = 184$; $n = 3$ gives the positive solution $x_3 = 48599, y_3 = 8460$.

The positive solutions x_n, y_n could also be obtained by finding (by Theorem 3-13) successive convergents to $\sqrt{33}$ and employing Theorem 8-21. For example, since $\sqrt{33} = [5, \overline{1, 2, 1, 10}]$, the first twelve convergents are

$$\frac{5}{1}, \frac{6}{1}, \frac{17}{3}, \frac{23}{4}, \frac{247}{43}, \frac{270}{47}, \frac{787}{137}, \frac{1057}{184}, \frac{11357}{1977}, \frac{12414}{2161}, \frac{36185}{6299}, \frac{48599}{8460}.$$

The fourth, eighth, and twelfth convergents give rise to the first three positive solutions of (8-59).

Example 2 Show that the equation

$$x^2 - 66y^2 = 5$$

has no solution in integers.

First Solution If there is a positive solution, we see, by Theorem 8-25, that it must be obtained from a convergent to $\sqrt{66}$. Let us compute the accompanying table, starting with $a_1 = [\sqrt{66}] = 8$ and $\xi = (0 + \sqrt{66})/1$. Hence $d = 66, e = 0, f = 1$. We employ the notation and results given in Theorem 8-21.

Table 8-2

j	1	2	3	4
e_j	8	8	8	8
f_j	2	1	2	1
ξ_j	$\dfrac{8 + \sqrt{66}}{2}$	$\dfrac{8 + \sqrt{66}}{1}$	$\dfrac{8 + \sqrt{66}}{2}$	$\dfrac{8 + \sqrt{66}}{1}$
a_j	8	8	16	8

Hence $\sqrt{66} = [8, \overline{8, 16}]$. By Theorem 8-21 (i), we see that $p_j^2 - 66q_j^2$ is either -2 or 1, but never 5. Hence there is no integral solution.

Pell's Equation

Second Solution From the proposed equation, we see that x must be odd. Hence, assuming that an integral solution exists, we have
$$1 - 2y^2 \equiv 5 \pmod{8}.$$
If y is odd, the congruence reduces to $-1 \equiv 5 \pmod 8$; if y is even, the congruence reduces to $1 \equiv 5 \pmod 8$. Hence a solution in integers is impossible.

Third Solution If there is a solution in integers, we would have $x^2 = 5 + 66y^2$; that is, $x^2 \equiv 5 \pmod{33}$. However, 5 is a quadratic nonresidue of 3. Consequently, there is no solution in integers.

8-12 Problems

1. Find the fundamental solution of $x^2 - 8y^2 = 1$ and, in addition, three other positive solutions.

2. Find the fundamental solution of each of the following equations: (a) $x^2 - 2y^2 = 1$; (b) $x^2 - 50y^2 = 1$; (c) $x^2 - 21y^2 = 1$.

3. Find the fundamental solution and two further positive solutions of each of the following equations: (a) $x^2 - 10y^2 = 1$; (b) $x^2 - 7y^2 = 1$; (c) $x^2 - 15y^2 = 1$.

4. If a is a positive integer, find all integral solutions of the following equations: (a) $x^2 - a^2 y^2 = 1$; (b) $x^2 - 9a^2 y^2 = -1$.

5. Find all positive solutions of the following equations for which $y < 5450$: (a) $x^2 - 11y^2 = 1$; (b) $x^2 - 12y^2 = 1$.

6. Exhibit all positive solutions of $x^2 - 45y^2 = 1$ for which $y < 7730$.

7. Find the fundamental solution of each of the following equations, and indicate a method of obtaining all positive solutions: (a) $x^2 - 18y^2 = 1$; (b) $x^2 - 72y^2 = 1$.

8. Show that each of the following equations has an infinite number of positive integral solutions; and, in each case, write down two such solutions: (a) $x^2 - 17y^2 = -1$; (b) $x^2 - 12y^2 = -3$.

9. If the positive integer d is of the form $4K + 3$, prove that there is no solution in integers of $x^2 - dy^2 = -1$.

10. Indicating two positive solutions of each equation, give a method of determining all positive solutions of the following equations: (a) $x^2 - 37y^2 = 1$; (b) $x^2 - 10y^2 = -1$.

11. Show that none of the following equations has a solution in integers: (a) $x^2 - 11y^2 = -1$; (b) $x^2 - 12y^2 = -1$; (c) $x^2 - 15y^2 = -1$.

12. Determine whether the following equations are solvable in integers:
    (a) $x^2 - 3y^2 = 2$;
    (b) $7y^2 - x^2 = -1$;
    (c) $x^2 - 86y^2 = 37$;
    (d) $x^2 + 17y^2 = 60$;
    (e) $x^2 - 19y^2 = -1$;
    (f) $x^2 - 145y^2 = 11$.

13. Determine the solvability in integers of the following equations:
    (a) $x^2 + 86y^2 = 37$;
    (b) $x^2 - 82y^2 = 79$;
    (c) $x^2 + 17y^2 = 61$;
    (d) $x^2 - 17y^2 = 61$;
    (e) $x^2 - 45y^2 = -1$;
    (f) $x^2 - 13y^2 = -1$;
    (g) $x^2 - 65y^2 = 7$;
    (h) $x^2 - 170y^2 = 4$.

14. Determine all positive solutions of the following equations:
    (a) $x^2 - 5y^2 = 1$;
    (b) $x^2 - 5y^2 = -1$;
    (c) $x^2 - 7y^2 = 4$;
    (d) $x^2 - 8y^2 = -4$;
    (e) $x^2 - 27y^2 = 9$.

15. Let d be a positive integer such that the irrational number $\sqrt{d}$ has a period of length 2; that is, let $\sqrt{d} = [a_1, \overline{a_2, a_3}]$. Using the notation of Theorem 8-21, establish the following results:
    (a) $e_j = a_1$ and $f_{2j} = f = 1$ for all $j \geq 1$;
    (b) $f_{2j-1} = d - a_1^2 \neq 1$ for all $j \geq 1$;
    (c) the even and the odd convergents p_j/q_j give rise to solutions $x = p_j$, $y = q_j$ of $x^2 - dy^2 = 1$ and of $x^2 - dy^2 = -f_1$, respectively;
    (d) $a_3 = 2a_1$.

8-13 Farey Sequences

In this section we shall be concerned with certain properties of the positive rational numbers, and the existence of rational approximations to real numbers.

The *Farey sequence of order n* is the sequence, in increasing order of magnitude, of the irreducible rational fractions between 0 and 1 (both inclusive) whose denominators do not exceed n. Thus, the Farey sequence of order 6 is:

$$\frac{0}{1}, \frac{1}{6}, \frac{1}{5}, \frac{1}{4}, \frac{1}{3}, \frac{2}{5}, \frac{1}{2}, \frac{3}{5}, \frac{2}{3}, \frac{3}{4}, \frac{4}{5}, \frac{5}{6}, \frac{1}{1}.$$

There are some simple well-known properties of Farey sequences which we shall now consider.

• **Theorem 8-29** If a/b and a'/b' are consecutive terms in a Farey sequence, then $a'b - ab' = 1$.

Farey Sequences

Proof[9] Since the first two terms of the Farey sequence are $0/1$ and $1/n$, the result is obviously true when a/b is the first term of the sequence or when $n = 1$. Next, let $a > 0$ (thus $n > 1$). Since the fractions in the sequence are in their lowest terms, $(a, b) = (a', b') = 1$. Consequently, by Theorem 2-6, there exists a solution $x = x_0, y = y_0$ of the equation

(8-60) $$bx + (-a)y = 1.$$

Where t is an arbitrary integer, all integral solutions (by Theorem 2-10) of (8-60) are given by $x = x_0 + at, y = y_0 + bt$. Since the set of integers w satisfying $n - b < w \leq n$ form a complete set of residues, modulo b, choose t so that $n - b < y_0 + bt \leq n$. Now, since $a, b,$ and y are all positive integers, we see from the equation $bx = 1 + ay$ that $x > 0$. Moreover, since $bx = 1 + ay \leq 1 + an$, we have

$$x \leq \frac{1 + an}{b} \leq \frac{1 + (b-1)n}{b} = \frac{bn - (n-1)}{b} = n - \frac{n-1}{b} < n.$$

Hence, since $(x, y) = 1, 0 \leq n - b < y \leq n$, and $0 < x < n, x/y$ is greater than a/b and is a fraction in the Farey sequence of order n. For, on dividing (8-60) through by by we have

$$\frac{x}{y} - \frac{a}{b} = \frac{1}{by} > 0,$$

and

$$x - y = \frac{1 + ay - by}{b} = \frac{1 - (b-a)y}{b} \leq \frac{1-y}{b} \leq 0.$$

If x/y is not the successor of a/b,

(8-61) $$\frac{x}{y} - \frac{a'}{b'} = \frac{b'x - a'y}{b'y} \geq \frac{1}{b'y}.$$

On the other hand,

(8-62) $$\frac{a'}{b'} - \frac{a}{b} = \frac{a'b - ab'}{bb'} \geq \frac{1}{bb'}.$$

Adding (8-61) and (8-62), we obtain

$$\frac{x}{y} - \frac{a}{b} \geq \frac{1}{b'y} + \frac{1}{bb'} = \frac{b+y}{bb'y} > \frac{n}{bb'y}.$$

However,

$$\frac{1}{by} = \frac{bx - ay}{by} = \frac{x}{y} - \frac{a}{b} > \frac{n}{bb'y} = \frac{1}{(b'/n) \cdot by} \geq \frac{1}{by}.$$

This gives us a contradiction. Therefore x/y must be a'/b'; consequently, $a'b - ab' = 1$.

- **Theorem 8-30** If $a/b, a'/b', a''/b''$ are three consecutive terms of a Farey sequence, then $a'/b' = (a + a'')/(b + b'')$.

    Proof This immediately follows from the preceding theorem on evaluating a'/b' from the equations $a'b - ab' = 1$ and $a''b' - a'b'' = 1$. It can be shown, also, that this present theorem implies the former.[10]

- **Theorem 8-31** If a/b and a'/b' are consecutive terms of a Farey sequence of order n, $b + b' > n$.

    Proof The *mediant* of a/b and a'/b' is defined as $(a + a')/(b + b')$ and lies between the two given consecutive terms. For, since $0 \leq a/b < a'/b'$, we have $ab + ab' < ab + a'b$ and $a'b' + ab' < a'b' + a'b$. From these last two inequalities we obtain $a/b < (a + a')/(b + b') < a'/b'$. If $b + b' \leq n$, then $(a + a')/(b + b')$ would be a term of the sequence of order n lying between the two consecutive terms a/b and a'/b'. Hence $b + b' > n$.

- **Theorem 8-32** When n exceeds unity, two consecutive terms of the Farey sequence of order n have different denominators.

    Proof Let a/b and a'/b' be consecutive terms of the sequence. If $n > 1$, there are at least three terms in the Farey sequence of order n. If a/b is the first term (namely, $0/1$), the next term is $1/n$ (and $n > 1$). If a'/b' is the last term of the sequence, a/b is $(n - 1)/n$ and $a'/b' = 1/1$. Assume, then, that $b > 1$. If $b' = b$, then $b > a' \geq a + 1$ and (since $a < a' \leq b - 1$)

$$\frac{a}{b} < \frac{a}{b - 1} < \frac{a + 1}{b} \leq \frac{a'}{b'}.$$

    Since $0 < a/(b - 1) < 1$, we have a term of the sequence between two consecutive terms of the sequence. This contradiction establishes the result.

- **Theorem 8-33** If n is a positive integer and ξ an arbitrary real quantity, there exists an irreducible fraction a/b such that $0 < b \leq n$ and

$$\left|\xi - \frac{a}{b}\right| \leq \frac{1}{b(n + 1)}.$$

    Proof Since $\xi = [\xi] + \xi_0$, where $0 \leq \xi_0 < 1$, we may assume, then, that the given ξ satisfies the condition $0 < \xi < 1$.

    Consider the Farey sequence of order n together with the mediants of each pair of consecutive terms of the Farey sequence. These terms (together with the mediants) will subdivide the interval $0 \leq \eta \leq 1$ into subintervals. Consequently, ξ will be in one of the closed intervals

Farey Sequences

formed by a member of the Farey sequence and one of its associated mediants. Therefore, where a/b and a'/b' are consecutive terms of the Farey sequence, either

$$\xi - \frac{a}{b} \leq \frac{a+a'}{b+b'} - \frac{a}{b} = \frac{1}{b(b+b')} \leq \frac{1}{b(n+1)}$$

or

$$\frac{a'}{b'} - \xi \leq \frac{a'}{b'} - \frac{a+a'}{b+b'} = \frac{1}{b'(b+b')} \leq \frac{1}{b'(n+1)}.$$

Hence there exists a rational number a/b such that

$$\left|\xi - \frac{a}{b}\right| \leq \frac{1}{b(n+1)}.$$

8-14 Problems

1. Determine the Farey sequences of order 7 and of order 9.

2. (a) Find a rational number a/b, $0 < b \leq 7$, differing from $\sqrt{33}$ by not more than $1/(8b)$.

   (b) Find a rational number a/b, $0 < b \leq 9$, differing from $\sqrt{43}$ by not more than $1/(10b)$.

3. Prove that the number of terms in the Farey sequence of order n is $1 + \phi(1) + \phi(2) + \ldots + \phi(n)$, where $\phi(k)$ denotes Euler's ϕ-function.

Certain Diophantine Equations and Sums of Squares 9

9-1 Introductory Remarks[1]

We have already given some consideration, in Section 2-12, to linear Diophantine equations, and have obtained an expression for all solutions of $ax + by = n$ when one solution is known (Theorem 2-10). We shall now be concerned with certain Diophantine equations of second and higher degrees. Since the history of Diophantine analysis has been that of special problems, general methods of treatment have been relatively few, and disconnected results and partial solutions of problems have often been given. The development of methods of general applicability, although very desirable, has, until recently, not been prominent in the treatment of Diophantine problems. The work of Axel Thue (1863–1922) and Th. Skolem (1887–1963), among others, has given some impetus to the search for general methods of investigation. Here special mention should be made of L. J. Mordell's *Diophantine Equations*.

A type of proof freely used by Fermat is often of great help in determining whether a given Diophantine equation is solvable or not. The procedure known as the *method of infinite descent* may be described as follows. If there exists a positive integer n having a specific property (for example, a positive integer n for which a certain proposition $P(n)$ is true), then it is proved that there must exist a smaller positive integer having this same property. In this way, starting from a definite positive integer, we obtain an infinite decreasing sequence of positive integers each of which has the stated property. Since, however, a non-empty set of positive integers contains a least positive integer (See Section 2-4, Fundamental Principle I), the above statement leads to a contradiction. Consequently, there exists no such positive integer possessing the designated property. This type of proof will be employed in Section 9-5.

9-2 The Pythagorean Equation

The well-known Pythagorean theorem states that in a right-angled triangle the area of the square constructed on the hypotenuse is equal to the sum of the areas of the squares on the two legs. The solutions given by the Pythagoreans assigned to the hypotenuse a value one greater than that of the larger leg. The first general solution of the Pythagorean problem of finding integral solutions of $x^2 + y^2 = z^2$ was given by Euclid (Elements, Book X).

We shall now obtain formulae giving all solutions in non-zero integers of the Pythagorean equation

$$x^2 + y^2 = z^2.$$

- **Theorem 9-1** The non-zero integral solutions, with y even, of the Pythagorean equation

(9-1) $$x^2 + y^2 = z^2$$

are given by

(9-2) $$x = k(r^2 - s^2), \quad y = 2krs, \quad z = k(r^2 + s^2),$$

where k is arbitrary but different from zero, and r and s are non-zero relatively prime integers of opposite parity.

> *Proof* First, we note that if any two of the three numbers x, y, z have a common divisor, this common divisor divides the third. Hence, without loss of generality, we may seek the *primitive* solutions of the equation; that is, as defined in Chapter 8, those solutions x, y, z for which the greatest common divisor of x, y, and z is unity. All non-zero

solutions can then be obtained by multiplying the primitive solutions by an arbitrary non-zero integer k.

Since the square of an odd integer is of the form $8M + 1$ and the square of an even integer is of the form $4M_1$, we see that not both x and y can be odd. For, if both were odd, we would have $8M_2 + 2 = z^2$. Let y, say, be even and x (and z) odd. Then

$$y^2 = 4y_1^2 = z^2 - x^2;$$

which implies that

$$y_1^2 = \frac{z+x}{2} \frac{z-x}{2}.$$

Any common divisor of $(z + x)/2$ and $(z - x)/2$ would divide their sum z and their difference x. Since $(z, x) = 1$, $(z + x)/2$ and $(z - x)/2$ are relatively prime and their product is a perfect square y_1^2. Hence, assuming each of x, y, z to be non-negative (that is, $x \geq 0$, $y \geq 0$, $z \geq 0$), we have

(9-3) $$\frac{z+x}{2} = r^2, \quad \frac{z-x}{2} = s^2,$$

where $(r, s) = 1$.

Subtracting and adding equations (9-3), we obtain²

(9-4) $$x = r^2 - s^2, \quad y = 2rs, \quad z = r^2 + s^2.$$

Since both x and z are to be odd, r and s are of opposite parity.

We have found that, if (9-1) has a primitive solution in non-negative integers, it is given by (9-4) with $r^2 > s^2$. Moreover, on substituting the values from (9-4) in equation (9-1), we see that (9-1) is satisfied for all integers r and s. Thus we have established the theorem.

9-3 The Diophantine Equation $x^2 + 2y^2 = z^2$

• **Theorem 9-2** The primitive solutions, in non-negative integers, of the Diophantine equation

(9-5) $$x^2 + 2y^2 = z^2$$

are given by

(9-6) $$x = \pm(r^2 - 2s^2), \quad y = 2rs, \quad \text{and} \quad z = r^2 + 2s^2,$$

where $x > 0$, and r and s are non-negative integers such that $(r, 2s) = 1$.

Proof Since all solutions can be obtained from primitive solutions, in which $(x, y, z) = 1$, by multiplying them by an arbitrary integer, consider possible non-negative primitive solutions. Since the other combinations are impossible, we assume that x (as well as z) is odd and y

even. If a solution in non-negative integers exists, let x, y, z denote such a non-negative primitive solution; thus $y = 2y_1$ and

$$2y_1^2 = \frac{z+x}{2} \cdot \frac{z-x}{2}.$$

Hence, since $(z+x)/2$ and $(z-x)/2$ are relatively prime integers, either

$$z + x = 2r^2 \quad \text{and} \quad z - x = 4s^2,$$

or

$$z - x = 2r^2 \quad \text{and} \quad z + x = 4s^2,$$

where r and s are non-negative integers such that $(r, s) = 1$.

In the former case, $x = r^2 - 2s^2$, $y = 2rs$, $z = r^2 + 2s^2$, where $(r, 2s) = 1$. In the latter case, $x = 2s^2 - r^2$, $y = 2rs$, $z = 2s^2 + r^2$. Hence, combining the two cases, we see that

(9-6) $x = \pm(r^2 - 2s^2), \quad y = 2rs, \quad \text{and} \quad z = r^2 + 2s^2,$

where $x > 0$ and $(r, 2s) = 1$. Substituting these values (9-6) in equation (9-5), we see that (9-5) is satisfied for all integral values of r and s. Hence, where r and $2s$ are relatively prime and where the sign is chosen to make x positive, we see that equations (9-6) yield all non-negative primitive solutions of (9-5). All solutions of (9-5) are obtained by multiplying the primitive solutions x, y, z by an arbitrary integer. If the sign of any one of x, y, z is changed, we still have a solution of (9-5).

Example 1 In a right-angled triangle with the lengths of the three sides integral, prove that the radius R of the inscribed circle is also an integer.

Proof

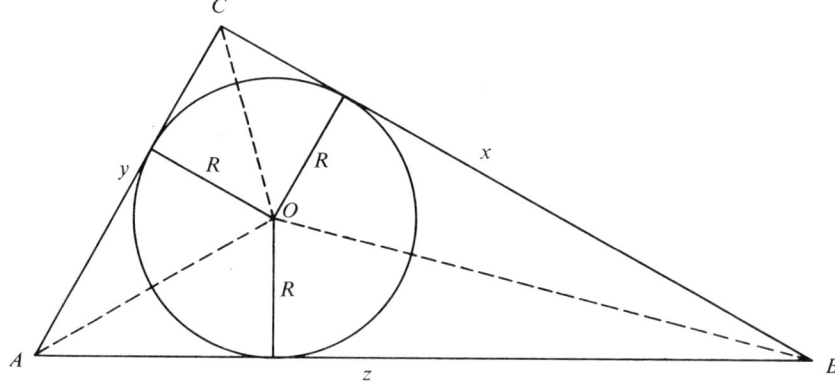

The Diophantine Equation $x^2 + 2y^2 = z^2$

Let O be the center of the circle inscribed in a right-angled triangle with hypotenuse of length z and legs of lengths x and y; therefore

(9-7) $$x^2 + y^2 = z^2.$$

Since the area of the right-angled triangle is equal to the sum of the areas of the three triangles with common vertex O, we have

$$\tfrac{1}{2}xy = \tfrac{1}{2}R(x + y + z).$$

Assuming y to be even, we have the solution of (9-7) given by

$$x = k(r^2 - s^2), \quad y = 2krs, \quad z = k(r^2 + s^2).$$

Hence, since $xy = R(x + y + z)$,

$$2k^2 rs(r^2 - s^2) = Rk(r^2 - s^2 + 2rs + r^2 + s^2);$$

consequently,

$$R = \frac{2k^2 rs(r^2 - s^2)}{2kr(r + s)} = ks(r - s),$$

an integer.

Example 2 Solve in positive integers the Diophantine equation

(9-8) $$x^2 + y^2 = 2z^2.$$

Solution Let x, y, z be integers satisfying (9-8). We see that a common divisor of any two of x, y, z divides the third. If $d = (x, y, z)$ and if $x = dx_0$, $y = dy_0$, $z = dz_0$, then x_0, y_0, z_0 are relatively prime in pairs and satisfy (9-8). Consider, then, primitive solutions x, y, z of (9-8). We now readily see that none of x, y, z can be even. Let x, y, z be positive odd integers satisfying (9-8) subject to the condition $(x, y, z) = 1$. Then $(x+y)/2$, $(x-y)/2$, z satisfy the equation

$$\left(\frac{x+y}{2}\right)^2 + \left(\frac{x-y}{2}\right)^2 = z^2.$$

If $x = y$, then $z = x = y = 1$ would be a primitive solution. Excluding this case, we may use the solution of the Pythagorean equation to yield: either

$$\frac{x+y}{2} = r^2 - s^2,$$

$$\frac{x-y}{2} = 2rs,$$

$$z = r^2 + s^2;$$

or

$$\frac{x-y}{2} = r^2 - s^2,$$

$$\frac{x+y}{2} = 2rs,$$

$$z = r^2 + s^2,$$

where the positive numbers r and s are relatively prime and of opposite parity. Hence, since $(x+y)/2$ and $(x-y)/2$ have the sum and difference x and y, respectively,

(9-9)
$$x = r^2 - s^2 + 2rs,$$
$$y = \pm(r^2 - s^2 - 2rs),$$
$$z = r^2 + s^2,$$

where the numerical value of each is to be taken in order to assure a positive solution. These values (9-9) satisfy (9-8) for all r and s.

9-4 Problems

1. List in tabular form all positive primitive solutions of $x^2 + y^2 = z^2$, with y even, for which $0 < z < 90$.

2. Obtain formulae for all primitive solutions of the Pythagorean equation $x^2 + y^2 = z^2$, with y even, in which z differs from x or y by one.

3. Obtain all primitive solutions of the Pythagorean equation $x^2 + y^2 = z^2$, with y even, in which the following condition is satisfied: (a) z differs from x or y by 2; (b) z differs from x or y by an odd prime p.

4. Find a complete solution in integers of $x^2 + 3y^2 = z^2$.

5. Obtain a complete solution in positive integers of $x^2 + 5y^2 = z^2$.

6. Obtain all solutions in positive integers of $x^2 + py^2 = z^2$ in the following cases: (a) when p is a prime of the form $4M + 3$; (b) when p is a prime of the form $4M + 1$.

9-5 Some Fourth Degree Diophantine Equations

We shall apply Fermat's method of "infinite descent" to establish the following result.

- **Theorem 9-3** There is no solution in positive integers of the Diophantine equation

(9-10)
$$x^4 + y^4 = z^2$$

Proof Since a prime factor of any two of the integers x, y, z would be a factor of the third, let x, y, z be assumed to be a specific primitive solution in positive integers. Then, since either x^2 or y^2 must be even, in

$$(x^2)^2 + (y^2)^2 = z^2,$$

Some Fourth Degree Diophantine Equations

we may take y to be even and, using the solution of the Pythagorean equation, set

(9-11)
$$x^2 = r^2 - s^2,$$
$$y^2 = 2rs,$$
$$z = r^2 + s^2,$$

where r and s ($r^2 > s^2$) are relatively prime, of the same sign, and of opposite parity.

First, we see that it is impossible to have r even and s odd. For, in that case, $r^2 - s^2$ is of the form $4M + 3$, and is therefore not a perfect square. Hence, we must have r odd and s even.

Since, in the equation
$$y^2 = 2rs,$$
$(r, 2s) = 1$ and 2 must occur to an even power in the right member, we have $r = \pm r_0^2$, $s = \pm 2s_0^2$, where r_0 and s_0 are positive integers. Substituting these values of r and s in the first of the equations (9-11), we obtain
$$x^2 = r_0^4 - 4s_0^4;$$
that is,

(9-12)
$$(r_0^2)^2 = (2s_0^2)^2 + x^2,$$

where $(r_0, 2s_0) = 1$. Hence, applying to (9-12) the solution of the Pythagorean equation, we take

(9-13)   and
$$r_0^2 = t^2 + u^2$$
$$s_0^2 = tu.$$

Since $(t, u) = 1$, $t = \pm v^2$ and $u = \pm w^2$, where v and w are relatively prime and positive. Therefore, from the first equation (9-13) we obtain
$$v^4 + w^4 = r_0^2.$$

We have, then, another primitive solution of (9-10) in which $r_0^2 = |r| \leq r^2 < z < z^2$. Consequently, the assumption that (9-10) has a primitive solution in positive integers x, y, z implies that it has another primitive solution in positive integers v, w, r_0 for which $r_0 < z$. Employing the method of infinite descent, we see that the initial assumption that equation (9-10) has a primitive solution in positive integers must be false. Consequently, equation (9-10) has no solution in positive integers.

• **Theorem 9-4** The area of a right-angled triangle with the lengths of the three sides integral cannot be a perfect (integral) square.

Proof Let the hypotenuse have the length c and the other two sides lengths a and b. Then $a^2 + b^2 = c^2$. We desire to show that $\frac{1}{2} ab = u^2$ is impossible for positive integral values of a, b, and u. Since, if any two of a, b, c have a common factor, this factor divides the third, we see that it is sufficient to prove the result for a primitive Pythagorean triangle. We shall assume, then, that for $(a, b, c) = 1$,

(9-14) $$ab = 2u^2.$$

Using the solution found for the Pythagorean equation (Theorem 9-1), we see that, on the assumption that b is even,

$$a = d^2 - e^2,$$
$$b = 2de,$$
$$c = d^2 + e^2,$$

where the positive numbers d and e ($d > e$) are relatively prime and of opposite parity. Hence, on substituting the above values in (9-14) we find that

(9-15) $$de(d - e)(d + e) = u^2.$$

Since d and e are relatively prime and of opposite parity, the four numbers d, e, $d - e$, $d + e$ are seen to be relatively prime in pairs. Hence, since the product of the four positive numbers is a perfect square, each must be a perfect square; that is,

(9-16)
$$d = g^2,$$
$$e = h^2,$$
$$d - e = g^2 - h^2 = k^2,$$
$$d + e = g^2 + h^2 = l^2,$$

where g, h, k, and l are relatively prime in pairs. Since d and e are of opposite parity, so also are g and h. We shall assume g, h, k, and l to be positive. On adding and subtracting the last two equations (9-16), we obtain

(9-17)
$$2g^2 = k^2 + l^2,$$
$$2h^2 = l^2 - k^2 = (l - k)(l + k).$$

From the last of the equations (9-17), we see that, since k and l are both odd, $l - k$ and $l + k$ are both even, and, consequently, so also is h. On substituting $h = 2m$ in $2h^2 = l^2 - k^2$, we obtain

(9-18) $$2m^2 = \frac{l - k}{2} \cdot \frac{l + k}{2}.$$

Since their sum and difference are relatively prime, we see that $(l - k)/2$ and $(l + k)/2$ are also relatively prime.

Some Fourth Degree Diophantine Equations

If we assume in (9-18) that $(l - k)/2$ contains 2 to an odd power, then

$$\frac{l-k}{2} = 2t^2,$$

(9-19) and

$$\frac{l+k}{2} = u^2,$$

where $(t, u) = 1$; therefore

(9-20) $$m^2 = t^2 u^2.$$

If, on the other hand, $(l + k)/2$ contains 2 to an odd power,

$$\frac{l-k}{2} = u^2,$$

(9-21) and

$$\frac{l+k}{2} = 2t^2,$$

where $(t, u) = 1$. In both instances, (9-20) remains valid. From (9-19) we obtain

(9-22)
$$k = u^2 - 2t^2,$$
$$l = u^2 + 2t^2.$$

Similarly, from (9-21) we obtain

(9-23)
$$k = 2t^2 - u^2,$$
$$l = 2t^2 + u^2.$$

When these values from (9-22) and (9-23) are substituted in the first of the equations (9-17), we find

$$2g^2 = (u^2 - 2t^2)^2 + (u^2 + 2t^2)^2$$
$$= 8t^4 + 2u^4;$$

that is,

$$g^2 = (u^2)^2 + (2t^2)^2.$$

The Pythagorean triangle with hypotenuse g and legs u^2 and $2t^2$ has an area equal to

$$\frac{1}{2} u^2 \cdot 2t^2 = u^2 t^2 = m^2.$$

Consequently, since $m^2 = \frac{1}{4}e$ and since, by (9-15), $e < u^2$, we now have a second Pythagorean triangle with length of sides integral whose area is again a perfect square smaller than the area u^2 of the first Pythagorean triangle. Since this process could be repeated starting with the second triangle to obtain a third of corresponding character with still smaller area, we see that (since the area is a positive integer) the method of infinite descent may be employed to show that the original assumption (9-14) is impossible. Thus the theorem is established.

- **Theorem 9-5** The Diophantine equation

(9-24) $$x^4 - y^4 = z^2$$

has no solution in positive integers.

Proof Assume that there are positive integers x, y, and z satisfying $x^4 - y^4 = z^2$. Then, if we set

$$a = (x^2)^2 - (y^2)^2,$$
$$b = 2x^2y^2,$$
$$c = (x^2)^2 + (y^2)^2,$$

we see that, since $a^2 + b^2 = c^2$, a, b, c are the integral sides of a Pythagorean triangle with hypotenuse c. The area of this triangle is

$$\frac{1}{2}ab = \frac{1}{2}(x^4 - y^4)2x^2y^2 = x^2y^2z^2,$$

an integral square. Since this result contradicts Theorem 9-4, there is no solution of (9-24) in positive integers.

- **Corollary 1** The only solution in positive integers, with $(x, y) = 1$, of $x^4 + y^4 = 2z^2$ is $x = y = z = 1$.

Proof If $x^4 + y^4 = 2z^2$ has a solution in positive integers x, y, z, then

$$(x^4 + y^4)^2 = 4z^4;$$

consequently,

$$(x^4 - y^4)^2 = 4z^4 - 4x^4y^4.$$

Hence,

$$\left(\frac{x^4 - y^4}{2}\right)^2 = z^4 - (xy)^4.$$

However, by this Theorem 9-5 there is no solution in positive integers of $X^4 - Y^4 = Z^2$. Hence $x^4 - y^4 = 0$ and $2(x^2)^2 = 2z^2$; thus $z^2 = x^4 = y^4$. Therefore, we obtain a primitive solution only when $z = 1$ and $x = y = 1$.

- **Corollary 2** The simultaneous Diophantine equations

$$x^2 + y^2 = z^2 \quad \text{and} \quad x^2 - y^2 = w^2$$

have no solution in positive integers x, y, z, w.

Some Fourth Degree Diophantine Equations

Proof If there were such a common solution x, y, z, w, then, on multiplication, we would have

$$x^4 - y^4 = (zw)^2,$$

a contradiction of Theorem 9-5.

- **Theorem 9-6** The Diophantine equation

(9-25) $$x^4 - 4y^4 = z^2$$

has no solution in positive integers.

Proof If x, y, z is a solution of (9-25) in which $(x, y) > 1$, let $(x, y) = p_1^{a_1} p_2^{a_2} \ldots p_r^{a_r}$, where the p's are distinct primes. Then, since $x = (x, y)x_1$ and $y = (x, y)y_1$, we have

$$p_1^{4a_1} p_2^{4a_2} \ldots p_r^{4a_r}(x_1^4 - 4y_1^4) = z^2.$$

Hence, $z = p_1^{2a_1} \ldots p_r^{2a_r} z_1$, and $x_1^4 - 4y_1^4 = z_1^2$, where $(x_1, y_1) = 1$.

Assuming, then, that x, y, z is a solution in positive integers of (9-25) with x and y relatively prime, we have

$$z^2 + (2y^2)^2 = (x^2)^2.$$

Employing the solution of the Pythagorean equation, we obtain

(9-26)
$$z = r^2 - s^2,$$
$$2y^2 = 2rs,$$
$$x^2 = r^2 + s^2,$$

where r and s are positive, relatively prime integers of opposite parity. Since $rs = y^2$, we have $r = t^2$, $s = u^2$, where t and u are positive and relatively prime. Hence, by the last equation in (9-26), we have $t^4 + u^4 = x^2$. Consequently, by Theorem 9-3, there can be no solution in positive integers of (9-25).

- **Corollary** There exists no right-angled triangle with integral sides whose area is double a square.

Proof Let the hypotenuse have length c and the legs lengths a and b, where $a \geq b > 0$. Then $a^2 + b^2 = c^2$. Let us assume the area of the triangle to be twice a square; that is, where $w > 0$, assume $\frac{1}{2}ab = 2w^2$; consequently,

$$ab = 4w^2.$$

Hence

$$(a + b)^2 = c^2 + 2ab = c^2 + 8w^2;$$

also

$$(a - b)^2 = c^2 - 8w^2.$$

Multiplying the last two equations, we obtain
$$(a^2 - b^2)^2 = c^4 - 4(2w)^4.$$
If we were to have $a^2 - b^2 = 0$, then $c^2 = 8w^2 = 2a^2$. Consequently, the exponent of the highest power of 2 dividing c^2 would be odd. Since this is an impossibility, we would have a positive solution
$$X = c,\ Y = 2w,\ Z = a^2 - b^2$$
of the equation
$$X^4 - 4Y^4 = Z^2,$$
a result contrary to our Theorem 9-6.

- **Theorem 9-7** The Diophantine equation
(9-27) $$x^4 + 2y^4 = z^2$$
has no solution in positive integers.

Proof Since a prime dividing any two of x, y, z divides the third, we shall assume that (9-27) has an explicit primitive solution in positive integers, namely, x, y, z, with $(x, y, z) = 1$.

If x were even, then z would be also. Consequently, x and z are odd and y even. Let $y = 2y_0$. From (9-27) we obtain
$$32y_0^4 = z^2 - x^4;$$
that is
$$8y_0^4 = \frac{z + x^2}{2} \cdot \frac{z - x^2}{2}.$$

Since $(z + x^2)/2$ and $(z - x^2)/2$ have the sum z and difference x^2, they are relatively prime. Thus, where u and v are positive, we must have either

(9-28) $$\frac{z + x^2}{2} = 8u^4 \quad \text{and} \quad \frac{z - x^2}{2} = v^4$$

or

(9-29) $$\frac{z + x^2}{2} = v^4 \quad \text{and} \quad \frac{z - x^2}{2} = 8u^4,$$

where $(2u, v) = 1$.

Since v is odd, we obtain from equations (9-28) $x^2 \equiv 8u^4 - v^4 \equiv 7$ (mod 8), a contradiction. On the other hand, from equations (9-29) we obtain

(9-30) $$\begin{aligned} x^2 &= v^4 - 8u^4, \\ y &= 2uv, \\ z &= v^4 + 8u^4. \end{aligned}$$

Some Fourth Degree Diophantine Equations

Hence, $x^2 + 8u^4 = v^4$; that is, $x^2 + 2(2u^2)^2 = (v^2)^2$. Using the solution of

(9-31) $$X^2 + 2Y^2 = Z^2$$

(Theorem 9-2), we set

(9-32) $\quad x = |r^2 - 2s^2|, \quad 2u^2 = 2rs, \quad \text{and} \quad v^2 = r^2 + 2s^2,$

where r and s are positive and $(r, 2s) = 1$.

From the second of the equations (9-32), since $(r, s) = 1$, we have $r = r_0^2$ and $s = s_0^2$. From the last of the equations (9-32), we have $r_0^4 + 2s_0^4 = v^2$; thus (9-27) has a solution in positive integers r_0, s_0, v, where, by the last of the equations (9-30), $v < v^4 < z$.

Consequently, we have shown that if (9-27) were to have a primitive solution in positive integers x, y, z, it would also have another primitive solution in positive integers s_0, r_0, v with v smaller than the original value z. This process can be repeated to get a primitive solution in positive integers with the right member of (9-27) still smaller than v^2. This implies that there is an infinite number of positive integers $< z$. Thus, the method of infinite descent shows that there can be no solution of (9-27) in positive integers.

Example 1 If p is a prime of the form $8M + 5$, prove that the equation
$$x^2 + 2y^2 = pz^2$$
has no solution in integers, except for the trivial case $x = y = z = 0$.

If there is a solution x, y, z in which $(x, y, z) = d > 1$, then, setting $x = dx_0, y = dy_0, z = dz_0$, we find that x_0, y_0, z_0 is a solution in which $(x_0, y_0, z_0) = 1$. Let, then, x, y, z be a primitive solution.

First Solution Since $(x, y, z) = 1$, we see that z cannot be even. Consequently $x, y,$ and z are all odd. Since the square of an odd number is of the form $8M + 1$, we have $x^2 + 2y^2 \equiv 3 \pmod{8}$, and $pz^2 \equiv 5 \pmod{8}$. Since $3 \not\equiv 5 \pmod 8$, there can be no solution in non-zero integers of $x^2 + 2y^2 = pz^2$.

Second Solution Since $(x, y, z) = 1$, we have $(y, p) = 1$. Also, an integer w exists such that $wy \equiv 1 \pmod p$. Hence, multiplying the given equation through by w^2, we obtain the congruence $(wx)^2 + 2(wy)^2 \equiv 0 \pmod p$; thus $(wx)^2 \equiv -2 \pmod p$. Consequently, the Legendre symbol $\left(\dfrac{-2}{p}\right) = 1$. However,

$$\left(\frac{-2}{p}\right) = \left(\frac{-1}{p}\right)\left(\frac{2}{p}\right) = 1 \cdot (-1) = -1.$$

This contradiction shows that there is no non-trivial solution in integers of the given equation.

Example 2 Show that there is no solution in positive integers of the Diophantine equation[3]

(9-33) $$x^4 - y^4 = 3z^2.$$

Proof We shall assume that there is a possible solution in positive integers x, y, z. Since a prime dividing any two of x, y, z must divide the third, we shall assume a primitive solution in which $(x, y) = 1$.

If z is odd, either x is even or y is even. Now, x cannot be even, since in that case $-1 \equiv 3 \pmod 8$. Moreover, y cannot be even, since that would imply that $1 \equiv 3 \pmod 8$.

Therefore, z is even; and, since both x and y must be odd, 16 divides $x^4 - y^4$; consequently, 4 divides z. Set $z = 4z_1$. Next, the G.C.D. of every pair of the three even values $x^2 + y^2$, $x + y$, $x - y$ is 2. Since

$$x^4 - y^4 = (x^2 + y^2)(x + y)(x - y) = 48z_1^2,$$

we have

(9-34) $$\frac{x^2 + y^2}{2} \cdot \frac{x + y}{2} \cdot \frac{x - y}{2} = 6z_1^2,$$

where $(x^2 + y^2)/2$, $(x + y)/2$, $(x - y)/2$ are relatively prime in pairs. Since each of x and y is of one of the forms $6M + 1, 6M + 3, 6M + 5$, each of the squares x^2 and y^2 is of one of the two forms $24N + 1$, $72N + 9$; and, consequently, since $(x, y) = 1$, $(x^2 + y^2)/2$ is either of the form $12K + 1$ or $12K + 5$.

We therefore have, from (9-34), the following possibilities: either

$$\frac{x^2 + y^2}{2} = u^2, \qquad \frac{x \pm y}{2} = 3v^2, \qquad \frac{x \mp y}{2} = 2w^2;$$

or

$$\frac{x^2 + y^2}{2} = u^2, \qquad \frac{x \pm y}{2} = v^2, \qquad \frac{x \mp y}{2} = 6w^2,$$

where the positive integers u, v, w are relatively prime in pairs. That is, *either*

(9-35) $\qquad x \pm y = 6v^2, \qquad x \mp y = 4w^2, \qquad x^2 + y^2 = 2u^2;$

or

(9-36) $\qquad x \pm y = 2v^2, \qquad x \mp y = 12w^2, \qquad x^2 + y^2 = 2u^2,$

where u, v, w are relatively prime in pairs and $(u, 6) = 1 = (v, 2)$. Adding and subtracting the first two equations (9-35) to obtain x and y in

Some Fourth Degree Diophantine Equations

terms of v and w, we have on substituting these values of x and y in the third equation of (9-35):

(9-37) $$9v^4 + 4w^4 = u^2.$$

Similarly, from equations (9-36), we obtain

(9-38) $$v^4 + 36w^4 = u^2.$$

Since the G.C.D. of the even numbers $u + 3v^2$ and $u - 3v^2$ divides their sum and their difference, we see that, when $(u, 6) = 1$, their G.C.D. is 2. From (9-37), we have

$$(u + 3v^2)(u - 3v^2) = 4w^4;$$

thus,

(9-39) $$u + 3v^2 = 2k^4,$$
$$u - 3v^2 = 2l^4,$$

where $k > 0$, $l > 0$, $w = kl$, and $(k, l) = 1$. Subtracting equations (9-39), we get $3v^2 = k^4 - l^4$, an equation of the same form as (9-33), in which v is odd. This contradicts our former result that z must be even in equation (9-33). Hence equations (9-35) are untenable.

Next, from equations (9-38) we obtain

$$(u + 6w^2)(u - 6w^2) = v^4.$$

Since $(u, w) = 1$ and $(u, 6) = 1$, we see that $u + 6w^2$ and $u - 6w^2$ are relatively prime, and, consequently, where $v = mn$ and $(m, n) = 1$, for suitable positive integers m and n,

$$u + 6w^2 = m^4 \quad \text{and} \quad u - 6w^2 = n^4.$$

Subtracting the last two equations, we obtain

$$3(2w)^2 = m^4 - n^4,$$

again an equation of the same form as the given equation (9-33). However, since from (9-36) we know that $x = v^2 + 6w^2$, $m \leq m^2n^2 < m^2n^2 + 6w^2 = x$.

Consequently, if there is a solution of (9-33) in positive integers, we have shown that there must exist another such solution with a smaller value of x. Applying Fermat's method of infinite descent, we see that there cannot exist a solution of (9-33) in integers for which $z > 0$.

9-6 Problems

1. Prove that there is no solution in positive integers x, y, z of the Diophantine equation $x^4 + y^4 = z^4$ (Fermat's Last Theorem for fourth powers).

2. Prove that if there is a solution in integers of $2u^4 - v^4 = w^2$ with $(u, v) = 1$, then w must be of the form $8M \pm 1$.

3. Prove that $x^4 + 4y^4 = z^2$ has no solution in positive integers.

4. Show that there is no solution in positive integers of $x^4 - 4y^4 = -z^2$.

5. When $k = 3, 4, 5,$ or 6, prove that $x^3 + y^3 = 9z + k$ has no solution in integers x, y, z.

6. When $|k| = 3, 4, 5, 9, 10,$ or 12, prove that there is no integral solution x, y, z of the Diophantine equation $x^5 + y^5 = 25z + k$.

7. If p is a prime of the form $6M + 5$, prove that, unless x, y, z are all zero, there is no solution in integers of the Diophantine equation $x^2 + 3y^2 - pz^2 = 0$.

8. If $p = 4M + 3$ and $q = 4pN + 1$ are primes, prove that $x^2 + qy^2 - pz^2 = 0$ has no solution in positive integers x, y, z.

9. If a and b are relatively prime integers of the same sign and if n is a positive integer, prove that $ax^n - by^n = 0$ is solvable in positive integers x and y when and only when both $|a|$ and $|b|$ are nth powers.

10. Obtain a complete solution in positive integers of the Diophantine equation $x^2 + y^2 = z^4$.

11. Show that the Diophantine equation $x^4 - y^4 = 2z^2$ has no solution in positive integers x, y, z.

12. Show that where p is a prime of one of the four forms $24M + 13$, $24M + 17$, $24M + 19$, $24M + 23$, there is no solution in positive integers of $2x^2 + 3y^2 - pz^2 = 0$.

13. Prove that the Diophantine equation $x^4 - 8y^4 = z^2$ has no solution in positive integers.

14. Prove that there is no common solution in positive integers of the simultaneous Diophantine equations $x^2 + y^2 = u^2$ and $x^2 + 2y^2 = v^2$.

9-7 Solution of the Equations $X^4 - 2Y^4 = \pm Z^2$

The solving[4] in positive integers of the equation

(9-40) $\qquad X^4 - 2Y^4 = Z^2$

and the solving of equation

(9-41) $\qquad 2y^4 - x^4 = z^2$

are intimately related problems.

Solution of the Equations $X^4 - 2Y^4 = \pm Z^2$

Let X, Y, Z be a solution in positive integers of (9-40). If $d = (X, Y)$, then $X = dX_0$, $Y = dY_0$, and $Z = d^2 Z_0$. Moreover, X_0, Y_0, Z_0, where $(X_0, Y_0) = 1$, is a solution in positive integers of (9-40). Consequently, we shall consider solutions of (9-40) in positive integers X, Y, Z, where $(X, Y) = 1$. Likewise, we shall consider solutions of (9-41) in positive integers x, y, z, where $(x, y) = 1$. Consequently, both X and Z are odd, as are also x and z.

From equation (9-40)

$$(X^2)^2 - Z^2 = 2(Y^2)^2.$$

Since X and Z are odd, Y must be even. Therefore,

$$(X^2 \pm Z)(X^2 \mp Z) = 2Y^4.$$

We see that the G.C.D. of $X^2 \pm Z$ and $X^2 \mp Z$ is 2, and that, therefore, one of $X^2 \pm Z$ must be double an odd number. Let

(9-42)
$$\begin{aligned} X^2 \pm Z &= 2z^4, \\ X^2 \mp Z &= (2w)^4, \\ Y &= 2zw, \end{aligned}$$

where $(z, 2w) = 1$. Adding the first two of equations (9-42), we obtain

(9-43) $$X^2 = z^4 + 8w^4.$$

From (9-43) we obtain

$$(X \pm z^2)(X \mp z^2) = 8w^4.$$

Since $(X, 2zw) = 1$, the G.C.D. of the even integers $X \pm z^2$ and $X \mp z^2$ is 2. Consequently, one of $X \pm z^2$ is double an odd, and

(9-44)
$$\begin{aligned} X \pm z^2 &= 2x^4, \\ X \mp z^2 &= 4y^4, \end{aligned}$$

where $w = xy$ and $(x, 2y) = 1$. On subtracting equations (9-44), we get

(9-45) $$\pm z^2 = x^4 - 2y^4.$$

On adding the two equations (9-44), substituting $w = xy$ in the last of equations (9-42), and subtracting the first two of equations (9-42), we obtain

(9-46)
$$\begin{aligned} X &= x^4 + 2y^4, \\ Y &= 2xyz, \\ Z &= \pm(z^4 - 8x^4 y^4), \end{aligned}$$

where $(x, 2y) = 1$ and $(z, 2xy) = 1$. These values X, Y, Z given in (9-46) satisfy (9-40) for all values of x, y, z satisfying either of the equations (9-45).

Hence, *from each solution x, y, z of either of the two equations* (9-45), *we obtain a solution of* (9-40). Note that in (9-45), the equation $z^2 = x^4 - 2y^4$ is the equation (9-40) itself, and the equation $-z^2 = x^4 - 2y^4$ is equation (9-41). Since $x_0 = y_0 = z_0 = 1$ is an immediate solution of equation (9-41), we can directly get an infinite sequence of solutions of (9-40).

Using $x_0 = y_0 = z_0 = 1$ in (9-41), we obtain from (9-46) a solution of (9-40):

$$X_1 = 3,$$
$$Y_1 = 2,$$
$$Z_1 = 7.$$

This is the solution of (9-40) with the smallest value of X. These last values X_1, Y_1, Z_1, used as x, y, z in equations (9-46), give rise to another solution of (9-40); namely,

$$X_2 = 3^4 + 2 \cdot 2^4 = 113,$$
$$Y_2 = 84,$$
$$Z_2 = -(7^4 - 8 \cdot 3^4 \cdot 2^4) = 7967.$$

Then, considering these values X_2, Y_2, Z_2 as the values of x, y, z in the first equation (9-45), we obtain from (9-46) a third solution of (9-40); namely,

$$X_3 = 113^4 + 2 \cdot 84^4 = 262621633,$$
$$Y_3 = 2 \cdot 113 \cdot 84 \cdot 7967 = 151245528,$$
$$Z_3 = \pm(7967^4 - 8 \cdot 113^4 \cdot 84^4) = 60912456065182847.$$

Thus, we can continue in like manner and, starting from the original solution $x_0 = y_0 = z_0 = 1$ of (9-41), can obtain an infinite sequence of solutions X_n, Y_n, Z_n of (9-40), in each of which the Y_n is equal to the product of 2^n by an odd integer.

There now remains the problem of determining all non-zero solutions of equation (9-41). When $(x, y) = 1$ in (9-41), we see that x, y, z are all odd. For the purpose of solving (9-41), set $2u = x^2 + z$ and $2v = x^2 - z$; thus $x^2 = u + v$, $z = u - v$. Consequently, we obtain from (9-41) the equation

(9-47) $$y^4 = u^2 + v^2,$$

in which u and v must have opposite parity.

The solution of equation (9-41) includes the solving of a problem proposed by Fermat. This problem consists in finding a right-angled triangle with integral sides such that the length of the hypotenuse and also the sum of the lengths of the legs about the right angle, be perfect squares. If the lengths of the legs be denoted by u and v and the length of the hypotenuse by y^2, it is required to find u, v, and y^2 such that $u + v = x^2$, say, and $u^2 + v^2 = (y^2)^2$. Consequently, setting $z = u - v$, we have $2y^4 - x^4 = (u - v)^2 = z^2$. This is again equation (9-41). Not every positive solution of (9-41), however, will give *positive* values u and v to satisfy Fermat's problem[5].

Since $u + v = x^2$ and $u^2 + v^2 = (y^2)^2$, where $(u, v) = 1$, we obtain from the solution of the Pythagorean equation that either

$$u = R^2 - S^2,$$
$$v = 2RS,$$
$$y^2 = R^2 + S^2;$$

or

$$u = 2RS,$$
$$v = R^2 - S^2,$$
$$y^2 = R^2 + S^2,$$

where $(R, S) = 1$ and R and S are of opposite parity. That is,

(9-48)
$$\frac{x^2 \pm z}{2} = R^2 - S^2,$$
$$\frac{x^2 \mp z}{2} = 2RS,$$
$$y^2 = R^2 + S^2,$$

where $(R, S) = 1$ and R and S are of opposite parity.

Let us assume that $y > 1$. Then it follows that $uv \neq 0$. For, if either $u = 0$ or $v = 0$, $z = \mp x^2$, and from (9-41) we have $y = x$. However, since $(x, y) = 1$, $y = x$ implies that $x = y = 1$, then also $z = 1$. Thus we have a contradiction of our assumption that $y > 1$. Moreover, since $uv \neq 0$, we see that neither R nor S can be zero. We see also that $R + S \neq 0$. For, if it were zero, then $(x^2 \pm z)/2 = 0$. Again this leads to the solution $x = y = z = 1$ of equation (9-41). If $R + S < 0$, we could replace in equations (9-48) R and S by $-R$ and $-S$, respectively. Hence we may assume that $R + S$ is positive.

Adding and subtracting the first two equations of (9-48), we obtain

(9-49)
$$x^2 = R^2 + 2RS - S^2,$$
$$z = \pm(R^2 - 2RS - S^2).$$

From the first of equations (9-49) we see that since x is odd and RS is even, we must have S even. Employing the solution of the Pythagorean equation, we obtain as a solution of the last equation in (9-48):

(9-50)
$$R = r^2 - s^2,$$
$$S = 2rs,$$
$$y = r^2 + s^2,$$

where $(r, s) = 1$ and r and s are non-zero of opposite parity.

Substituting the values of R and S in the first equation of (9-49), we obtain

$$x^2 = (r^2 - s^2)^2 + 4rs(r^2 - s^2) - (2rs)^2$$
$$= r^4 + 4r^3s - 6r^2s^2 - 4rs^3 + s^4.$$

Hence

(9-51)
$$(r^2 + 2rs - s^2)^2 - x^2 = 8r^2s^2.$$

Now, since $r^2 + 2rs - s^2 > 0$ as a result of $R + S$ being positive,

$$(r^2 + 2rs - s^2 \pm x)(r^2 + 2rs - s^2 \mp x) = 8r^2 s^2.$$

Since r and s are of opposite parity and x is odd, we see that the number in each of the parentheses on the left is even, and

(9-52) $$\left(\frac{r^2 + 2rs - s^2 \pm x}{2}\right)\left(\frac{r^2 + 2rs - s^2 \mp x}{2}\right) = 2r^2 s^2.$$

Since the sum and difference of the numbers in the two parentheses on the left are $r^2 + 2rs - s^2$ and $\pm x$, respectively, we see that a prime p dividing each of these parentheses must divide the odd number x and also $r^2 + 2rs - s^2$. Consequently, by (9-51), p divides rs. If $p|r$ and $p|(r^2 + 2rs - s^2)$, then $p|s$. This contradicts the fact that $(r, s) = 1$; likewise, if $p|s$. Hence the two parentheses in the left member of (9-52) are relatively prime. Therefore,

(9-53)
$$\frac{r^2 + 2rs - s^2 \pm x}{2} = f^2$$
$$\frac{r^2 + 2rs - s^2 \mp x}{2} = 2g^2,$$

where $(fg)^2 = (rs)^2$ and $(f, g) = 1$. Adding and subtracting equations (9-53), we obtain

$$r^2 + 2rs - s^2 = f^2 + 2g^2,$$

(9-54) and

$$x = \pm(f^2 - 2g^2).$$

Since we may take f and g so that $fg = rs \neq 0$, we may write $r/f = g/s = K/L$, where $(K, L) = 1$; consequently, where α, β, K, L are all non-zero integers,

(9-55) $$r = \alpha K, \ s = \beta L, \ f = \alpha L, \ g = \beta K.$$

Substituting these values in the first equation of (9-54), we have

$$\alpha^2 K^2 + 2\alpha\beta KL - \beta^2 L^2 = \alpha^2 L^2 + 2\beta^2 K^2;$$

thus,

$$(L^2 + 2K^2)\left(\frac{\beta}{\alpha}\right)^2 - 2KL\left(\frac{\beta}{\alpha}\right) + (L^2 - K^2) = 0.$$

Hence, where β/α is in its lowest terms,

(9-56)
$$\frac{\beta}{\alpha} = \frac{KL \pm \sqrt{K^2 L^2 - (L^2 + 2K^2)(L^2 - K^2)}}{L^2 + 2K^2}$$
$$= \frac{KL \pm \sqrt{2K^4 - L^4}}{L^2 + 2K^2}.$$

Since the left member of the equation

$$\frac{\beta}{\alpha}(L^2 + 2K^2) - KL = \pm\sqrt{2K^4 - L^4}$$

is rational, $2K^4 - L^4$ must be a perfect square, V^2; that is, if we set $y = K$, $x = L$, and $z = V$, we obtain a solution of (9-41). Since from (9-50) we have

(9-57) $$y = r^2 + s^2 = \alpha^2 K^2 + \beta^2 L^2,$$

it follows, since $y > 1$, that

$$|K| = \frac{\sqrt{y - \beta^2 L^2}}{|\alpha|} < \sqrt{y} < y.$$

Thus, assuming y, x, z to be a solution in positive integers, we have found another solution of (9-41) in which $|K|$, $|L|$, $|V|$ replace y, x and z, respectively. That is, in view of (9-57), we have found another solution in positive integers with smaller y.

Conversely, if we begin with a solution $y = K \geq 1$, $x = L$, $z = V$, we determine from (9-56) two relatively prime integers α, β, which when applied in (9-55) give two sets of integers r, s, f, g. We obtain the x from (9-54), the y from (9-50), and the z from (9-50) and (9-49). Since, from (9-54) and (9-53)

$$x^2 = (f^2 - 2g^2)^2 = (f^2 + 2g^2)^2 - 8f^2g^2$$
$$= (r^2 + 2rs - s^2)^2 - 8r^2s^2,$$

these values for y, x and z satisfy (9-41) for arbitrary values of r and s.

For example, $K = L = V = 1$ is a solution of (9-41) since these values satisfy $2K^4 - L^4 = V^2$. Using $K = L = 1$ in (9-56), we obtain $\beta/\alpha = (1 \pm 1)/3$. Hence $\beta/\alpha = 0/3$ or $\beta/\alpha = 2/3$. Since $\beta = 0$ is inadmissible, take $\alpha = 3$, $\beta = 2$ and obtain $r = 3$, $s = 2$, $f = 3$, $g = 2$. Hence, $y = r^2 + s^2 = 13$, $x = \pm(f^2 - 2g^2) = 1$, $z = 239$. The primitive solution $y = 13$, $x = 1$, $z = 239$ is that one with the least value of y exceeding 1. Starting from this last solution of (9-41), we obtain from (9-56) $\beta/\alpha = (13 \pm 239)/339$. Therefore $\beta/\alpha = -226/339 = -2/3$ or $\beta/\alpha = 252/339 = 84/113$. This gives two possibilities: $\alpha = 3$, $\beta = -2$ and $\alpha = 113$, $\beta = 84$. Using $K = 13$, $L = 1$, $\alpha = 3$, $\beta = -2$, we obtain $r = 39$, $s = -2$, $f = 3$, $g = -26$; thus $y = 1525$, $x = 1343$, $z = 2750257$. Using $K = 13$, $L = 1$, $\alpha = 113$, $\beta = 84$, we obtain $y = 2165017$, $x = 2372159$, $z = 3503833734241$.

This procedure can be repeated to obtain successively the positive solutions of (9-41). The method employed here may be regarded as a modification of Fermat's method of infinite descent.

9-8 Sum of Two Squares

We shall now consider, for a given integer n, conditions under which a Diophantine equation of the type

$$n = x^2 + y^2$$

has a solution in integers x and y. We shall first, in this connection, prove a useful result of A. Thue.

- **Theorem 9-8** *If $m > 1$ and if k denotes the least integer $> \sqrt{m}$ (that is, $k - 1 = [\sqrt{m}]$), then for an integer a relatively prime to m there exist positive integers x and y, each not exceeding $k - 1$, such that either*

$$ay \equiv x \pmod{m}$$

or

$$ay \equiv -x \pmod{m}.$$

Proof Consider the numbers of the form $ay + x$, where x and y are subject to the restrictions $0 \leq x \leq k - 1$ and $0 \leq y \leq k - 1$. We note that $(k - 1)^2 \leq m < k^2$. Moreover, $k = 1 + [\sqrt{2}] = 2$ when $m = 2$, $k = 1 + [\sqrt{3}] = 2$ when $m = 3$, and $k \leq (k - 1)^2$ when $k \geq 3$ (that is, when $m \geq 4$). Consequently, $k \leq m$ for $m \geq 2$.

Since there are altogether $k^2 > m$ combinations of x with y, we have more than m numbers $ay + x$. Hence at least two of them must belong to the same residue class, modulo m. Suppose

$$ay_1 + x_1 \equiv ay_2 + x_2 \pmod{m}.$$

We then have

(9-58) $$a(y_1 - y_2) \equiv x_2 - x_1 \pmod{m}.$$

If $x_2 - x_1 = 0$, then $y_1 = y_2$, and conversely. Since $y_1 \not\equiv y_2 \pmod{m}$, set $y = |y_1 - y_2|$ and $x = |x_2 - x_1|$, where $1 \leq y \leq k - 1$ and $1 \leq x \leq k - 1$. Then from (9-58) we see that we always have solutions x and y of either $ay \equiv x \pmod{m}$, when $y_1 - y_2$ and $x_2 - x_1$ have the same sign, or $ay \equiv -x \pmod{m}$, when $y_1 - y_2$ and $x_2 - x_1$ have opposite signs.

- **Theorem 9-9** *If m and n are each a sum of two squares, then their product mn is also a sum of two squares.*

Proof In view of the identity

(9-59) $$(x^2 + y^2)(u^2 + v^2) = (xu + yv)^2 + (xv - yu)^2,$$

we see that, if $m = x^2 + y^2$ and $n = u^2 + v^2$, then $mn = (xu + yv)^2 + (xv - yu)^2$.

Sum of Two Squares

Corollary If each of $n_1, n_2, \ldots, n_s$, where $s \geq 2$, is a sum of two squares, then $n_1 n_2 \ldots n_s$ is also a sum of two squares.

The following result is often referred to as Girard's Theorem.

• **Theorem 9-10** Every prime p of the form $4m + 1$ can be expressed as the sum of two squares; that is, $p = x^2 + y^2$.

Proof Since -1 is a quadratic residue of $p = 4m + 1$ (by Theorem 6-7 (iv)),

(9-60) $$w^2 + 1 \equiv 0 \pmod{p}.$$

is solvable. By Thue's Theorem 9-8 (with a replaced by w and m by p), there exist positive integers x and y, each less than $\sqrt{p}$, such that

$$wy \equiv \pm x \pmod{p}.$$

That is, from (9-60)

$$w^2 y^2 + y^2 \equiv 0 \pmod{p};$$

consequently,

$$x^2 + y^2 \equiv 0 \pmod{p}.$$

Moreover,

$$x^2 + y^2 = pq,$$

where $q \geq 1$.

But, since $x^2 + y^2 < 2p$,

$$p = x^2 + y^2.$$

• **Theorem 9-11** Let the positive integer n be written in the form $n = lm^2$, where l is not divisible by the square of a prime. Then n can be written as a sum of two squares if and only if l contains no prime factor of the form $4M + 3$.

In other words, a positive integer n is a sum of two squares when and only when the exponent of the highest power of every prime factor of the form $4M + 3$ is even; that is to say, when and only when n has the form $2^e a^2 b$, where e is non-negative, a is odd, and every prime factor of b is of the form $4M + 1$.

For example, $20 = 5 \cdot 2^2 = 4^2 + 2^2$, but $12 = 3 \cdot 2^2$ cannot be written as the sum of two squares. However, $90 = 2 \cdot 3^2 \cdot 5 = 9^2 + 3^2$.

Proof Since $u^2 \equiv 0 \pmod 4$ when u is even, and $u^2 \equiv 1 \pmod 4$ when u is odd, it is evident that for every x and every y, $x^2 + y^2 \not\equiv 3 \pmod 4$. Hence, no prime of the form $4M + 3$ can be written as the sum of two squares. Moreover, since $2 = 1^2 + 1^2$, every prime not of the form $4M + 3$ can be written as the sum of two squares.

We shall prove that, where l contains no square of a prime and no prime factor of the form $4M + 3$, $n = lm^2$ is a sum of two squares. In the case where $l = 1$, we have $n = m^2 + 0^2$. If $l > 1$, let $l = p_1 p_2 \cdots p_s$ be the canonical decomposition of l as a product of distinct primes. Each of these primes, being either 2 or of the form $4M + 1$, is a sum of two squares. Hence, by Theorem 9-9, Corollary, l is a sum of two squares, $l = g^2 + h^2$. Therefore, since $n = lm^2 = (gm)^2 + (hm)^2$, n is a sum of two squares.

Conversely, if $n = lm^2$ is a sum of two squares, we shall see that l cannot have a prime factor of the form $4M + 3$. Since this statement is obviously true if $l = 1$ or $l = 2$, let us assume that $l > 2$. Let $n = lm^2 = a^2 + b^2$, where $ab \neq 0$, $d = (a, b)$, $a = da_0$, $b = db_0$, $(a_0, b_0) = 1$.

If $d > 1$, let q be an arbitrary prime factor of d so that $d = q^r d_1$, where $r \geq 1$ and $(d_1, q) = 1$. Since $d^2 | n$, $q | m$ and $m = q^s m_1$, where $(m_1, q) = 1$. If $r > s$, then $2r \geq 2s + 2$. Since the highest power of q dividing lm^2 is not greater than $2s + 1$, $2r \leq 2s + 1$. This results in a contradiction. Hence, since $d^2 | n$ and $r \leq s$, we see that $d^2 | m^2$. Write $m^2 = d^2 m_0^2$. Consequently, since

$$lm_0^2 = \frac{a^2 + b^2}{d^2} = a_0^2 + b_0^2,$$

we have $a_0^2 + b_0^2 \equiv 0 \pmod{l}$. Next, let p be an odd prime factor of l. Since $(a_0, b_0) = 1$, $(a_0 b_0, p) = 1$. Let c satisfy the congruence $a_0 c \equiv 1 \pmod{p}$. Then, since $a_0^2 + b_0^2 \equiv 0 \pmod{p}$,

$$(a_0 c)^2 + (b_0 c)^2 \equiv 0 \pmod{p};$$

therefore $(b_0 c)^2 \equiv -1 \pmod{p}$. Now, since -1 is a quadratic residue of p, p must be of the form $4M + 1$. Consequently, the Diophantine equation $n = X^2 + Y^2$ is solvable in integers when and only when the positive integer n has the property stated in the theorem.

• **Theorem 9-12** If a prime is the sum of two squares and divides a sum of two relatively prime squares, then the quotient is likewise a sum of two relatively prime squares.[6]

Proof Let the prime p, represented as

$$p = a^2 + b^2,$$

be a divisor of $A^2 + B^2$, where $(A, B) = 1$. Then

(9-61) $$pP = (a^2 + b^2)P = A^2 + B^2.$$

Since, obviously, $(a, b) = 1$, there is an infinitude of solutions x, y of the equation

(9-62) $$ax + by = A$$

given by the formulae (Theorem 2-10)

(9-63) $$x = x_0 + tb,$$
(9-64) $$y = y_0 - ta,$$

where x_0, y_0 denotes any particular solution and t is arbitrary.

If we set $C_0 = bx_0 - ay_0$ and $C = bx - ay$, then

(9-65) $$bx - ay = C = C_0 + pt.$$

Since x_0, y_0 is a solution of (9-62), we have the following identities:

(9-66) $$(a^2 + b^2)(x_0^2 + y_0^2) = A^2 + C_0^2,$$
(9-67) $$(a^2 + b^2)(x^2 + y^2) = A^2 + C^2.$$

Subtracting (9-66) from (9-61), we obtain

$$B^2 - C_0^2 \equiv (B - C_0)(B + C_0) \equiv 0 \pmod{p}.$$

Hence, the prime $p = a^2 + b^2$ divides at least one of $B - C_0, B + C_0$. If p divides $-(B + C_0) = -B - C_0$, then we could replace the original B in (9-61) by its negative B' and then have $p|(B' - C_0)$. Let us assume, then, that p divides $B - C_0$; consequently

(9-68) $$B - C_0 = pm.$$

Taking $t = m$ in formulae (9-63), (9-64), and (9-65), we derive from (9-65) and (9-68) that $C = B$; then it follows, from (9-67), that

$$\frac{A^2 + B^2}{a^2 + b^2} = x^2 + y^2.$$

In view of (9-62) and $bx - ay = B$, we see that a common divisor of x and y must divide both of the relatively prime integers A and B. Hence x and y are relatively prime, and the theorem is established.

The preceding theorem yields another method of proof[7] of Theorem 9-10.

9-9 Sum of Three Squares

The problem of determining what positive numbers can be represented as a sum of three squares is somewhat difficult. We can, however, readily prove that no integer of the form $4^a(8m + 7)$ can be a sum of three squares. Conversely, it can be shown[8] by employing the theory of ternary quadratic forms that only numbers of this form cannot be represented as a sum of three squares.

In other words, *the positive integer n can be represented as a sum of three squares when and only when* $n \neq 4^a(8m + 7)$.

• **Theorem 9-13** If a positive integer n can be represented as a sum of three squares, it is not of the form $4^a(8m + 7)$, where a and m are non-negative integers.

Proof First, we see that the integer $8m + 7$ cannot be represented as a sum of three squares. For, if $8m + 7 = x_1^2 + x_2^2 + x_3^2$, then exactly one of the x's is odd or exactly three are odd. But, since for an arbitrary square x^2, we have $x^2 \equiv 0 \pmod{4}$ or $x^2 \equiv 1 \pmod{8}$, such a representation is impossible.

Next, if $4^a(8m + 7)$, where $a \geq 1$, were represented as

$$4^a(8m + 7) = x_1^2 + x_2^2 + x_3^2,$$

then $x_1^2 + x_2^2 + x_3^2 \equiv 0 \pmod{4}$; that is, each of x_1, x_2, x_3 must be even. Consequently,

$$4^{a-1}(8m + 7) = \left(\frac{x_1}{2}\right)^2 + \left(\frac{x_2}{2}\right)^2 + \left(\frac{x_3}{2}\right)^2.$$

If $a - 1 > 0$, the process could be repeated until we eventually reached the false conclusion that $8m + 7$ is a sum of three squares. Hence, no integer of the form $4^a(8m + 7)$ can be a sum of three squares.

Example Where $Z^2 = z^2$, exhibit an explicit one-to-one correspondence between the integral solutions of

(9-69) $$x^2 - 2y^2 = z^2$$

and

(9-70) $$2X^2 - Y^2 = Z^2$$

Solution Neither the solution $x = 3, y = -2, z = 1$ nor the solution $x = -3, y = 2, z = -1$ is to be regarded as a solution different from the solution $x = 3, y = 2, z = 1$. In other words, we shall consider solutions where the values of the unknowns are non-negative. Moreover, since all solutions of each equation may be obtained from its primitive solutions, we may consider $(x, y) = (X, Y) = 1$.

Using the identities

(9-71) $$x^2 - 2y^2 = 2(x + y)^2 - (x + 2y)^2$$

and

(9-72) $$2X^2 - Y^2 = (2X - Y)^2 - 2(Y - X)^2,$$

Sum of Three Squares

we see that every primitive solution x, y, z of (9-69) yields a primitive solution $X = x + y$, $Y = x + 2y$, $Z = z$ of (9-70); and, conversely, every primitive solution X, Y, Z of (9-70) gives rise to a primitive solution of (9-69).

If, for two non-negative solutions x_0, y_0, z_0 and x_1, y_1, z_1 of (9-69), we were to have $x_1 + y_1 = x_0 + y_0$ and $x_1 + 2y_1 = x_0 + 2y_0$, then $x_1 = x_0$ and $y_1 = y_0$. Hence, in view of (9-71), we see that different solutions of (9-69) yield different solutions of (9-70), and conversely (cf. Theorem 9-2).

9-10 Problems

1. If p is a prime of the form $4M + 3$, show that the equation $x^2 + y^2 - pz^2 = 0$ has no solution in positive integers.

2. (a) Give a proof independent of Theorem 9-10 that every prime $p = 4M + 1$ is a divisor of a sum of certain two relatively prime squares.

    (b) Prove that every prime $p = 4M + 1$ is a divisor of a sum of certain two relatively prime squares where each square exceeds 3. (Hint: consider an odd primitive root of p.)

3. Show that the positive integers n and $2n$ have the same number of different representations as a sum of two squares. (One of the squares may be zero, and $(\pm a)^2 + (\pm b)^2$ and $(\pm b)^2 + (\pm a)^2$ are not to be regarded as representations different from $a^2 + b^2$.)

4. Using the identity $(x^2 + y^2 + z^2)^2 = (2xz)^2 + (2yz)^2 + (x^2 + y^2 - z^2)^2$, find three primes p and three composite numbers n, each a sum of three squares, such that p^2 and n^2 are sums of three squares. Also, find these three squares in each of the six cases.

5. Prove that the product of two numbers each of the form $X^2 + mY^2$ is again a number of this same form.

6. Prove that a number having a prime factor $p = 4n + 3$ is not a sum of two relatively prime squares.

7. Using the identity $(a^2 + b^2 + c^2)(a_1^2 + b_1^2 + c_1^2) = (ab_1 - ba_1)^2 + (bc_1 + cb_1)^2 + (ca_1 + ac_1)^2 + (aa_1 + bb_1 - cc_1)^2$ and the identity given in Problem 4, prove that the cube of a sum of three squares is itself a sum of three squares.

8. Prove that every positive divisor of a sum of two relatively prime squares is a sum of two squares.

9-11 Sum of Four Squares

In commenting on the work of Diophantus on the sum of squares, C.G. Bachet remarked that in his opinion, Diophantus had known that a positive integer is either a square or the sum of two, three, or four squares. Bachet verified this proposition for numbers up to 325. This result is often referred to as Bachet's Theorem. Fermat stated that he had a proof that every positive number is a sum of at most four squares. Like Bachet, he believed that Diophantus apparently was aware of this theorem. L. Euler gave much time and thought to the problem, but he was not the first to obtain a complete proof.

In 1770, utilizing ideas from Euler, J. L. Lagrange gave a proof[9] of Bachet's Theorem. Later proofs, including one by Euler, considerably simplified Lagrange's original proof.

• **Theorem 9-14** (Euler's Identity) The product of two integers each a sum of four squares is again a sum of four squares.

Proof The identity[10] given by Euler is as follows:

(9-73)
$$\begin{aligned}(x_1^2 + x_2^2 + x_3^2 + x_4^2)&(y_1^2 + y_2^2 + y_3^2 + y_4^2) \\ = (x_1 y_1 + x_2 y_2 &+ x_3 y_3 + x_4 y_4)^2 \\ + (x_1 y_2 - x_2 y_1 &+ x_3 y_4 - x_4 y_3)^2 \\ + (x_1 y_3 - x_3 y_1 &+ x_4 y_2 - x_2 y_4)^2 \\ + (x_1 y_4 - x_4 y_1 &+ x_2 y_3 - x_3 y_2)^2.\end{aligned}$$

This result may be obtained by multiplying, row by column, the following determinants (with elements complex numbers):

$$\begin{vmatrix} x_1 + ix_2 & -(x_3 + ix_4) \\ x_3 - ix_4 & x_1 - ix_2 \end{vmatrix} \cdot \begin{vmatrix} y_1 - iy_2 & y_3 + iy_4 \\ -(y_3 - iy_4) & y_1 + iy_2 \end{vmatrix}$$

$$= \begin{vmatrix} z_1 - iz_2 & z_3 + iz_4 \\ -(z_3 - iz_4) & z_1 + iz_2 \end{vmatrix},$$

where

(9-74)
$$\begin{aligned} z_1 &= x_1 y_1 + x_2 y_2 + x_3 y_3 + x_4 y_4, \\ z_2 &= x_1 y_2 - x_2 y_1 + x_3 y_4 - x_4 y_3, \\ z_3 &= x_1 y_3 - x_3 y_1 + x_4 y_2 - x_2 y_4, \\ z_4 &= x_1 y_4 - x_4 y_1 + x_2 y_3 - x_3 y_2. \end{aligned}$$

• **Theorem 9-15** For every prime p, there is a positive integer m less than p such that mp is a sum of four squares.

Proof For the prime 2, we must have $m = 1$, and obviously $1 \cdot 2 = 1^2 + 1^2 + 0^2 + 0^2$. Assume, then, that p is an odd prime.

No two of the $(p + 1)/2$ integers x^2, where $0 \leq x \leq (p - 1)/2$, are congruent, modulo p. For, if $x_1^2 \equiv x_2^2 \pmod{p}$, then $p|(x_1 - x_2)(x_1 + x_2)$. This, however, is impossible when $x_1 \neq x_2$. Likewise, no two of the $(p + 1)/2$ integers $-1 - y^2$, where $0 \leq y \leq (p - 1)/2$, are congruent, modulo p. Hence, since there are $p + 1$ integers in the two sets and only p residue classes, modulo p, two of these numbers must be in the same residue class; that is, there is a pair of values x, y such that

$$x^2 \equiv -1 - y^2 \pmod{p},$$

where $0 \leq x < p/2$ and $0 \leq y < p/2$.

From this it follows that

$$x^2 + y^2 + 1^2 + 0^2 = mp < \frac{p^2}{4} + \frac{p^2}{4} + 1 < p^2;$$

consequently $0 < m < p$.

We shall next prove a crucial result, which when combined with Euler's identity, will establish the four-square theorem.[11]

- **Theorem 9-16** Every prime p is representable as a sum of four integral squares.

Proof For $p = 2$, we have $2 = 1^2 + 1^2 + 0^2 + 0^2$. Assume the prime p to be *odd*. Let m be the least positive integer such that mp is a sum of four squares; that is,

(9-75) $$mp = x_1^2 + x_2^2 + x_3^2 + x_4^2.$$

We shall first prove that this least positive m is odd, and finally that $m = 1$.

If m were even,

$$x_1 + x_2 + x_3 + x_4 \equiv x_1^2 + x_2^2 + x_3^2 + x_4^2 \equiv 0 \pmod{2}.$$

Then exactly none, two, or four of the x's are even. Assume, therefore, that $x_1 + x_2$ and $x_3 + x_4$ are even. Consequently,

$$\frac{m}{2}p = \left(\frac{x_1 + x_2}{2}\right)^2 + \left(\frac{x_1 - x_2}{2}\right)^2 + \left(\frac{x_3 + x_4}{2}\right)^2 + \left(\frac{x_3 - x_4}{2}\right)^2.$$

This contradicts the definition of m as the least positive number for which mp is a sum of four squares.

Let us now assume that m is odd and greater than unity. Choose integers y_1, y_2, y_3, y_4 so that

(9-76) $$y_i \equiv x_i \pmod{m},$$

where $|y_i| < m/2$, $i = 1, 2, 3, 4$. Then

$$y_1^2 + y_2^2 + y_3^2 + y_4^2 \equiv x_1^2 + x_2^2 + x_3^2 + x_4^2 \equiv mp \equiv 0 \pmod{m}.$$

Write

(9-77) $$y_1^2 + y_2^2 + y_3^2 + y_4^2 = mn.$$

Now, $n > 0$. For, if $n = 0$, each y is zero; consequently, m divides each x and $m^2 | mp$. Since p is a prime, this contradicts the statement that $1 < m < p$. Moreover, $n < m$ since by (9-76), $mn < 4(m^2/4)$.

From (9-75) and (9-77) we obtain, on multiplying corresponding sides,

(9-78) $$m^2 np = z_1^2 + z_2^2 + z_3^2 + z_4^2,$$

where the z's are given by (9-74). Since, by (9-76),

$$z_1 \equiv x_1 y_1 + x_2 y_2 + x_3 y_3 + x_4 y_4$$
$$\equiv x_1^2 + x_2^2 + x_3^2 + x_4^2 \equiv 0 \pmod{m}$$

and

$$x_i y_j - x_j y_i \equiv x_i x_j - x_j x_i \equiv 0 \pmod{m},$$

we see that z_2, z_3, z_4 can be written as a sum of two parts, each of which is now divisible by m. Hence, from (9-78), we note that each term in the right member is divisible by m^2. Consequently, np is a sum of four squares. Again, this contradicts the minimal character of m. Hence, the supposition that $m > 1$ is false, and $m = 1$.

• **Theorem 9-17** (Lagrange) Every positive integer n is a sum of four squares.

Proof Since $1 = 1^2 + 0^2 + 0^2 + 0^2$, the result is obvious for $n = 1$. Assume $n > 1$ and let $n = p_1 p_2 \cdots p_k$, where the p's are primes, not necessarily distinct. Since each p_i is a sum of four squares, then by Euler's identity (Theorem 9-14), the product of any two primes is again a sum of four squares. Hence, by mathematical induction, we see that n is a sum of four squares.

9-12 Remarks on Waring's Problem

Waring's Problem considers the representation of positive integers as sums of a specific number of non-negative kth powers. Waring stated without proof that every number is a sum of four squares, of 9 cubes, and of 19 biquadrates. The implication seems to be that he believed there was a certain least positive number, depending only on k, such that every positive integer can be represented as a sum of that many kth powers.

The question, by no means obvious, is whether there does exist a fixed number s, depending only on k, such that for every positive integer n,

Remarks on Waring's Problem

$$n = x_1^k + x_2^k + \cdots + x_s^k$$

is solvable in non-negative integers $x_1, x_2, \cdots, x_s$.

If there is such a number s for a fixed k, there must be a *least* such number. This least number is denoted by $g(k)$. In the preceding section we treated the case $k = 2$, and proved that $g(2) = 4$. Hilbert was the first to prove[12] (in 1909) the existence of such a positive number, depending only on k, for which it can be asserted that every positive integer is representable as a sum of that number of kth powers.

It may happen that for large values of n, fewer than $g(k)$ kth powers suffice to represent n. That is, there may be only certain specific numbers that actually require as many as $g(k)$ kth powers for their representation. For example, it is known that $g(3) = 9$, but for n sufficiently large it has been proved that fewer than 9 cubes suffice to represent n. Consequently, $G(k)$ is defined as the least number of kth powers for which it can be said that all sufficiently large numbers (that is, all numbers with not more than a finite number of exceptions) are representable by $G(k)$ kth powers. It is known[13] that, for $k \geq 2$, $k + 1 \leq G(k) \leq g(k)$. For squares, we have $G(2) = g(2) = 4$. For cubes, we know that $4 \leq G(3) \leq 7$ and $g(3) = 9$. It seems probable that the value of $G(3)$ will be found to be 4 or 5. For biquadrates, it is known that $G(4) = 16$ and $19 \leq g(4) \leq 35$. The numbers 79, 159, 239, 319, 399, 479, and 559 all require 19 biquadrates for their representation.

Since Hilbert's result of 1909, much work has been done on this problem and its generalization. For an interesting and informative summary of progress in the solution of Waring's Problem, reference may be made to the notes at the end of Chapter 21 of Hardy and Wright's *Introduction to the Theory of Numbers*.

9-13 Problems

1. If $(x, y) = 1$, prove that every positive divisor of $x^2 + y^2$ is the sum of two relatively prime squares.

2. Where x and y are relatively prime, show that, for $a \geq 0$, $x^2 + y^2 = (4a + 3)z$ has no solution in positive integers x, y, z.

3. Show that $x^2 = y^3 + 7$ has no solution in integers x and y.

4. Where $n > 0$ and $q = [(3/2)^n]$, prove that all positive integers not exceeding $2^n q$ are sums of $2^n + q - 2$ integral nth powers of non-negative integers. Moreover, show that $2^n q - 1$ is not a sum of $2^n + q - 3$ non-negative nth powers.

Notes

Chapter 1

1. *Nature*, XLII (1890), 467.
2. The reference here is to C. F. Gauss, *Disquisitiones Arithmeticae* (Leipzig, 1801). This is contained in the first volume of Gauss's *Werke* (Göttingen, 1863). An English translation of the *Disquisitiones Arithmeticae* prepared by Arthur A. Clarke, S. J., was published in 1966 by Yale University Press.
3. Report on the Theory of Numbers, British Association, 1859-1865. This Report is found in The Collected Mathematical Papers of Henry John Stephen Smith, Edited by J. W. L. Glaisher (Oxford, 1894), I, 38-364. The quotation is from Vol. I, pp. 38-39.
4. See the Preface of L. E. Dickson's *Introduction to the Theory of Numbers* (Chicago, 1929).
5. British Association for the Advancement of Science. Report of the ninetieth

meeting (Hull, 1922). Address to Section A (Mathematics and Physics) on "The Theory of Numbers" by Professor G. H. Hardy, President of the Section, pp. 16-24. Quotations are from pp. 16 and 21.

6. Before the American Mathematical Society on December 28, 1928, Professor Hardy began his address with the following opening remarks:

> The theory of numbers has always occupied a peculiar position among the purely mathematical sciences. It has the reputation of great difficulty and mystery among many who should be competent to judge; I suppose that there is no mathematical theory of which so many well-qualified mathematicians are so much afraid. At the same time it is unique among mathematical theories in its appeal to the uninstructed imagination and in its fascination for the amateur.

G. H. Hardy, "An Introduction to the Theory of Numbers," *Bulletin of the American Mathematical Society*, XXXV, No. 6 (1929), 778-818. The quotation is from p. 778.

7. Marshall Stone, "The Revolution in Mathematics," *The American Mathematical Monthly*, LXVIII, No. 8 (Oct., 1961), 715-734. The quotation is from page 723.

8. This result is sometimes referred to as Bachet's Theorem, since C. G. Bachet verified this result for integers up to 325 but lacked a proof of the general result. He believed that Diophantus had assumed that every positive integer can be written as a sum of not more than four squares. It was, however, J. L. Lagrange who, utilizing ideas of L. Euler, gave the first complete proof of Bachet's Theorem.

9. Arthur Wieferich in May, 1908 (*Mathematische Annalen*, LXVI (1909), 95-101) proved—except for a case he overlooked—that nine cubes suffice. Aubrey Kempner completed the proof by considering the case overlooked by Wieferich. (*Mathematische Annalen*, LXXII (1912), 387-399, particularly pp. 392-395.)

10. This is often known as "Goldbach's Theorem," but more accurately, "Goldbach's Problem." For an historical survey of this problem, see Ralph G. Archibald, "Goldbach's Theorem," *Scripta Mathematica*, III (1935), 44-50 and 153-161. See also a more recent paper by R. D. James, "Recent Progress in the Goldbach Problem," *Bulletin of the American Mathematical Society*, LV (1949), 246-260. It might be added that R. Descartes stated, without proof, that every even number is a sum of 1, 2, or 3 primes. Moreover, E. Waring stated that every odd number is either a prime or is a sum of three primes. See L. E. Dickson, *History of the Theory of Numbers* (Washington: Carnegie Institution of Washington, 1919, Publication No. 256) I, 421.

11. In 1959, only twenty perfect numbers were known; three more were found four years later. See G. H. Hardy and E. M. Wright, *An Introduction to the Theory of Numbers*, Fourth Ed. (1960), pp. 16, 240. An even number is perfect when and only when $2^p - 1$ is a prime and the even number has the form $2^{p-1}(2^p - 1)$. Primes of the form $2^p - 1$ are known as *Mersenne primes*. The known Mersenne primes are given by $p = 2, 3, 5, 7, 13, 17, 19, 31, 61, 89, 107, 127, 521, 607, 1279, 2203, 2281, 3217, 4253, 4423, 9689, 9941, 11213$. See Donald B. Gillies, "Three New Mersenne Primes and a Statistical Theory," *Mathematics of Computation*

(Washington, D. C.), XVIII, No. 85 (Jan., 1964), 93-97. The last three Mersenne Primes listed above were discovered by Illiac II at the Digital Computer Laboratory of the University of Illinois. The computing times for these three new Mersenne primes were, respectively, 1 hr. 23 min., 1 hr. 30 min., 2 hr. 15 min. Computations were checked by repetition. Thus the largest known prime is $2^{11213} - 1$, a number of 3376 digits. The twenty-third perfect number is $2^{11212}(2^{11213} - 1)$, a number of 6751 digits.

12. By a non-zero integral solution we mean one in which none of the unknowns x, y, z is zero: $0^n + a^n = a^n$ and $0^n + 0^n = 0^n$ would be trivial solutions.

13. The relationship between Fermat's Last Theorem and the theory of algebraic numbers is lucidly given in L. E. Dickson's paper, "Fermat's Last Theorem and the Origin and Nature of the Theory of Algebraic Numbers," *Annals of Mathematics*, XVIII, No. 4 (June, 1917), 161-187.

Chapter 2

1. See E. Landau, *Grundlagen der Analysis* (Leipzig, 1930), 148 pp. This work has been published in an English translation: *Foundations of Analysis* (Chelsea, New York, 1951), 148 pp. Other useful and helpful works are: Solomon Feferman, *The Number Systems, Foundations of Algebra and Analysis* (Reading, Mass.: Addison Wesley, 1964), 430 pp.; L. Henkin, W. N. Smith, V. J. Varineau, and M. J. Walsh, *Retracing Elementary Mathematics* (New York: Macmillan, 1962), 436 pp. (especially Chapters 2, 8, and 11); G. Birkhoff and S. Mac Lane, *A Survey of Modern Algebra* (Third Ed.; New York: Macmillan, 1965), 448 pp. (especially Chapter 1).

2. Results (i), (ii), and (iii) follow immediately from the definition of numerical value. If either a or b is zero, or if both are of the same sign, the left member in (iv) is clearly equal to the right.

Next, let $ab < 0$. There are the two possibilities: $a > 0$ and $b < 0$, and $a < 0$ and $b > 0$. Because of the symmetry of the relation in a and b, we need to consider only the case $a > 0$ and $b = -b_1 < 0$. If $a + b \geq 0$, then $|a + b| = |a - b_1| = a - b_1 < a < |a| + |b|$. Moreover, if $a + b < 0$, $|a + b| = |a - b_1| = |-(b_1 - a)| = b_1 - a < |b| < |a| + |b|$. Hence, (iv) is established.

Now, using (iv) and (ii), we have $|a| = |(a + b) + (-b)| \leq |a + b| + |-b| = |a + b| + |b|$; consequently, $|a| - |b| \leq |a + b|$. Thus, (v) is proved.

3. We shall prove that *Principles* I, II, IIa, and III *are equivalent statements*. First, we shall prove that Principle II (Mathematical Induction) implies Principle IIa (Complete Mathematical Induction).

Let S be the set of positive integers n having the property that, if all integers m with $0 < a \leq m \leq n$ are in S, then $n + 1$ is in S. Suppose also that a is a member of S. In order to establish our desired result, we now wish to prove that S contains all integers $\geq a$.

Let T be the set of integers n such that all integers m satisfying $a \leq m \leq n$ are in S. Then T contains a since the only integer m satisfying $a \leq m \leq a$ is a, and

a is in S. Suppose now that n is a member of T. This means that all integers m with $a \leq m \leq n$ are in S. Hence, by our definition of S, $n+1$ is in S. Thus all m with $a \leq m \leq n+1$ are in S. Hence, $n+1$ is in T. Therefore, by mathematical induction (Principle II), T contains all integers $\geq a$. From the definition of T, it follows that, if n is in T, then n is in S. Hence, S contains all integers $\geq a$.

Secondly, we wish to prove that Principle IIa (Complete Mathematical Induction) implies Principle I (Well-ordering). Assuming the principle of complete mathematical induction, let E be a non-empty set of positive integers. Suppose E has no smallest integer. Now, since every positive integer ≥ 1, 1 is not a member of E. Let S be the set of all integers not in E. Then 1 is an element of S. Moreover, if all m with $1 \leq m \leq n$ are in S, then $n+1$ is in S. For, if not, $n+1$ would be in E, and thus would be the smallest integer in E. This contradicts our assumption that E has no smallest integer. Therefore, by complete mathematical induction (using Principle IIa with $a = 1$), S must contain all positive integers. Hence E is empty—contrary to the hypothesis regarding E. Consequently, the statement that E has no smallest integer is false; and thus every non-empty set E does contain a smallest integer (Principle I).

Thirdly, we desire to prove that Principle I (Well-ordering) implies Principle II (Mathematical Induction). Let S be a set of positive integers containing integer a and having the property that, if l is in the set, then $l+1$ is also in the set. Assume that S does not contain all positive integers $\geq a$. Let T be the set of all positive integers $\geq a$ which are not in S. By our present assumption, then, T is not empty. By the Well-ordering principle (Principle I) T contains a least element (which is $> a$ since a is in S). Therefore, the least element of T may be written $t+1$, where $t > 0$. This implies that t is in S and that $t+1$ is not in S. This contradicts our hypothesis regarding S. Hence, the assumption that S does not contain all integers $\geq a$ is false, and the theorem is established.

These proofs show that Principles II, IIa, and I are equivalent.

We shall now prove that Principle II implies Principle III. Let E be a non-empty set of positive integers all $\leq n$. We desire to prove that E contains a greatest integer.

Let S be the set of values of n for which the theorem holds. Then 1 is in S because a set of positive integers cannot contain an integer less than 1. Now suppose n is in S. We shall prove that $n+1$ is in S, and thus establish the theorem by mathematical induction.

Let E be a non-empty set of positive integers all $\leq n+1$. If $n+1$ is in E, then $n+1$ is clearly the largest member of E; and in this case E does have a largest member. If $n+1$ is not in E, then all members of E are $\leq n$. By the induction hypothesis, n is in S. Therefore E has a largest member. We consequently conclude that $n+1$ is in S. Hence, by mathematical induction (Principle II with $a=1$), we see that S contains all positive integers, and the theorem is true.

We shall next prove that Principle III implies Principle I (and hence implies Principle II).

Let S be a non-empty set of positive integers. We desire to prove that S has a smallest integer. Let n be an element of S. Where y is in S and $y \leq n$, let T be

Chapter 2

the set of all integers x such that $x = n + 1 - y$. Since T contains $n + 1 - n = 1$, T is non-empty. The elements of T are positive integers not exceeding n. By Principle III, T has a greatest integer g. Let m be the corresponding value of y; that is, $g = n + 1 - m$, where m is in S and $m \leq n$. Since g is the greatest such integer, m is the smallest such integer in S which is $\leq n$. Since n is in S, m is the smallest integer in S. That is, Principle III implies Principle I.

Consequently, we see that Principles I, II, IIa, and III are equivalent.

Next, we shall see that Principle II implies Principle IV. In other words, using mathematical induction, we shall prove that, given positive integers a and b, there exists a positive integer n such that $na > b$.

Let a be a fixed positive integer. We shall show that for every b, n exists such that $na > b$.

Let S be the set of values of b for which such an n exists. Since $a > 0$, $2a \geq 2 > 1$. Hence 1 is in S. Assuming b to be in S, we shall prove that $b + 1$ is in S. Since b is in S, there is an integer q such that $qa > b$. Since $a \geq 1$, by addition we get $qa + a > b + 1$; that is, $(q + 1)a > b + 1$. Hence the condition is satisfied for $b + 1$ with $n = q + 1$. Thus $b + 1$ is an element of S. Therefore, using Principle II with $a = 1$, we see that S contains all positive integers, and the theorem is established.

4. If $a > 0$ and $b \geq 0$, then $b + na > 0$ when $n > 0$; if $a > 0$ and $b < 0$, then by Archimedes' Principle (Principle IV) there is a positive integer n such that $na > -b$, and $b + na > 0$ (when $n > -b/a$); if $a < 0$ and $b \geq 0$, then $b + na > 0$ if $n < 0$; if $a < 0$ and $b < 0$, then by Archimedes' Principle there is a positive integer, $-n$ say, such that $(-n)(-a) > -b$, and $b + na > 0$ (if $n < -b/a$).

5. Following E. Landau (*Vorlesungen über Zahlentheorie*, I, 5), we shall first prove, by complete mathematical induction on n, the existence of such a representation and, after that, its uniqueness.

For $n = 1$, we obviously have such a representation for $s = 0$, $a_0 = 1$, $0 < a_0 < g$. Let $n > 1$ and assume the existence of a representation for $1, 2, \ldots, n - 1$. Now, n belongs to one of the following non-overlapping intervals:
$1 \leq n < g$, $g \leq n < g^2$, $g^2 \leq n < g^3$, $\ldots$. There is, then, an integer s such that $g^s \leq n < g^{s+1}$. Consequently, $n = a_s g^s + r$, where $0 \leq r < g^s$. Now, $a_s > 0$ since $a_s g^s = n - r > g^s - g^s = 0$; moreover, $a_s < g$ since $a_s g^s \leq n < g^{s+1}$. If $r = 0$, then we have the desired representation: $n = a_s g^s + 0 \cdot g^{s-1} + \ldots + 0 \cdot g + 0$, where $0 < a_s < g$. If $r > 0$, then, since $r < g^s \leq n$, we have, by the hypothesis of the induction, $r = b_t g^t + b_{t-1} g^{t-1} + \ldots + b_1 g + b_0$, where $t \geq 0$, $b_t > 0$, $0 \leq b_i < g$ for $0 \leq i \leq t$. Since $g^s > r \geq b_t g^t \geq g^t$, we see that $t < s$ and $n = a_s g^s + 0 \cdot g^{s-1} + \ldots + 0 \cdot g^{t+1} + b_t g^t + \ldots + b_1 g + b_0$. Hence, such a representation exists for n, and the induction is complete.

We now consider the question of uniqueness of the representation. Assume two possible representations: $n = a_s g^s + \ldots + a_1 g + a_0 = c_u g^u + \ldots + c_1 g + c_0$, where $s \geq 0$, $a_s > 0$, $0 \leq a_i < g$ for $0 \leq i \leq s$, $u \geq 0$, $c_u > 0$, $0 \leq c_j < g$ for $0 \leq j \leq u$. We desire to prove that $s = u$ and $a_i = c_i$ for $0 \leq i \leq s$. For, if this were not the case, we obtain on subtracting the two representations of n, $0 = d_v g^v + \cdots + d_1 g + d_0, v > 0, d_v \neq 0, -g < d_i < g$ for $0 \leq i \leq v$; thus

$$g^v \leq |d_v g^v| = |d_{v-1} g^{v-1} + \cdots + d_1 g + d_0|$$
$$\leq (g-1)(g^{v-1} + g^{v-2} + \cdots + g + 1)$$
$$= (g-1) \frac{g^v - 1}{g - 1} = g^v - 1.$$

This contradiction establishes the uniqueness of the representation.

6. See Euclid's Elements, IX, 20.
7. *Annales fac. sc. de Toulouse*, IV (1890), 14, final paper.
8. "Das Fortschreitungsgesetz der Primzahlen durch eine transcendente Gleichung exakt dargestellt," *Wiss. Beilage Jahresbericht*, Gymn., Trier, 1899, 96 pp.
9. Monatsber. *Akad. Wiss. Berlin für 1878*, 1879, 777-8.
10. This proof was mentioned by Professor Harold N. Shapiro in the course of an address on March 18, 1967.

Chapter 3

1. An ancient Chinese manuscript of approximately 2500 years ago erroneously stated that $2^{n-1} - 1$ is not divisible by n if n is not a prime number. Also, G. W. Leibniz (in 1680 and in 1681), employing a false argument, concurred in this erroneous statement. See L. E. Dickson, *History of the Theory of Numbers*, I, 91.

2. The first even pseudoprime (namely, 161038) was found by D. H. Lehmer. N. G. W. H. Beeger exhibited three more such numbers and proved that there exist infinitely many even pseudoprimes: "On even numbers m dividing $2^m - 2$," *American Mathematical Monthly*, LVIII, No. 8 (1951), 553-555.

3. André Rotkiewicz, "Sur les nombres pseudopremiers de la forme $ax + b$," *Comptes rendus hebdomadaires des Séances de l'Académie des Sciences* (Paris), CCLVII, No. 18 (1963), 2601-2604.

4. Being the first to publish, in 1770, the theorem that $(p - 1)! + 1$ is divisible by the prime p, E. Waring attributed it to Sir John Wilson (1741-1793). However, in 1682, G. W. Leibniz had stated the equivalent of Wilson's Theorem. J. L. Lagrange was the first to publish a proof of this theorem; he also proved its converse. (See *Oeuvres de Lagrange* (Paris, 1869), III, 425-433.)

5. The Chinese Remainder Theorem had a restriction on the moduli, but none on the a's. In this generalized statement of Theorem 3-17, there is no restriction on the m's, but definite restrictions on the a's.

It is readily seen that if there is a common solution and if M is the least common multiple of $m_1, m_2, \ldots, m_r$, all solutions are congruent to each other, modulo M. For, if x_1 and x_2 are common solutions, $x_1 \equiv a_i \equiv x_2 \pmod{m_i}$, $i = 1, 2, \ldots, r$; consequently, $x_1 \equiv x_2 \pmod{M}$.

Chapter 3

If the given set of congruences has a common solution, it necessarily follows that

$$a_1 - a_2, a_1 - a_3, \ldots, a_1 - a_r,$$
$$a_2 - a_3, \ldots, a_2 - a_r,$$
$$\ldots \qquad \ldots,$$
$$a_{r-1} - a_r$$

must be divisible, respectively, by

$$d_{12}, d_{13}, \ldots, d_{1r},$$
$$d_{23}, \ldots, d_{2r},$$
$$\ldots \qquad \ldots,$$
$$d_{r-1\,r}.$$

We could write these $(r-1)r/2$ congruences more briefly as $a_i - a_j \equiv 0 \pmod{d_{ij}}$, where $i = 1, 2, \ldots, r-1$ and $j = 2, \ldots, r$, with $i < j$.

The proof that the given conditions are sufficient for a common solution will be given by induction on r, the number of given congruences. First, consider $r = 2$:

$$x \equiv a_1 \pmod{m_1},$$
$$x \equiv a_2 \pmod{m_2}.$$

For an arbitrary integer y, the first congruence is satisfied by $a_1 + ym_1$, and every x satisfying the first congruence is of this form. Substituting this for x in the second congruence, we get $ym_1 \equiv a_2 - a_1 \pmod{m_2}$, which has a solution $y \equiv k_1 \pmod{m_2}$, say, when and only when $d_{12} = (m_1, m_2)$ divides $a_1 - a_2$; and hence, if M_2 is the L.C.M. of m_1 and m_2, $x \equiv a_1 + k_1 m_1 \pmod{M_2}$ is a solution of the two congruences. Since $x_0 = a_1 + k_1 m_1$ satisfies the two congruences, integers satisfying these two congruences are exclusively of the form $x_0 + kM_2$, k being an arbitrary integer. This establishes the truth of the theorem for the case of two congruences ($r = 2$).

Let us assume, then, that the conditions are sufficient for a common solution in the case of s (where $s \geq 2$) congruences. We desire to show that the conditions are sufficient in the case of $s + 1$ congruences. By hypothesis, then, when

(1) $a_i - a_j \equiv 0 \pmod{d_{ij}}, i = 1, 2, \ldots, s-1, j = 1, 2, \ldots, s, i < j,$

there exists a common solution $x = x_0$ of the congruences

$$x \equiv a_1 \pmod{m_1},$$
$$x \equiv a_2 \pmod{m_2},$$

(2) $\qquad \ldots$

$$x \equiv a_s \pmod{m_s}.$$

Where M_s is the L.C.M. of $m_1, \ldots, m_s$, every x satisfying these s congruences is of the form $x \equiv x_0 \pmod{M_s}$; and conversely, every x of this form satisfies these s congruences.

We now desire to show, under the hypothesis of the induction, that if

(3) $a_i \equiv a_j \pmod{d_{ij}}, i = 1, \ldots, s; j = 1, \ldots, s+1, i < j,$

then there is a common solution of the set of congruences (2) *and* the congruence

(4) $$x \equiv a_{s+1} \pmod{m_{s+1}}.$$

Since conditions (3) imply conditions (1), we know there is a solution x_0 of congruences (2). Hence, under conditions (3), we desire to prove that there is a common solution of

(5) $$x \equiv x_0 \pmod{M_s} \quad \text{and} \quad x \equiv a_{s+1} \pmod{m_{s+1}}.$$

We readily see that (M_s, m_{s+1}) is equal to the L.C.M. of $(m_1, m_{s+1}), (m_2, m_{s+1}),$ $\ldots, (m_s, m_{s+1})$. If $(M_s, m_{s+1}) = 1$, the statement is evidently true. Obviously, since each (m_i, m_{s+1}), $i = 1, 2, \ldots, s$, divides $d = (M_s, m_{s+1})$, this L.C.M. divides (M_s, m_{s+1}). Let, then, $(M_s, m_{s+1}) = d > 1$, and let p be an arbitrary prime factor of d. Let p^A be the highest power of p occurring in any one of $m_1, m_2, \ldots, m_s$, and let p^B be the highest power of p occurring in m_{s+1}. Then, if C is the smaller of A and B ($C = A$ if $A = B$), p^C is the highest power of p occurring not only in $d = (M_s, m_{s+1})$ but also in the L.C.M. of $(m_1, m_{s+1}), \ldots, (m_s, m_{s+1})$. Hence, (M_s, m_{s+1}) is equal to the L.C.M. of $(m_1, m_{s+1}), \ldots, (m_s, m_{s+1})$.

Since the main theorem has been established for two congruences ($r = 2$), there is a common solution of the two congruences (5) if and only if

(6) $$x_0 \equiv a_{s+1} \pmod{(M_s, m_{s+1})}.$$

By the hypothesis of the induction,

$$a_1 \equiv a_{s+1} \pmod{(m_1, m_{s+1})},$$
$$a_2 \equiv a_{s+1} \pmod{(m_2, m_{s+1})},$$
$$\cdots \quad \cdots$$
$$a_s \equiv a_{s+1} \pmod{(m_s, m_{s+1})}.$$

Moreover, since $x = x_0$ satisfies congruences (2), we have

$$x_0 \equiv a_1 \equiv a_{s+1} \pmod{(m_1, m_{s+1})},$$
$$x_0 \equiv a_2 \equiv a_{s+1} \pmod{(m_2, m_{s+1})},$$
$$\cdots \quad \cdots$$
$$x_0 \equiv a_s \equiv a_{s+1} \pmod{(m_s, m_{s+1})}.$$

Hence $x_0 \equiv a_{s+1} \pmod{(m_i, m_{s+1})}$ for $i = 1, 2, \ldots, s$. Since the L.C.M. of $(m_1, m_{s+1}), \ldots, (m_s, m_{s+1})$ is equal to (M_s, m_{s+1}), we have

$$x_0 \equiv a_{s+1} \pmod{(M_s, m_{s+1})}.$$

This, we saw, was the sufficient condition (6) for a common solution of the two congruences (5). This completes the proof by induction on r that the given conditions are sufficient for a common solution of the given congruences.

Assuming the conditions are met for a common solution of a set of congruences of the type given in the theorem, we may write the L.C.M. of $m_1, m_2, \ldots, m_r$ as $p_1^{\alpha_1} \cdots p_t^{\alpha_t}$, where the p's are distinct primes and the α's are positive integers. There will be at least one congruence $x \equiv a_{i_k} \pmod{m_{i_k}}$ such that $p_k^{\alpha_k}$ divides m_{i_k}; that is, $x \equiv a_{i_k} \pmod{p_k^{\alpha_k}}$. If we retain only those t congruences whose moduli $m_{i_1}, m_{i_2}, \ldots, m_{i_t}$ are divisible by $p_1^{\alpha_1}, p_2^{\alpha_2}, \ldots, p_t^{\alpha_t}$, respectively, and take the moduli of these t congruences to be $p_1^{\alpha_1}, p_2^{\alpha_2}, \ldots, p_t^{\alpha_t}$, respectively, we will have a set of t congruences with moduli relatively prime in pairs. The Chinese Remainder Theorem may be used to obtain a common solution of these t congruences. Since the conditions for solvability were satisfied, the solution found for these t congruences will satisfy the given r congruences. For, the common solution of the t congruences is unique modulo $p_1^{\alpha_1} \cdots p_t^{\alpha_t}$ and must, therefore, be the unique solution modulo $p_1^{\alpha_1} \cdots p_t^{\alpha_t}$ of the original r congruences.

For example, suppose the following congruences are submitted for a possible common solution:

(7)
$$x \equiv 21 \pmod{2^3 \cdot 3 \cdot 5},$$
$$x \equiv 381 \pmod{2 \cdot 3^3 \cdot 5^2},$$
$$x \equiv 7 \pmod{11},$$
$$x \equiv 45 \pmod{2^3 \cdot 3 \cdot 7},$$
$$x \equiv 13 \pmod{2^4}.$$

The L.C.M. of the moduli is $831600 = 2^4 \cdot 3^3 \cdot 5^2 \cdot 7 \cdot 11$. We see that the sufficient conditions for a common solution are met. The t ($t = 5$) congruences we use are:

(8)
$$x \equiv 13 \pmod{2^4},$$
$$x \equiv 381 \equiv 3 \pmod{3^3},$$
$$x \equiv 381 \equiv 6 \pmod{5^2},$$
$$x \equiv 45 \equiv 3 \pmod{7},$$
$$x \equiv 7 \pmod{11}.$$

The common solution of congruences (8) is $x \equiv 381 \pmod{831600}$. This is, of course, the common solution of congruences (7).

6. See H. Griffin, *Elementary Theory of Numbers* (New York: McGraw-Hill, 1954), pp. 68–70.

7. In an extended sense, we may say that a single congruence and a set of simultaneous congruences are "equivalent." Thus, if $m_1, m_2, \ldots, m_r$ are relatively prime in pairs and if $m = m_1 m_2 \cdots m_r$, then we may say that the conditional congruence

(1) $$f(x) \equiv 0 \pmod{m}$$

is "equivalent" to the set of simultaneous conditional congruences

(2) $$f(x) \equiv 0 \pmod{m_i}, \, i = 1, 2, \ldots, r;$$

and, conversely, the set of simultaneous congruences (2) is "equivalent" to the single congruence (1). This "equivalence" is used in the sense that every x satisfying (1) obviously satisfies the set of simultaneous congruences (2), and conversely. For example, if $f(x) = x^2 + 11x - 12$, the congruence $f(x) \equiv 0 \pmod{2310}$ is "equivalent" to the set of simultaneous congruences

$$f(x) \equiv 0 \pmod{6},$$
$$f(x) \equiv 0 \pmod{11},$$
$$f(x) \equiv 0 \pmod{35},$$

a solution of which is $x \equiv 43 \pmod{2310}$.

8. In Chapter 4 (Theorem 4-7, Corollary), it will be explicitly proved that the product of r consecutive positive integers is divisible by $r!$.

9. See Section 7-7, footnote 4.
10. This will be proved in Chapter 4 (Theorem 4-7, Corollary).

Chapter 4

1. Theorem 4-5 and Theorem 4-6 were proved by A. M. Legendre. See his *Théorie des nombres* (2nd ed., 1808), p. 8; (3rd ed., 1830), I, 10.

2. A proof by induction on n (given by L. J. Mordell) can be obtained by using the fact that, where $0! = 1! = 1$,

$$\frac{n!}{n_1! n_2! \cdots n_r!} = \frac{(n_1 + n_2 + \cdots + n_r)!}{n_1! n_2! \cdots n_r!} = \frac{(n-1)!}{(n_1 - 1)! n_2! \cdots n_r!} + \cdots + \frac{(n-1)!}{n_1! n_2! \cdots (n_r - 1)!}.$$

3. A Calendar Problem.

To correct errors of the Julian calendar, Pope Gregory XIII introduced, in 1582, the Gregorian calendar which is now used almost universally. In the Gregorian calendar the common year consists of 365 days and the leap year of 366 days. Those years divisible by 4 but not by 100 are leap years, as are also those years divisible by 400; all other years are common years of 365 days. For example, 1904 and 1968 were leap years, as well as 1600. However, 1800, 1900, and 1966 were common years. Various methods involving tabulations have been given to determine on what day of the week a certain date would fall. Following the exposition of Uspensky and Heaslet given in their *Elementary Number Theory*, we shall develop a simple formula to determine the day of the week of any date beyond the leap year 1600.

To determine the number l of leap years n such that $1600 < n \leq N$, we proceed as follows. Those years $4n_1$ in this interval which are divisible by 4 satisfy the condition $1600 < 4n_1 \leq N$; that is, $400 < n_1 \leq N/4$. The number of such years $4n_1$ is

(1) $$\left\lfloor \frac{N}{4} \right\rfloor - 400.$$

From this number we deduct the number of years $100n_2$ (only some of which may be leap years) such that $1600 < 100n_2 \leq N$, and add the number of (leap) years $400n_3$ such that $1600 < 400n_3 \leq N$. That is, we subtract from (1) the number $[N/100] - 16$ and add to this last result $[N/400] - 4$. Consequently, the number of leap years n such that $1600 < n \leq N$ is given by

(2) $$\begin{aligned} l &= \left\lfloor \frac{N}{4} \right\rfloor - 400 - \left\lfloor \frac{N}{100} \right\rfloor + 16 + \left\lfloor \frac{N}{400} \right\rfloor - 4 \\ &= \left\lfloor \frac{N}{4} \right\rfloor - \left\lfloor \frac{N}{100} \right\rfloor + \left\lfloor \frac{N}{400} \right\rfloor - 388. \end{aligned}$$

Dividing the year N by 100 we obtain $N = 100c + d$, where $0 \leq d < 100$. Hence, the number of leap years l greater than 1600 and not exceeding N may be written as follows:

$$l = 24c + \left[\frac{d}{4}\right] + \left[\frac{c}{4}\right] - 388.$$

Since the month of February has 28 days in a common year and 29 days in a leap year, it is convenient for computation to proceed as if the year N begins with March 1st. Then, March and April will be considered as the first and second months of the year N and January and February of the year $N + 1$ as the eleventh and twelfth months of the year N. Consequently, January and February of the year N will be considered as the eleventh and twelfth months of the year $N - 1$. Denoting the days of the week from Sunday to Saturday by the numbers 0, 1, ..., 6, respectively, and assuming that the first of March of the year 1600 had the number a, we see that, since $365 = 7 \cdot 52 + 1$, March 1st 1601 would have its non-negative number congruent, modulo 7, to $a + 1$. March 1st 1602 and March 1st 1603 would have numbers congruent, modulo 7, to $a + 2$ and $a + 3$, respectively; but, since 1604 was a leap year and since there are $366 = 7 \cdot 52 + 2$ days between March 1st 1603 and March 1st 1604, March 1st 1604 would have a number congruent, modulo 7, to $a + 5$.

Consequently, every time we pass a common year, the number of March 1st, modulo 7, is increased by one, and every time we come to a leap year, the number of March 1st is increased, modulo 7, by two. Hence, to find the number e of March 1st of the year N, we must add to a the number of all years exceeding 1600 and not exceeding N, and also add the number of leap years in this same interval, and finally reduce this sum modulo 7. Thus,

$$e \equiv a + 100c + d - 1600 + 24c + \left[\frac{d}{4}\right] + \left[\frac{c}{4}\right] - 388 \pmod{7}$$

and

(3) $$e \equiv a + d - 2c + \left[\frac{d}{4}\right] + \left[\frac{c}{4}\right] \pmod{7}.$$

Since March 1st 1969 is a Saturday, this day has the number $e = 6$. Hence, since in this instance, $c = 19$, $d = 69$, and $e = 6$, we have from (3)

$$6 \equiv a + 69 - 38 + 17 + 4 \equiv a + 3 \pmod{7}.$$

Therefore, $a = 3$ and March 1st 1600 was a Wednesday. Thus, the congruence

(4) $$e \equiv 3 + d - 2c + \left[\frac{d}{4}\right] + \left[\frac{c}{4}\right] \pmod{7}$$

determines the day of the week on which March 1st falls in every year $N \geq 1600$.

In order to determine the day of the week on which the first of all other months of the year would fall, we note that (since $31 = 7 \cdot 4 + 3$) the number of April 1st is 3 units greater than that of March 1st; that of May 1st is 2 units greater than that of April 1st; that of June 1st is 3 units greater than that of May 1st; that of July 1st is 2 units greater than that of June 1st; that of August 1st is 3 units greater than that of July 1st; that of September 1st is 3 units greater than that of August 1st; that of October 1st is 2 units greater than that of September 1st; that of November 1st is 3 units greater than that of October 1st; that of December 1st is 2 units greater than that of November 1st; that of January 1st is 3 units greater than that of December 1st; that of February 1st is 3 units greater than that of January 1st.

It has been observed by Chr. Zeller that for the function $f(m) = [2.6m - 0.2]$, the increments $f(m + 1) - f(m)$ for $m = 1, 2, \ldots, 11$ give precisely the eleven increments listed in the above tabulation. Consequently, since $f(1) = [2.6 - 0.2] = 2$, congruence (4) for March 1st may be written

$$e = 1 + [2.6 - 0.2] + d - 2c + \left[\frac{d}{4}\right] + \left[\frac{c}{4}\right] \pmod{7}.$$

The first day of the mth month has, therefore, the number given by

$$e' \equiv 1 + [2.6m - 0.2] + d - 2c + \left[\frac{d}{4}\right] + \left[\frac{c}{4}\right] \pmod{7}.$$

Finally, the number f of the day of the week corresponding to the sth day of the mth month of the year $N = 100c + d \geq 1600$ is determined by the congruence

(5) $\qquad f \equiv s + [2.6m - 0.2] + d - 2c + \left[\frac{d}{4}\right] + \left[\frac{c}{4}\right] \pmod{7}.$

For example, for May 23rd 1971, we have $s = 23, m = 3, c = 19$, and $d = 71$. Then $f \equiv 23 + 7 + 71 - 38 + 17 + 4 \equiv 0 \pmod{7}$. Consequently, this date falls on Sunday.

In order to determine what dates fall on a specified day of the week, we may restate (5) in a more convenient form. Let $N \equiv 100c + d \equiv a \pmod 4$, where $0 \leq a < 4$, and $N \equiv 100c + d \equiv b \pmod 7$, where $0 \leq b < 7$. From these congruences we have $d \equiv a \pmod 4$ and $d \equiv b - 2c \pmod 7$. By the Chinese Remainder Theorem, we have $d \equiv 21a + 8b - 16c \pmod{28}$; consequently, $[d/4] \equiv [5a + (1/4)a] + 2b - 4c \equiv 5a + 2b - 4c \pmod 7$, and $[d/4] + d \equiv 5a + 3b + c \pmod 7$. Therefore, (5) may now be written as follows:

$$s \equiv f - [2.6m - 0.2] + 2a + 4b + c - \left[\frac{c}{4}\right] \pmod 7.$$

For example, to determine on what dates Saturdays occur in May, we take $f = 6, m = 3$, to obtain $s \equiv 6 - 7 + 2a + 4b + c - [c/4] \pmod 7$. For the year 1922, we see that $s \equiv 6 \pmod 7$, and, consequently, May 6th, 13th, 20th, and 27th fell on Saturday. Similarly, using $m = 12, a = 1, b = 3, c = 19$, we see that in 1922, February 4th, 11th, 18th, and 25th fell on Saturday.

4. See E. Lucas, *Théorie des nombres* (Paris, 1891), I, 402; also P. Bachmann, *Niedere Zahlentheorie* (Leipzig, 1902), I, 91-94.

5. See W. Sierpiński, *Elementary Theory of Numbers* (New York, 1964), pp. 179-180.

6. See Hans-Joachim Kanold, "Über mehrfach vollkommene Zahlen, II" *Journal für die reine und angewandte Mathematik* (Berlin), CXCVII (1957), 82-96, especially p. 90. See also Jacques Touchard, "On Prime Numbers and Perfect Numbers," *Scripta Mathematica*, XIX (1953), 35-39. Reference may be made also to the expository paper of Paul J. McCarthy, "Odd Perfect Numbers," *Scripta Mathematica*, XXIII (1957), 43-47. In addition, there is the paper of Karl K. Norton, "Remarks on the number of factors of an odd perfect number," *Acta arithmetica*, VI·4 (1961), 365-374.

Chapter 4

7. This simple proof was given by L. E. Dickson, *American Mathematical Monthly*, XVIII (1911), 109.

8. See G. H. Hardy and E. M. Wright, *An Introduction to the Theory of Numbers* (4th Ed., Oxford University Press, 1960), p. 237.

9. J. Liouville, "Sur quelques fonctions numériques", *Journal de mathématiques pures et appliquées*, Ser. 2, II (1857), 244-248. (See especially p. 246.)

10. See *Journal de mathématiques pures et appliquées*, Ser. 2, II (1857), 141-144, 244-248, 377-384.

11. See H. Siebeck, *Journal für die reine und angewandte Mathematik*, XXXIII (1846), 71-77.

12. With the convention that $\binom{0}{0} = 1$, we shall prove that:

$$x_n = \sum_{k=0}^{(n-1)/2} \binom{n-k-1}{k} a^{n-2k-1} c^k \quad \text{if } n \text{ is odd}, n \geq 1;$$

$$x_n = \sum_{k=0}^{(n-2)/2} \binom{n-k-1}{k} a^{n-2k-1} c^k \quad \text{if } n \text{ is even}, n \geq 2.$$

Proof The proof is by complete mathematical induction on n. For $n = 1$,

$$\sum_{k=0}^{0} \binom{-k}{k} a^{-2k} c^k = \binom{0}{0} a^0 c^0 = 1 = x_1.$$

For $n = 2$,

$$\sum_{k=0}^{0} \binom{1-k}{k} a^{1-2k} c^k = \binom{1}{0} a^1 c^0 = a = x_2.$$

Assume the theorem to be true for all n less than some fixed integer $r \geq 3$.

Case I. Let r be odd. Then

$$x_{r-2} = \sum_{k=0}^{(r-3)/2} \binom{r-k-3}{k} a^{r-2k-3} c^k,$$

$$x_{r-1} = \sum_{k=0}^{(r-3)/2} \binom{r-k-2}{k} a^{r-2k-2} c^k.$$

Therefore,

$$x_r = ax_{r-1} + cx_{r-2} = \sum_{k=0}^{(r-3)/2} \binom{r-k-2}{k} a^{r-2k-1} c^k$$

$$+ \sum_{k=0}^{(r-3)/2} \binom{r-k-3}{k} a^{r-2k-3} c^{k+1}$$

$$= a^{r-1} + \sum_{k=1}^{(r-3)/2} \binom{r-k-2}{k} a^{r-2k-1} c^k + c^{(r-1)/2}$$

$$+ \sum_{k=0}^{(r-5)/2} \binom{r-k-3}{k} a^{r-2k-3} c^{k+1}$$

$$= a^{r-1} + c^{(r-1)/2} + \sum_{k=1}^{(r-3)/2} \binom{r-k-2}{k} a^{r-2k-1} c^k$$

$$+ \sum_{k=1}^{(r-3)/2} \binom{r-k-2}{k-1} a^{r-2k-1} c^k$$

$$= a^{r-1} + c^{(r-1)/2} + \sum_{k=1}^{(r-3)/2} \left[\binom{r-k-2}{k} + \binom{r-k-2}{k-1}\right] a^{r-2k-1} c^k$$

$$= a^{r-1} + c^{(r-1)/2} + \sum_{k=1}^{(r-3)/2} \binom{r-k-1}{k} a^{r-2k-1} c^k$$

$$= \sum_{k=0}^{(r-1)/2} \binom{r-k-1}{k} a^{r-2k-1} c^k.$$

Case II. Let r be even. Then

$$x_{r-2} = \sum_{k=0}^{(r-4)/2} \binom{r-k-3}{k} a^{r-2k-3} c^k,$$

$$x_{r-1} = \sum_{k=0}^{(r-2)/2} \binom{r-k-2}{k} a^{r-2k-2} c^k.$$

Therefore,

$$\begin{aligned}
x_r &= ax_{r-1} + cx_{r-2} \\
&= \sum_{k=0}^{(r-2)/2} \binom{r-k-2}{k} a^{r-2k-1} c^k \\
&\quad + \sum_{k=0}^{(r-4)/2} \binom{r-k-3}{k} a^{r-2k-3} c^{k+1} \\
&= a^{r-1} + \sum_{k=1}^{(r-2)/2} \binom{r-k-2}{k} a^{r-2k-1} c^k \\
&\quad + \sum_{k=1}^{(r-2)/2} \binom{r-k-2}{k-1} a^{r-2k-1} c^k \\
&= a^{r-1} + \sum_{k=1}^{(r-2)/2} \left[\binom{r-k-2}{k} + \binom{r-k-2}{k-1}\right] a^{r-2k-1} c^k \\
&= \sum_{k=0}^{(r-2)/2} \binom{r-k-1}{k} a^{r-2k-1} c^k.
\end{aligned}$$

Hence the theorem is true for all $n \geq 1$.

13. Since α and β are the solutions of the equation $x^2 - ax - c = 0$, $\alpha = \beta$ when and only when $a^2 + 4c = 0$. Hence there are the two cases: $a = 2, c = -1$; and $a = -2, c = -1$. In the former case ($a = 2$), it is readily seen that $x_k = k$; consequently, the theorem reduces to the identity $m + n = (-1)m(n-1) + (m+1)n$. In the latter case ($a = -2$), we have $x_k = (-1)^{k+1}k$; thus the theorem becomes $(-1)^{m+n+1}(m+n) = (-1)(-1)^{m+1}m(-1)^n(n-1) + (-1)^{m+2}(m+1)$ $\cdot (-1)^{n+1}n$. Hence the theorem is true when $\alpha = \beta$.

14. This proof was given by Ronald M. Solomon.

15. Leonardo Pisano, better known by his nickname Fibonacci (an abbreviation of *filius bonacci*) wrote a book in 1202 entitled *Liber Abacci*, containing nearly all the arithmetic and algebra known at that time. A second version was brought out in 1228. It was through this work that Europeans became acquainted with the Arabic numerals. He illustrated the theory by many examples, one of which was the famous rabbit problem. In discussing the number of offspring of a pair of rabbits, he obtained the sequence of the Fibonacci numbers.

16. To prove that $y_n = \alpha^n + \beta^n, n \geq 0$, let $z_n = \alpha^n + \beta^n, n \geq 0$, where $\alpha + \beta = a$ and $\alpha\beta = -c$. Obviously, $z_n = y_n$ for $n = 0, 1, 2$. If it be assumed that $z_n = y_n$ for $n = 0, 1, 2, \ldots, k$, where $k \geq 1$, then $y_{k+1} = ay_k + cy_{k-1} = az_k + cz_{k-1} = a(\alpha^k + \beta^k) + c(\alpha^{k-1} + \beta^{k-1}) = (\alpha + \beta)(\alpha^k + \beta^k) - \alpha\beta(\alpha^{k-1} + \beta^{k-1})$ $= \alpha^{k+1} + \beta^{k+1} = z_{k+1}$. Hence, by complete induction on n, we have $z_n = y_n$ for $n \geq 0$.

Chapter 5

1. Here we follow L. E. Dickson, *Introduction to the Theory of Numbers* (Chicago, 1929), p. 19.

2. This theorem and the next are generalizations of results of F. Arndt, *Journal für die reine und angewandte Mathematik* (Berlin), XXXI (1846), 259-268.

Theorem 5-11 is not true for $p = 2$. For example, if $a = 3$, $p = 2$, $e = l = 1$, we see that when $n = 3$, although $ep^{n-l} = 1 \cdot 2^{3-1} = 4$, 3 belongs to exponent 2, modulo 2^3.

3. See A. Cunningham, 'Haupt-exponents of 2,' *The Quarterly Journal of pure and applied Mathematics*, XXXVII (1906), 122-145. See footnote on page 124 of his paper.

4. See A. Cunningham and H. J. Woodall, 'On Haupt-exponents of 2,' *The Quarterly Journal of pure and applied Mathematics*, XLV (1914), 114-125. Note page 115.

5. Erna H. Pearson, "On the congruences $(p - 1)! \equiv -1$ and $2^{p-1} \equiv 1$ (mod p^2)," *Mathematics of Computation*, XVII (1963), 194-195. Those primes p satisfying $(p - 1)! \equiv -1 \pmod{p^2}$ are known as Wilson primes. It has been shown that the only Wilson primes not exceeding 200183 are 5, 13, and 563. The computations by Erna Pearson were carried out on the Control Data 1604 Computer (a binary computer) at the University of Texas.

6. See C. F. Gauss, *Disquisitiones Arithmeticae*, Section 73.

7. See M. A. Stern, *Journal für die reine und angewandte Mathematik*, VI (1830), 147-153, esp. pp. 148-149.

8. We here follow the proof of L. E. Dickson, *Introduction to the Theory of Numbers* (Chicago, 1929), pp. 19-20.

9. *Decimal Fractions* As is well known, some rational fractions, such as 2/5 and 3/8, give rise to terminating decimals; others, such as 1/3 and 4/21, give rise to repeating decimals; while still others, such as 5/6 and 2/15, give rise to a mixed decimal, a non-repeating part followed by a repeating part. In order to examine the reasons for these differences, let us briefly outline an arithmetical theory of decimal fractions.

Every positive rational fraction (not an integer) can be expressed as the sum of an integer, positive or zero, and a positive proper fraction r/h in which $0 < r < h$ and $(r, h) = 1$. Let us write $h = h_1 h_2$, where $h_1 = 2^b 5^c$, $b \geq 0, c \geq 0, h_2 \geq 1$, and $(h_2, 10) = 1$.

We can determine integers $a_1, a_2, \ldots,$ and the corresponding integers $r_1, r_2, \ldots$ such that

(1)
$$10r = a_1 h + r_1, \ 0 \leq r_1 < h,$$
$$10r_1 = a_2 h + r_2, \ 0 \leq r_2 < h,$$
$$\cdots \qquad \cdots$$
$$\cdots \qquad \cdots$$
$$\cdots \qquad \cdots$$

$$10r_{i-1} = a_i h + r_i, \ 0 \leq r_i < h,$$
$$\cdots \qquad \cdots$$
$$\cdots \qquad \cdots$$
$$\cdots \qquad \cdots .$$

Moreover, since $r < h$, each $a_i \geq 0$ and < 10. Rewriting the above equations as congruences, modulo h, we have $10 r_{i-1} \equiv r_i \pmod{h}$, where $r_0 = r$ and $i = 1, 2, \ldots$; that is,

(2) $\qquad 10^m r_{i-1} \equiv 10^{m-1} r_i \equiv 10^{m-2} r_{i+1} \equiv \cdots \equiv r_{m+i-1} \pmod{h},$

where $m \geq 1$ and $i = 1, 2, \ldots$. For $i = 1$, we obtain from (2)

(3) $\qquad\qquad\qquad 10^m r \equiv r_m \pmod{h},$

where $m \geq 0$. The sequence of equations (1) terminates when a zero remainder is reached; that is, for some value of m, we obtain $10^m r \equiv 0 \pmod{h}$. This occurs when and only when 10^m is divisible by h; that is, when and only when $h = 2^b 5^c$ and $h_2 = 1$. Then, if s is the least positive integer such that $10^s \equiv 0 \pmod{h}$, there are s equations in (1); and, moreover, s is equal to the greater of the two numbers b and c. In this case we can, however, with slight modification, write equations (1) as an infinite sequence of equations by rewriting the sth equation in the form

$$10 r_{s-1} = (a_s - 1)h + r_s, \ r_s = h,$$

and continuing as follows:

$$10 r_s = 9h + r_{s+1}, \ r_{s+1} = h,$$
$$10 r_{s+1} = 9h + r_{s+2}, \ r_{s+2} = h,$$
$$\cdots \qquad \cdots$$
$$\cdots \qquad \cdots$$
$$\cdots \qquad \cdots .$$

Then, however great the i may be, we have in all cases:

$$\frac{r}{h} = \frac{a_1}{10} + \frac{a_2}{10^2} + \frac{a_3}{10^3} + \cdots + \frac{a_i}{10^i} + \frac{r_i}{10^i h},$$

where it is understood that in the case where the equations (1) terminate with the sth equation (with a zero remainder), a_s is to be replaced by $a_s - 1$ and $a_{s+1}, a_{s+2}, \ldots$ are each equal to 9. Since $r_i/(10^i h) \leq 1/10^i$, we obtain the following convergent series for the proper fraction r/h:

$$\frac{r}{h} = \frac{a_1}{10} + \frac{a_2}{10^2} + \cdots + \frac{a_i}{10^i} + \cdots.$$

This infinite series, as we shall see, has the sequence of integers $a_1, a_2, \ldots,$ periodic. The r_i's in (1) cannot all be different since each r_i is not greater than h. Let r_k ($r_0 = r$) be the first one which is equal to a later one, and let r_{k+l} be the first one after r_k equal to r_k. Hence, k and l are the smallest integers, $k \geq 0, l > 0$, for which $r_{k+l} = r_k$. Since $r_{k+l} = r_k$, we see from equations (1) that

$$r_{k+l+1} = r_{k+1},$$
$$r_{k+l+2} = r_{k+2},$$
$$\ldots$$
$$\ldots$$
$$\ldots$$
$$r_{k+2l} = r_{k+l} = r_k.$$

Since the remainders $r_k, r_{k+1}, \ldots, r_{k+l-1}$, and therefore the sequence of numbers $a_{k+1}, \ldots, a_{k+l}$, repeat as a period, the entire sequence of a's consists of k terms a_1, $a_2, \ldots, a_k$ at the beginning of the repeating period of l terms $a_{k+1}, a_{k+2}, \ldots, a_{k+l}$. If $k = 0$, the decimal begins with $a_1, a_2, \ldots, a_l$.

The numbers k and l are determined as follows. In view of (3), we see from the equality $r_{k+l} = r_k$ that

$$10^{k+l} r \equiv 10^k r \pmod{h},$$

and, since $(r, h) = 1$,

(4) $$10^k(10^l - 1) \equiv 0 \pmod{h}.$$

Since, conversely, the equality $r_{k+l} = r_k$ follows from the congruence (4), we see that k and l are the least integers, the first positive or zero, the second positive, for which (4) holds. In other words, $k \geq 0$ and $l > 0$ are the least numbers such that

(5) $$10^k \equiv 0 \pmod{2^b 5^c} \quad \text{and} \quad 10^l \equiv 1 \pmod{h_2}.$$

If, therefore, $h = 2^b 5^c h_2$ contains either 2 or 5 as a prime factor and if $h_2 > 1$, we get a mixed periodic decimal. Here k is the greater of the two numbers b and c, and l is the exponent to which 10 belongs, modulo h_2. For example, for the fraction $3/22$, $h = 2^1 5^0 h_2$, $h_2 = 11$. Hence, $k = 1$ and $l = 2$. In fact, $3/22 = 0.1\overline{3636}\ldots$.

If $h = 2^b 5^c h_2$, $h_2 = 1$, then from (5) it is observed that $l = 1$. The infinite decimal representation of r/h would again be mixed periodic, containing k terms at the beginning and a period of a single term, namely, 9. For example, for the fraction $3/20$ we have $h = 2^2 5$ and $k = 2, l = 1$. In fact, $3/20 = 0.149999\ldots = 0.15$.

If, finally, h is relatively prime to 10, $h = 2^0 5^0 h_2$, then $k = 0$. In this case the decimal is purely periodic, and the number of digits in the period is the exponent to which 10 belongs, modulo h. For example, for $3/7$ we have $k = 0, l = 6$, and $3/7 = 0.\overline{428571}\ldots$.

It is to be noted that the type of decimal is determined by the value of h and not by the value of r in the fraction r/h.

The above treatment given for the scale 10 can readily be modified to treat rational fractions with respect to any scale of notation (greater than one). For such a treatment see P. Bachmann, *Niedere Zahlentheorie*, I (1902), 351-360.

Chapter 6

1. More generally, for $s \geq 2$, D is an *sth power residue modulo m* (or briefly, an *sth power residue of m*) if, where $(D, m) = 1$, there exists a solution of $x^s \equiv D$

(mod m). We use the term *quadratic, cubic, biquadratic* (or *quartic*) *residue modulo m* according as $s = 2$, 3, or 4, respectively.

2. In 1901 P. Bachmann listed 49 proofs of the Quadratic Reciprocity Law in his *Niedere Zahlentheorie* (Leipzig: B. G. Teubner, 1902), I, 203-204. Thirty years later this number was increased to 59 (P. Bachmann and R. Haussner, *Grundlehren der Neuren Zahlentheorie* (3rd ed., 1931), pp. 47-48).

3. See Chr. Zeller, *Monatsberichte der Königlich Preussischen Akademie der Wissenschaften zu Berlin* aus dem Jahre 1872, pp. 846-847; G. Frobenius, "Über das quadratische Reziprozitätsgesetz" *Sitzungsberichte der Königlich Preussischen Akademie der Wissenschaften* (Berlin), (Jan.-June, 1914), pp. 335-349; R. Dedekind, Festschrift Heinrich Weber zu seinem siebzigsten Geburtstag am 5. März 1912 gewidmet von Freunden und Schülern (Leipzig and Berlin), 1912, pp. 23-36; R. Dedekind, *Gesammelte Mathematische Werke*, II (1931), 340-352.

4. We shall here consider Gauss's *first proof* of the Quadratic Reciprocity Law. The proof is rather long and is by complete mathematical induction. Gauss considered, in fact, eight different cases. Some of these, however, can be combined to give rise to two mutually exclusive cases. In the exposition given here, we have followed the presentation given by Dirichlet (P. G. Lejeune Dirichlet and R. Dedekind, *Vorlesungen über Zahlentheorie* (4th rev. edn., 1894), 112-121; also, the exposition by P. Bachmann in *Niedere Zahlentheorie*, I, 206-212). We shall refer to the Quadratic Reciprocity Law (Theorem 6-9) as the *theorem*, and to the Generalized Quadratic Reciprocity Law (Theorem 6-12) as the *generalized theorem*.

The procedure is as follows. If the theorem is assumed true for all pairs of distinct odd primes each less than a specific prime q then it will be shown that this theorem must be true also for every combination of q with an arbitrary odd prime less than q. Since $\left(\frac{3}{5}\right) = \left(\frac{5}{3}\right) = -1$, we note that the theorem is true for the two smallest odd primes 3 and 5, the only odd primes less than the prime $q = 7$. Consequently, the theorem would then be true for every combination of 7 with each smaller odd prime, namely, with 3 and 5. Since now the theorem is true for every pair of odd primes less than 11, the theorem would be true for every combination of 11 with each smaller odd prime, namely, with 3, 5, and 7. Hence, by complete mathematical induction, the theorem would be true for every pair of distinct odd primes.

In the course of the proof we utilize the fact that, on the assumption that the theorem is true for every pair of odd primes $< q$, the validity of the generalized theorem will necessarily follow for the relatively prime positive odd integers P and Q, provided that all prime factors of both P and Q are less than q; that is,

(1) $$\left(\frac{P}{Q}\right)\left(\frac{Q}{P}\right) = (-1)^{\frac{P-1}{2} \frac{Q-1}{2}}.$$

That this last statement is true follows from the fact that the proof of the generalized theorem is based exclusively on the truth of the original theorem for all pairs of odd primes one of which is a factor of P and the other of Q.

Let us assume, then, that the theorem is true for every pair of distinct odd primes less than the prime q. We wish to prove that the theorem is true for every combination of q with an odd prime p less than q. We shall consider the two mutually exclusive and exhaustive cases:

(i) at least one of $\left(\dfrac{p}{q}\right)$ and $\left(\dfrac{-p}{q}\right)$ is 1;

(ii) q is of the form $4M+1$ and $\left(\dfrac{p}{q}\right) = -1$.

Since $\left(\dfrac{-1}{q}\right) = (-1)^{(q-1)/2}$, we note that case (i) includes each of the three possibilities:

q is of the form $4M+1$ and $\left(\dfrac{p}{q}\right) = 1$;

q is of the form $4M+3$ and $\left(\dfrac{p}{q}\right) = 1$;

q is of the form $4M+3$ and $\left(\dfrac{p}{q}\right) = -1$.

For, if $q = 4M+3$, then

$$\left(\dfrac{-p}{q}\right) = (-1)^{(q-1)/2}\left(\dfrac{p}{q}\right) = -\left(\dfrac{p}{q}\right).$$

Therefore, in this case, either $\left(\dfrac{p}{q}\right) = 1$ or $\left(\dfrac{-p}{q}\right) = 1$. Moreover, if $q = 4M+1$,

$$\left(\dfrac{-p}{q}\right) = (-1)^{(q-1)/2}\left(\dfrac{p}{q}\right) = \left(\dfrac{p}{q}\right);$$

and, consequently, either $\left(\dfrac{p}{q}\right) = -1$ or $\left(\dfrac{p}{q}\right) = \left(\dfrac{-p}{q}\right) = 1$. On the other hand, if $\left(\dfrac{p}{q}\right) = \left(\dfrac{-p}{q}\right) = -1$, then $(-1)^{(q-1)/2} = 1$ and $q = 4M+1$.

In the more difficult case (ii) we desire to prove that, when p is a quadratic nonresidue of q, then q also is a quadratic nonresidue of p. For, then

$$\left(\dfrac{p}{q}\right)\left(\dfrac{q}{p}\right) = (-1)^2 = 1 = (-1)^{[(p-1)/2]\cdot[(q-1)/2]}.$$

Consider now case (i): either $\left(\dfrac{p}{q}\right) = 1$ or $\left(\dfrac{-p}{q}\right) = 1$ (or $\left(\dfrac{p}{q}\right) = \left(\dfrac{-p}{q}\right) = 1$).

In order to prove the theorem in case (i), we desire to show that

(2) $$\left(\dfrac{q}{p}\right) = (-1)^{[(p-1)/2]\cdot[(q-1)/2]}\left(\dfrac{p}{q}\right).$$

Let w be that one of the two integers $p, -p$ for which $\left(\dfrac{w}{q}\right) = 1$. There are now two distinct solutions x of $x^2 \equiv w \pmod{q}$. We choose these solutions to be positive and $< q$. If the positive integer x_0 is one such solution, the other is $q - x_0$. Let e be that one of the two solutions which is even. Then, where $0 < e < q$,

(3) $$e^2 - w = fq.$$

Now, f cannot be negative; for, if this were so, w would be positive, $w = p$. Then

$p - e^2$ would be a positive number divisible by q. This, however, is impossible since $p - e^2 < p$ and $p < q$. We now consider two possibilities.

First, if f is not divisible by p, we see from (3) that $\left(\dfrac{w}{f}\right) = 1$; and, moreover, since fq is a quadratic residue of p, we have

$$1 = \left(\frac{fq}{|w|}\right) = \left(\frac{f}{|w|}\right)\left(\frac{q}{|w|}\right).$$

Hence,

(4) $$\left(\frac{f}{|w|}\right) = \left(\frac{q}{|w|}\right).$$

The positive number f must be less than q. This is obviously the case if $w = p > 0$. If $w = -p$, then

$$\begin{aligned}fq &= e^2 - w \\ &\leq (q-1)^2 + p \\ &= q^2 - 2q + p + 1 \\ &< q^2 - 2q + q \\ &= q(q-1).\end{aligned}$$

Hence $f < q - 1$. Since the odd numbers f and $|w|$ are relatively prime and since both numbers are also positive and $< q$, we have by the generalized theorem

(5) $$\left(\frac{f}{|w|}\right)\left(\frac{|w|}{f}\right) = (-1)^{[(|w|-1)/2]\cdot[(f-1)/2]}.$$

Now, when $w = p$, we have

$$\left(\frac{|w|}{f}\right) = \left(\frac{w}{f}\right) = 1 \text{ and } \left(\frac{f}{|w|}\right) = (-1)^{[(w-1)/2]\cdot[(f-1)/2]}.$$

Moreover, if $w = -p$, we have from (5)

$$\begin{aligned}\left(\frac{f}{|w|}\right) &= (-1)^{[(-w-1)/2]\cdot[(f-1)/2]}\left(\frac{-w}{f}\right) \\ &= (-1)^{[-(w+1)/2]\cdot[(f-1)/2]}(-1)^{(f-1)/2}\left(\frac{w}{f}\right) \\ &= (-1)^{[(-w+1)/2]\cdot[(f-1)/2]} \\ &= (-1)^{[-(w-1)/2]\cdot[(f-1)/2]} \\ &= (-1)^{[(w-1)/2]\cdot[(f-1)/2]}.\end{aligned}$$

Hence, in both instances

$$\left(\frac{f}{|w|}\right) = (-1)^{[(w-1)/2]\cdot[(f-1)/2]}.$$

Since e is even, we have from (3), $-w - 1 \equiv fq - 1 \pmod 4$; therefore,

$$-\frac{w+1}{2} \equiv \frac{fq-1}{2} \pmod 2.$$

Setting $f' = (f-1)/2$ and $q' = (q-1)/2$, we have

$$fq - 1 = (2f'+1)(2q'+1) - 1 = 4f'q' + 2(f'+q').$$

Chapter 6

Hence

$$\frac{fq-1}{2} = 2f'q' + f' + q'.$$

Therefore,

$$-\frac{w+1}{2} \equiv \frac{f-1}{2} + \frac{q-1}{2} \pmod{2}.$$

Multiplying this last congruence through by $(w-1)/2$, we obtain

$$-\frac{w-1}{2}\frac{w+1}{2} \equiv \frac{w-1}{2}\frac{f-1}{2} + \frac{w-1}{2}\frac{q-1}{2} \pmod{2}.$$

Since $(w-1)/2$ and $(w+1)/2$ are consecutive integers, their product is even and

$$\frac{w-1}{2}\frac{f-1}{2} \equiv \frac{w-1}{2}\frac{q-1}{2} \pmod{2}.$$

Therefore,

(6) $$\left(\frac{f}{|w|}\right) = (-1)^{[(w-1)/2]\cdot[(q-1)/2]},$$

whether $w = p$ or whether $w = -p$. The theorem will be established for p and q if we show that (2) is valid.

In the case where $w = p$, we know from (4) and (6) that

$$\left(\frac{q}{p}\right) = \left(\frac{f}{p}\right) = (-1)^{[(p-1)/2]\cdot[(q-1)/2]}.$$

Since $1 = \left(\frac{w}{q}\right) = \left(\frac{p}{q}\right)$, (2) is established for p and q when $w = p$. Next, if $w = -p$, then, since $1 = \left(\frac{-p}{q}\right) = \left(\frac{-1}{q}\right)\left(\frac{p}{q}\right)$, we see that $\left(\frac{p}{q}\right) = \left(\frac{-1}{q}\right)$. Then, using (4) and (6), we obtain

$$\left(\frac{q}{p}\right) = \left(\frac{f}{p}\right) = (-1)^{[(-p-1)/2]\cdot[(q-1)/2]} = (-1)^{[-(p+1)/2]\cdot[(q-1)/2]}$$
$$= (-1)^{[(p+1)/2]\cdot[(q-1)/2]} = (-1)^{[(p-1)/2]\cdot[(q-1)/2]}\cdot(-1)^{(q-1)/2}$$
$$= (-1)^{[(p-1)/2]\cdot[(q-1)/2]}\left(\frac{-1}{q}\right) = (-1)^{[(p-1)/2]\cdot[(q-1)/2]}\left(\frac{p}{q}\right).$$

This establishes the desired result (2) when $w = -p$. This completes the proof of the theorem in case (i) in the event that $(f, p) = 1$.

Next, considering the second possibility in case (i), let us assume that f is divisible by p: $f = wf_1$ where f_1 denotes a suitable odd number numerically $< q$. Then, from (3) we see that e is divisible by w: $e = we_1$, where e_1 is a suitable even number. From (3) we obtain

(7) $$e_1^2 w - 1 = f_1 q.$$

Consequently, f_1 cannot be divisible by w. Also, $(f_1, e_1) = 1$, and, from (7), we have

(8) $$1 = \left(\frac{-f_1 q}{|w|}\right) \text{ and } 1 = \left(\frac{e_1^2 w}{|f_1|}\right) = \left(\frac{w}{|f_1|}\right).$$

Now, $|w|$ and $|f_1|$ are relatively prime positive odd numbers, each $< q$. We may, therefore, assume the truth of the generalized theorem for $|w|$ and $|f_1|$. That is, we assume (1) with P and Q replaced by $|f_1|$ and $|w|$, respectively; namely,

$$\left(\frac{|f_1|}{|w|}\right)\left(\frac{|w|}{|f_1|}\right) = (-1)^{[(|f_1|-1)/2]\cdot[(|w|-1)/2]}.$$

In the case where $w = p$, we have $f_1 > 0$. Hence,

$$\left(\frac{f_1}{p}\right)\left(\frac{p}{f_1}\right) = (-1)^{[(f_1-1)/2]\cdot[(p-1)/2]}.$$

But then, from (8), we have $\left(\frac{-q}{p}\right) = \left(\frac{f_1}{p}\right)$ and $\left(\frac{p}{f_1}\right) = 1$. Hence,

$$\left(\frac{-1}{p}\right)\left(\frac{q}{p}\right) = (-1)^{[(f_1-1)/2]\cdot[(p-1)/2]};$$

consequently,

$$\left(\frac{q}{p}\right) = (-1)^{[(f_1-1)/2]\cdot[(p-1)/2]} \cdot (-1)^{(p-1)/2} = (-1)^{[(f_1+1)/2]\cdot[(p-1)/2]}.$$

From (7) we have $f_1 q + 1 \equiv 0 \pmod{4}$. Since f_1 and q are both odd, both cannot be of the form $4M + 1$ nor can both be of the form $4M + 3$. Hence one is of the form $4M + 1$ and the other of the form $4M + 3$. In both possibilities we see that

$$\frac{f_1 + 1}{2} \equiv \frac{q-1}{2} \pmod{2}.$$

Hence, since $\left(\frac{p}{q}\right) = 1$, we see that (2) is valid; that is,

$$\left(\frac{q}{p}\right) = (-1)^{[(p-1)/2]\cdot[(q-1)/2]} \cdot \left(\frac{p}{q}\right).$$

If $w = -p$, then $f_1 < 0$. Hence, since $|f_1|$ and p are both $< q$, we assume the truth of the generalized theorem for these numbers: thus,

$$\left(\frac{|f_1|}{p}\right)\left(\frac{p}{|f_1|}\right) = (-1)^{[(-f_1-1)/2]\cdot[(p-1)/2]}$$
$$= (-1)^{[-(f_1+1)/2]\cdot[(p-1)/2]}$$
$$= (-1)^{[(f_1+1)/2]\cdot[(p-1)/2]}.$$

Since, from (8) $\left(\frac{-f_1}{p}\right) = \left(\frac{q}{p}\right)$ and $1 = \left(\frac{-p}{|f_1|}\right) = \left(\frac{-1}{|f_1|}\right)\left(\frac{p}{|f_1|}\right)$, we have

$$\left(\frac{|f_1|}{p}\right) = \left(\frac{-f_1}{p}\right) = \left(\frac{q}{p}\right) = (-1)^{[(f_1+1)/2]\cdot[(p-1)/2]}\left(\frac{p}{|f_1|}\right)$$
$$= (-1)^{[(f_1+1)/2]\cdot[(p-1)/2]}\left(\frac{-1}{|f_1|}\right) = (-1)^{[(f_1+1)/2]\cdot[(p-1)/2]}\cdot(-1)^{(-f-1)/2}.$$
$$= (-1)^{[(f_1+1)/2]\cdot[(p-1)/2]+(f_1+1)/2}.$$

Since $1 = \left(\frac{w}{q}\right) = \left(\frac{-1}{q}\right)\left(\frac{p}{q}\right), \left(\frac{p}{q}\right) = \left(\frac{-1}{q}\right) = (-1)^{(q-1)/2}$.
Just as above, we see that

$$\frac{f_1 + 1}{2} \equiv \frac{q-1}{2} \pmod{2}.$$

Chapter 6

Hence,

$$\left(\frac{|f_1|}{p}\right) = \left(\frac{q}{p}\right) = (-1)^{[(p-1)/2]\cdot[(q-1)/2]}\cdot(-1)^{(q-1)/2}$$
$$= (-1)^{[(p-1)/2]\cdot[(q-1)/2]}\left(\frac{-1}{q}\right).$$

Thus,

$$\left(\frac{q}{p}\right) = (-1)^{[(p-1)/2]\cdot[(q-1)/2]}\left(\frac{p}{q}\right).$$

This completes the proof of (2), and thus establishes the theorem for case (i).

We now consider the distinctly more difficult case (ii). Let q be a prime of the form $4N + 1$, and let p be an odd prime $< q$ such the $\left(\frac{p}{q}\right) = -1$. We shall now give a lemma with Gauss's proof, the complete form of which had long eluded him.

• *Lemma* If q is a prime of the form $4N + 1$, there exists an odd prime $p' < q$ for which $\left(\frac{q}{p'}\right) = -1$.

Proof This result is quite simply established when q is of the form $8N + 5$; for then $(q + 1)/2$ is an integer of the form $4N + 3$. Consequently, since not all its prime factors can be of the form $4N + 1$, $(q + 1)/2$ must have a prime factor p' of the form $4N + 3$; that is, $q + 1$ is divisible by p'. In other words, $q \equiv -1$ (mod p') and $\left(\frac{q}{p'}\right) = \left(\frac{-1}{p'}\right) = -1$.

We shall next consider the more difficult case where q is of the form $8N + 1$. If, in this instance, q were a quadratic residue of every prime which does not exceed an odd number $2m + 1 < q$, then, since it has the form $8N + 1$, q would be a quadratic residue also of every positive integer which is a product of numbers each $\leq 2m + 1$ (See Theorem 3-21 and Theorem 3-22 in conjunction with Theorem 3-16); and, accordingly, if $M = 1\cdot 2\cdot 3\cdot\ldots 2m(2m+1)$, the congruence $x^2 \equiv q$ (mod M) would be solvable. Let $x = k$ be one of its solutions. Then k, as well as q, is relatively prime to M; moreover,

$$k(q-1^2)(q-2^2)\cdots(q-m^2)$$
$$\equiv k(k^2-1^2)(k^2-2^2)\cdots(k^2-m^2)$$
$$\equiv (k+m)(k+m-1)(k+m-2)\cdots(k+1)k(k-1)\cdots$$
$$\cdot (k-m+2)(k-m+1)(k-m)$$
$$\equiv 0 \pmod{M},$$

since the product of the $2m + 1$ consecutive integers in the right member is divisible by $M = (2m + 1)!$ (See Theorem 4-7, Corollary). Therefore, since $(k, M) = 1$ and since

$$M = (m+1)\{(m+1)^2 - 1^2\}\{(m+1)^2 - 2^2\}\cdots\{(m+1)^2 - m^2\},$$

the product

$$\frac{1}{m+1}\frac{q-1^2}{(m+1)^2-1^2}\frac{q-2^2}{(m+1)^2-2^2}\cdots\frac{q-m^2}{(m+1)^2-m^2}$$

is an integer. On the other hand, this product is certainly not an integer if we choose $m = [\sqrt{q}]$. For, if $m < \sqrt{q} < m + 1$, then $m^2 < q < (m + 1)^2$ and all the factors of this product are proper fractions. Moreover, since $q \geq 17$, we obviously have $8 < (q-3)^2$ and $4q < q^2 - 2q + 1 = (q-1)^2$; thus, $2\sqrt{q} < q - 1$, $2\sqrt{q} + 1 < q$, and (when $m = [\sqrt{q}]$) $2m + 1 < q$. Hence, for this number m, the assumption that q is a quadratic residue of every prime which does not exceed the odd number $2m + 1 < q$ is untenable. Consequently, we have proved that if q is a prime of the form $8N + 1$, there exists a prime $p' < 2\sqrt{q} + 1$, and consequently $<q$, for which $\left(\frac{q}{p'}\right) = -1$. This establishes the truth of the Lemma.

If it were true that $\left(\frac{p'}{q}\right) = 1$, then case (i), which we have already established, would be applicable; and hence Theorem 6-9 would be true for primes p' and q; namely,

$$\left(\frac{q}{p'}\right) = (-1)^{[(p'-1)/2] \cdot [(q-1)/2]} = 1.$$

This, however, would contradict the assumption that $\left(\frac{q}{p'}\right) = -1$. Hence $\left(\frac{p'}{q}\right) = -1$, and Theorem 6-9 is then valid for the primes p' and q; that is,

$$\left(\frac{p'}{q}\right)\left(\frac{q}{p'}\right) = 1 = (-1)^{[(p'-1)/2] \cdot [(q-1)/2]}.$$

In order to complete the proof of Theorem 6-9, it remains to be shown that, if there is another odd prime $p < q$, different from the existing prime p', such that $\left(\frac{p}{q}\right) = -1$, then also $\left(\frac{q}{p}\right) = -1$; in other words, we desire to establish that

(9) $$\left(\frac{q}{pp'}\right) = 1.$$

Since $\left(\frac{pp'}{q}\right) = \left(\frac{p}{q}\right)\left(\frac{p'}{q}\right) = 1$, then pp' is a quadratic residue of q; and there exists a positive even integer e satisfying an equation of the form

(10) $$e^2 - pp' = fq.$$

Here f is an odd integer numerically $<q$. We shall consider three possibilities.

First, suppose that $(f, pp') = 1$. Then from (10) we have $\left(\frac{pp'}{|f|}\right) = 1$. Likewise, since by (10) fq is a quadratic residue of pp', we have $\left(\frac{fq}{pp'}\right) = 1$. Since the generalized theorem is obviously applicable to the odd numbers $|f|$ and pp', we have, when $f > 0$,

$$\left(\frac{q}{pp'}\right) = \left(\frac{f}{pp'}\right) = (-1)^{[(f-1)/2] \cdot [(pp'-1)/2]}.$$

Moreover, when $f < 0$, we have

Chapter 6

$$\left(\frac{q}{pp'}\right) = \left(\frac{f}{pp'}\right) = \left(\frac{-1}{pp'}\right)\left(\frac{|f|}{pp'}\right) = (-1)^{(pp'-1)/2} \cdot (-1)^{[(-f-1)/2] \cdot [(pp'-1)/2]}$$
$$= (-1)^{[(-f+1)/2] \cdot [(pp'-1)/2]} = (-1)^{[-(f-1)/2] \cdot [(pp'-1)/2]}$$
$$= (-1)^{[(f-1)/2] \cdot [(pp'-1)/2]}.$$

But, from (10) we see that $-pp' \equiv fq \pmod{4}$. Since $q \equiv 1 \pmod{4}$, $f \equiv -pp' \pmod{4}$ and

$$\frac{f-1}{2} \equiv -\frac{pp'+1}{2} \equiv \frac{pp'+1}{2} \pmod{2}.$$

Since $(pp'-1)/2$ and $(pp'+1)/2$ are consecutive integers, their product is even. Consequently,

$$\frac{q-1}{2}\frac{pp'-1}{2} \equiv 0 \equiv \frac{pp'+1}{2}\frac{pp'-1}{2} \equiv \frac{f-1}{2}\frac{pp'-1}{2} \pmod{2}.$$

Therefore,

$$\left(\frac{q}{pp'}\right) = (-1)^{[(q-1)/2] \cdot [(pp'-1)/2]} = 1.$$

In this instance (9) is true.

Second, suppose that $p'|f$ and $(f, p) = 1$; thus $f = p'f_1$, where $(f_1, p) = 1$. Then, of course, we must have $e = p'e_1$, and equation (10) takes on the form

(11) $$p'e_1^2 - p = f_1 q.$$

From equation (11) we see that $(f_1, p'e_1) = 1$ and $(e_1, p) = 1$. Consequently, since $p'e_1^2 \equiv p \pmod{|f_1|}$, we obtain from (11) the following relations:

$$\left(\frac{p'e_1^2}{|f_1|}\right) = \left(\frac{p'}{|f_1|}\right) = \left(\frac{p}{|f_1|}\right);$$

that is,

$$\left(\frac{pp'}{|f_1|}\right) = 1; \quad \left(\frac{f_1 q}{p}\right) = \left(\frac{p'}{p}\right); \quad \left(\frac{f_1 q}{p'}\right) = \left(\frac{-p}{p'}\right).$$

Now,

$$\left(\frac{q}{pp'}\right) = \left(\frac{|f_1|}{pp'}\right)^2 \left(\frac{q}{pp'}\right) = \left(\frac{|f_1|^2 q}{pp'}\right) = \left(\frac{|f_1|}{pp'}\right)\left(\frac{|f_1| q}{pp'}\right).$$

Since the generalized theorem is valid for the positive numbers $|f_1|$ and pp', we have

$$\left(\frac{|f_1|}{pp'}\right) = (-1)^{[(|f_1|-1)/2] \cdot [(pp'-1)/2]}\left(\frac{pp'}{|f_1|}\right) = (-1)^{[(|f_1|-1)/2] \cdot [(pp'-1)/2]}.$$

Hence,

$$\left(\frac{q}{pp'}\right) = (-1)^{[(|f_1|-1)/2] \cdot [(pp'-1)/2]}\left(\frac{|f_1| q}{p}\right)\left(\frac{|f_1| q}{p'}\right).$$

If $f_1 > 0$, we have:

$$\left(\frac{q}{pp'}\right) = (-1)^{[(f_1-1)/2] \cdot [(pp'-1)/2]}\left(\frac{p'}{p}\right)\left(\frac{-p}{p'}\right)$$
$$= (-1)^{[(f_1-1)/2] \cdot [(pp'-1)/2] + (p'-1)/2}\left(\frac{p'}{p}\right)\left(\frac{p}{p'}\right).$$

Since the theorem is valid for p and p',

$$\left(\frac{q}{pp'}\right) = (-1)^{[(f_1-1)/2]\cdot[(pp'-1)/2]+[(p'-1)/2]\cdot[(p+1)/2]}.$$

If $f_1 < 0$, we have:

$$\left(\frac{q}{pp'}\right) = (-1)^{[(-f_1-1)/2]\cdot[(pp'-1)/2]}\left(\frac{-1}{p}\right)\left(\frac{-1}{p'}\right)\left(\frac{f_1q}{p}\right)\left(\frac{f_1q}{p'}\right)$$

$$= (-1)^{[(-f_1-1)/2]\cdot[(pp'-1)/2]+(p-1)/2+(p'-1)/2}\left(\frac{p'}{p}\right)\left(\frac{-p}{p'}\right)$$

$$= (-1)^{[(-f_1-1)/2]\cdot[(pp'-1)/2]+(p-1)/2+[(p-1)/2]\cdot[(p'-1)/2]}$$

$$= (-1)^{[(f_1+1)/2]\cdot[(pp'-1)/2]+[(p-1)/2]\cdot[(p'+1)/2]}.$$

From (11), we know that $-p \equiv f_1 q \pmod 4$; that is, $f_1 \equiv -p \pmod 4$ and

$$\frac{f_1 - 1}{2} \equiv \frac{p+1}{2} \pmod 2.$$

Therefore,

$$\frac{f_1-1}{2}\frac{pp'-1}{2} + \frac{p'-1}{2}\frac{p+1}{2}$$

$$\equiv \frac{p+1}{2}\left\{\frac{pp'-1}{2} + \frac{p'-1}{2}\right\} = \frac{p+1}{2}\left\{p'\cdot\frac{p+1}{2} - 1\right\}$$

$$\equiv \frac{p+1}{2}\cdot\frac{p-1}{2} \equiv 0 \pmod 2;$$

moreover, since $(f_1 + 1)/2 \equiv (p-1)/2 \pmod 2$,

$$\frac{f_1+1}{2}\frac{pp'-1}{2} + \frac{p-1}{2}\frac{p'+1}{2} \equiv \frac{p-1}{2}\frac{pp'-1}{2} + \frac{p-1}{2}\frac{p'+1}{2}$$

$$\equiv \frac{p-1}{2}\left\{\frac{pp'+p'}{2}\right\} \equiv \frac{p-1}{2}\left\{p'\frac{p+1}{2}\right\}$$

$$\equiv \frac{p-1}{2}\frac{p+1}{2} \equiv 0 \pmod 2.$$

Hence in both cases ($f_1 \gtrless 0$), we see that (9) is true; that is,

$$\left(\frac{q}{pp'}\right) = 1.$$

Since, in this proof, we have not made use of any special property of the prime p', such as the fact that $\left(\frac{q}{p'}\right) = -1$, we see that by simply interchanging p and p' the above proof is valid for the case where f is divisible by p but not divisible by p'.

Finally, let f be divisible by pp'; that is, $f = pp'f_2$. Consequently, $e = pp'e_2$ and

(12) $$pp'e_2^2 - 1 = f_2 q.$$

From this we see that

$$1 = \left(\frac{pp'e_2^2}{|f_2|}\right) = \left(\frac{pp'}{|f_2|}\right) \quad \text{and} \quad \left(\frac{-f_2 q}{pp'}\right) = 1;$$

consequently,

$$\left(\frac{q}{pp'}\right) = \left(\frac{-f_2}{pp'}\right) = \left(\frac{-1}{pp'}\right)\left(\frac{f_2}{pp'}\right).$$

The generalized theorem being valid for the positive odd numbers $|f_2|$ and pp', we have

$$\left(\frac{|f_2|}{pp'}\right) = \left(\frac{pp'}{|f_2|}\right)(-1)^{[(|f_2|-1)/2] \cdot [(pp'-1)/2]}$$

Therefore, when $f_2 > 0$,

$$\left(\frac{q}{pp'}\right) = (-1)^{(pp'-1)/2 + [(f_2-1)/2] \cdot [(pp'-1)/2]} = (-1)^{[(f_2+1)/2] \cdot [(pp'-1)/2]}$$

when $f_2 < 0$,

$$\left(\frac{q}{pp'}\right) = (-1)^{(pp'-1)/2 + (pp'-1)/2 + [(-f_2-1)/2] \cdot [(pp'-1)/2]} = (-1)^{[(f_2+1)/2] \cdot [(pp'-1)/2]}.$$

From (12) we see that, since $f_2 q \equiv f_2 \equiv -1 \pmod 4$, $(f_2 + 1)/2 \equiv 0 \pmod 2$; and therefore

$$\frac{f_2 + 1}{2} \frac{pp' - 1}{2} \equiv 0 \pmod 2.$$

Consequently, (9) is true; that is, $\left(\dfrac{q}{pp'}\right) = 1$.

This concludes Gauss's first proof of the Quadratic Reciprocity Law (Theorem 6-9).

Chapter 7

1. See P. Bachmann, *Die analytische Zahlentheorie* (Leipzig: B. G. Teubner, 1894), p. 401; E. Landau, *Handbuch der Lehre von der Verteilung der Primzahlen* (Leipzig and Berlin: B. G. Teubner, 1909), I, 59-62.

2. In connection with a problem in the theory of groups, J. Bertrand wanted to be able to state that for every integer $n > 6$, there should be at least one prime p belonging to the interval $n/2 < p \leq n - 2$ (*Journal de l'Ecole royale polytechnique*, XVIII, Pt. 30, (1845), 123-140). He verified the statement for numbers n less than six million.

However, P. L. Tchebychef was the first to prove, in 1850, "Bertrand's Postulate" (*Mémoires présentés á l'Academie Impériale des Sciences de St. Pétersbourg*, VII (1854), 15-33; *Journal de mathématiques pures et appliquées* (Paris) XVII (1852), 366-390). In fact, Tchebychef proved that for $\epsilon > 1/5$ there is a real number ζ such that for $n \geq \zeta$ there exists at least one prime between n and $(1 + \epsilon)n$. Several, including J. J. Sylvester, reduced the size of the ϵ in Tchebychef's result. E. Cahen reputedly proved the statement of T. J. Stieltjes that ϵ may be taken to be any positive value, however small (*Comptes rendus hebdomadaires des séances de l'Académie des Sciences*, CXVI (1893), 490). In other words, for an arbitrary positive quantity ϵ, however small, there exists a positive quantity ξ (depending on the ϵ) such that for $n \geq \xi$ there exists at least one prime p in the interval $n < p \leq (1 + \epsilon)n$. From this it immediately follows that, where p_i denotes the ith prime,

$$\lim_{i \to \infty} \frac{p_{i+1}}{p_i} = 1.$$

In our treatment of Bertrand's Postulate, we follow the method of proof due to Paul Erdös (Acta litterarum ac scientiarum Regiae Universitatis Hungaricae Francisco-Josephinae, sectio scientiarum mathematicarum, Szeged, V (1932), 194-198).

3. P. L. Tchebychef proved that for sufficiently large values of η in Theorem 7-7, we may take $\alpha = 0.921\ldots$ and $\beta = 1.10555\ldots$ (*Journal de mathématiques pures et appliquées* (Paris), XVII (1852), 366-390).

4. Although no simple usable formula has been found for $\pi(x)$, the following has been given:

$$\pi(x) = 1 + \sum_{2 \leq n \leq x} \left\{ 1 + \lim_{m \to \infty} \left[1 - \prod_{k=2}^{n-1} \left(\sin \frac{\pi n}{k} \right)^2 \right]^m \right\}.$$

Moreover, where p_n is the nth prime and $\alpha = \sum_{n=1}^{\infty} p_n 10^{-2^n}$, a formula giving all the primes is the following:

$$p_n = [10^{2^n} \alpha] - 10^{2^{n-1}}[10^{2^{n-1}} \alpha]$$

(See W. Sierpiński, "Sur une formule donnant tous les nombres premiers," *Comptes rendus hebdomadaires des séances de l'Académie des Sciences* (Paris), CCXXXV (1952), 1078-1079).

A function $f(x)$ which assumes prime number values for positive integral values of x is called a *prime-representing function*. Moreover, if there is an infinite number of distinct primes among these prime values, $f(x)$ is known as a *proper prime-representing function*. It is known that there exists a real number α such that $[\alpha^{3^x}]$ yields only primes for positive integral values of x. See W. H. Mills, "A prime-representing function," *Bulletin of the American-Mathematical Society*, LIII, No. 6 (1947), 604. See also R. C. Buck, *The American Mathematical Monthly*, LIII, No. 5 (1946), 265.

Moreover, there exists a quantity α_0 such that, if $\alpha_{n+1} = 2^{\alpha_n}$, $[2^{\alpha_0}]$, $[2^{\alpha_1}]$, ..., $[2^{\alpha_n}]$, ... all yield prime numbers (E. M. Wright, "A prime-representing function," *The American Mathematical Monthly*, LVIII, No. 9 (1951), 616-618).

In addition, reference may be made to a summary of results given by Underwood Dudley, "History of a Formula for Primes," *The American Mathematical Monthly*, LXXVI, No. 1 (1969), 23-28.

In connection with the distribution of primes, further information may be found in the illuminating notes at the end of the second chapter of R. Ayoub, *An Introduction to the Analytic Theory of Numbers* (Providence, R.I.: The American Mathematical Society, 1963), pp. 129-134.

If $x = p_n$, the nth prime, then $\pi(x) = n$; and the prime number theorem yields the result that $p_n \sim n \log n$. It is now known that for all positive integers n we have $p_n > n \log n$ (Barkley Rosser, "The nth prime is greater than $n \log n$," *Proceedings of the London Mathematical Society*, (2), XLV (1939), 21-44).

5. J. Hadamard, "Sur la distribution des zéros de la fonction ζ (s) et ses conséquences arithmétiques," *Bulletin de la Société mathématique de France* (Paris), XXIV (1896), 199-220.

C. de la Vallée Poussin, "Recherches analytiques sur la théorie des nombres premiers," *Annales de la Société scientifique de Bruxelles* (Louvain), XX_2 (1896), 183-256, 281-362, and 363-397.

6. Atle Selberg, "An elementary proof of the prime number theorem," *Annals of Mathematics*, (2) L, No. 2 (1949), 305-313.

Paul Erdös, "On a new method in elementary number theory which leads to an elementary proof of the prime number theorem," *Proceedings of the National Academy of Sciences* (U. S. A.), XXXV (1949), 374-384.

An analytic proof of the prime number theorem is given in R. Ayoub, *An Introduction to the Analytic Theory of Numbers*, pp. 37-71, 72-101. Reference may be made also to G. H. Hardy and E. M. Wright, *An Introduction to the Theory of Numbers* (Fourth Edition, 1960), Chapter 22, pp. 340-374.

For an exposition of an "elementary" proof, reference may be made to T. Nagell, *Introduction to Number Theory* (New York: Wiley, 1951), Chapter 8, pp. 275-297; to Karl Prachar, *Primzahlverteilung* (Berlin: Springer, 1957), pp. 55-70, 81-92; and to E. M. Wright, "The Elementary Proof of the Prime Number Theorem," *Proceedings of the Royal Society of Edinburgh*, Section A, LXIII, Part III, No. 18 (1950-51), 257-267, as well as to Hardy and Wright, *An Introduction to the Theory of Numbers*.

7. Bericht über die zur Bekanntmachung geeigneten Verhandlungen der *Königl. Preuss. Akademie der Wissenschaften zu Berlin*, zweiter Jahrgang 1837, pp. 108-110; Mathematische Abhandlungen der *Königlichen Akademie der Wissenschaften zu Berlin*, aus dem Jahre 1837 (Berlin, 1839), pp. 45-71; P. G. L. Dirichlet, *Vorlesungen über Zahlentheorie* (4th ed. (rev. by R. Dedekind), 1894), Sections 132-137, pp. 342-359.

8. This result was proved by de la Vallée Poussin, *Annales de la Société scientifique de Bruxelles*, XX_2 (1896), 281-362.

9. S. Ramanujan, "Highly Composite Numbers," *Proceedings of the London Mathematical Society*, Ser. 2, XIV (1915), 347-409.

10. For $m \geqq 3$, we see that $p_{m+1} < p_{m-2}^3$. It is true (for $m = 3$) that $p_4 < p_1^3$; that is, $7 < 2^3$. By Bertrand's Postulate, for $m \geqq 4$, $p_{m+1} < 2p_m < 2^2 p_{m-1} < 2^3 p_{m-2} < p_{m-2}^3$ since $p_{m-2}^2 > 2^3$. Hence $p_{m+1} < p_{m-2}^3$ for $m \geqq 3$.

11. Ralph G. Archibald, "Concerning Highly Composite Numbers," *Transactions of the Royal Society of Canada*, Third Series, XXVI, Sec. III (1932), 111-118. See. pp. 111-112.

12. G. H. Hardy and S. Ramanujan, "Asymptotic formulae for the distribution of integers of various types," *Proceedings of the London Mathematical Society*, Ser. 2, XVI (1917), 112-132.

13. Ralph G. Archibald, "Relatively Highly Composite Numbers," *Transactions of the Royal Society of Canada*, Third Series, XXXIII, Sec. III (1939), 11-27.

14. Ralph G. Archibald, "Highly Composite Ideals," *Transactions of the Royal Society of Canada*, Third Series, XXX, Sec. III (1936), 41-47.

Chapter 8

1. Other common notations for this fraction are:

$$a_1 + \frac{1}{a_2 +}\frac{1}{a_3 +} \cdots \frac{1}{+ a_n},$$

$$<a_1, a_2, \ldots, a_n>,$$

$$\{a_1, a_2, \ldots, a_n\},$$

and sometimes

$$(a_1, a_2, \ldots, a_n).$$

2. Probably the most complete and useful treatment of continued fractions is found in O. Perron, *Die Lehre von den Kettenbrüchen* (Leipzig and Berlin: Teubner, 1913); 2nd Rev. Edn. (1929); 3rd Rev. Edn., Vol. I (1954), Vol. II (1957). In the latter part of the present chapter, especially in Sections 8-8, 8-9, and 8-11, our exposition has been modelled after that given in Neal H. McCoy, *The Theory of Numbers* (New York: Macmillan, 1965).

3. See Hardy and Wright, *An Introduction to the Theory of Numbers*, Section 10.15.

4. The following results are from A. Hurwitz (*Mathematische Annalen*, XXXIX (1891), 279-284). Every irrational number ξ has an infinite number of rational approximations p/q which satisfy the condition

$$\left|\xi - \frac{p}{q}\right| < \frac{1}{\sqrt{5}\,q^2}.$$

Moreover, the quantity $\sqrt{5}$ occurring in the right member of the inequality is the best possible value that can be employed; that is, the statement made regarding this inequality is not true for every irrational number ξ if $\sqrt{5}$ were replaced by a larger value.

É. Borel, employing a different approach, established a result which implies that of Hurwitz: Borel proved the result that of any three consecutive convergents to ξ at least one satisfies the inequality given above (*Journal de mathématiques pures et appliquées*, 5th Ser., IX (1903), 329-375).

See Hardy and Wright, *An Introduction to the Theory of Numbers* (Fourth Ed., 1960), Sections 11.8 and 11.9.

5. It can be shown that the last term in the period (namely, a_{m+1}) of the expansion of $\sqrt{d}$ is always $2a_1$.

6. Here we follow the exposition of Uspensky and Heaslet, *Elementary Theory of Numbers*, pp. 332-335.

7. If $d < 0$, there is obviously only a finite number of solutions. If $d = c^2$, a perfect square, we have $(x - cy)(x + cy) = m$. Since m has only a finite number of divisors, this last equation has only a finite number of integral solutions—there may, of course, be no solution, as is the case for $x^2 - 9y^2 = 6$. There remains the interesting case where d is positive and not a perfect square.

8. In this proof we follow the exposition in James E. Shockley, *Introduction to Number Theory*, p. 204.

9. Here we follow the proof given in Landau, *Vorlesungen über Zahlentheorie*, I, 98-99.

10. For, if $a/b, a'/b', a''/b''$ are *any* three consecutive terms in a Farey sequence,

$$\frac{a'}{b'} = \frac{a + a''}{b + b''}$$

implies that $a'b - ab' = a''b' - a'b''$. In other words, if a_j/b_j denotes the jth term of the sequence, we have for all possible values of $j \geqq 1$, $a_{j+1}b_j - a_j b_{j+1} = a_{j+2}b_{j+1} - a_{j+1}b_{j+2}$. Consequently, $a_{j+1}b_j - a_j b_{j+1}$ has the same value for all possible values of j. Since we have $a_1/b_1 = 0/1$ and $a_2/b_2 = 1/n$, its value for $j = 1$ is 1.

Chapter 9

1. See R. D. Carmichael, *Diophantine Analysis* (New York: Wiley, 1915); also Th. Skolem, "Diophantische Gleichungen," *Ergebnisse der Mathematik und ihrer Grenzgebiete*, Vol. 5 (Berlin: Springer, 1938).

2. For any set of non-zero values x, y, z satisfying (9-1), $\pm x$, $\pm y$, $\pm z$ provide eight solutions when we take all combinations of signs. We shall, however, consider positive values for each of x, y, z.

3. A more general statement in which 3 is replaced by a prime of the form $8M + 3$, is considered in T. Nagell, *Introduction to Number Theory* (New York: Wiley, 1951), pp. 230-232.

4. Our treatment of these equations follows that given by É. Lucas, the latter part being modified along the lines of T. Nagell. É. Lucas, "Recherches sur l'analyse indéterminée et l'arithmétique de Diophante," *Bulletin de la société d'émulation du Département de l'Allier* (Sciences, Arts et Belles-Lettres) (Moulins), XII (1873), 441-532. See pp. 465-472. T. Nagell, *Introduction to Number Theory* (New York: Wiley, 1951), pp. 232-235.

An extensive and complete treatment of the equation $2x^4 - y^4 = z^2$ was given by J. L. Lagrange, "Sur quelques problèmes de l'analyse de Diophante", *Nouveaux Mémoires de l'Académie royale des Sciences et Belles-Lettres de Berlin*, année 1777, p. 140. See his *Oeuvres*, Paris, Vol. 4 (1869), pp. 377-398.

5. Fermat stated, and later investigators proved, that the smallest such triangle is given by $u = 4\ 565\ 486\ 027\ 761$, $v = 1\ 061\ 652\ 293\ 520$, $u + v = 2372159^2$; that is, $x = 2372159$. Also $u^2 + v^2 = 20\ 843\ 662\ 669\ 680\ 914\ 462\ 673\ 121 + 1\ 127\ 105\ 592\ 336\ 276\ 233\ 990\ 400 = (4\ 687\ 298\ 610\ 289)^2 = 2165017^4$; consequently, $y = 2165017$.

6. M. Chalaux, "Tout nombre premier de la forme $4n + 1$ est une somme de deux carrés," *Nouvelles annales de mathématiques*, sec. 4, t. 17 (1917), 305-308.

7. For, since -1 is a quadratic residue of every prime p of the form $p = 4M + 1$ but of no prime of the form $4M + 3$, there exists an integer X such that p divides $X^2 + 1^2$. Assume, then, that every prime of the form $4M + 1$ less than p can be written as a sum of two squares. Now, we know that for a suitable

positive integer X less than p, p divides $X^2 + 1$. If $p/2 < X < p$, then $0 < -(X - p) < p/2$; and, setting $x = -(X - p)$, we see that $x^2 + 1$ is divisible by p. Thus, p divides a sum of two relatively prime squares $A^2 + B^2$, where $0 < A < p/2$ and $0 < B < p/2$. Let $pP = A^2 + B^2$, where $A^2 + B^2 < p^2/2$. Hence $0 < P < p/2$. Since P divides a sum of two relatively prime squares, P can have as possible prime factors merely 2 (to the first power only) and primes of the form $4M + 1$. Let p' be a prime factor of P of the form $4M + 1$. Write $P = P'p'$; thus $pP'p' = A^2 + B^2$. The prime number p' is, by hypothesis, a sum of two squares. Consequently, by Theorem 9-12 just proved, pP' is a sum of two relatively prime squares.

If P' still contains one or more prime factors of the form $4M + 1$, we could repeat the foregoing procedure until we would finally arrive at the result that p or $2p$ is a sum of two squares. In view of the identity

$$\frac{c^2 + d^2}{2} = \left(\frac{c+d}{2}\right)^2 + \left(\frac{c-d}{2}\right)^2,$$

we see that if $2p$ is a sum of two squares $c^2 + d^2$, then so is p a sum of two squares.

Consequently, since 5 is a sum of two squares, by complete mathematical induction on a positive integer q, we see that when q is a prime of the form $4M + 1$, q is a sum of two squares.

8. See L. E. Dickson, *Modern Elementary Theory of Numbers* (Chicago, 1939), pp. 93-96.

9. J. L. Lagrange, *Nouveaux Mémoires de l'Académie royale des Sciences et Belles-Lettres de Berlin*, année 1770 (Berlin, 1772), pp. 123-133; also his *Oeuvres*, Vol. 3 (1869), 189-201.

10. There are other identities expressing the product of sums of four squares as a sum of four squares. Consider, for example, the following:

$$(x_1^2 + x_2^2 + x_3^2 + x_4^2)(y_1^2 + y_2^2 + y_3^2 + y_4^2)$$
$$= (x_1y_1 - x_2y_2 \mp x_3y_3 \pm x_4y_4)^2$$
$$+ (x_1y_2 + x_2y_1 \mp x_3y_4 \pm x_4y_3)^2$$
$$+ (x_1y_3 + x_2y_4 \pm x_3y_1 \pm x_4y_2)^2$$
$$+ (x_1y_4 - x_2y_3 \pm x_3y_2 \mp x_4y_1)^2;$$

$$(x_1^2 + x_2^2 + x_3^2 + x_4^2)(y_1^2 + y_2^2 + y_3^2 + y_4^2)$$
$$= (x_1y_1 - x_2y_2 \mp x_3y_3 \mp x_4y_4)^2$$
$$+ (x_1y_2 + x_2y_1 \pm x_3y_4 \mp x_4y_3)^2$$
$$+ (x_1y_3 - x_2y_4 \pm x_3y_1 \pm x_4y_2)^2$$
$$+ (x_1y_4 + x_2y_3 \mp x_3y_2 \pm x_4y_1)^2.$$

The former identity is readily obtained by taking the norm of the product (which is the product of their norms) of the two quaternions $x_1 + x_2i + x_3j + x_4k$ and $y_1 + y_2i \pm y_3j \mp y_4k$; the latter identity is obtained by taking the norm of the product of the two quaternions $x_1 + x_2i + x_3j + x_4k$ and $y_1 + y_2i \pm y_3j \pm y_4k$. Euler's identity is obtained by taking the norm of the product of the two quaternions $x_1 + x_2i + x_3j + x_4k$ and $y_1 - y_2i - y_3j - y_4k$.

Chapter 9

11. In the proof given here, we follow the exposition given by E. Landau, *Vorlesungen über Zahlentheorie*, I, 108-109.

12. See *Göttinger Nachrichten*, 1909, pp. 17-36; *Mathematische Annalen*, LXVII (1909), 281-300. See also pp. 301-305. Here D. Hilbert proved that every positive integer P is a sum of N_k integral k^{th} powers $\geqq 0$, where N_k is a finite (but undetermined) number depending on k but not on P. The motivations at the basis of his proof are not readily perceived or understood. Analytical methods were employed involving multiple integrals. Subsequently, several other proofs were developed. Especially significant was the "elementary" (but not by any means simple) proof given in 1942 by Y. V. Linnik. His treatment of the problem was based upon important concepts of L. G. Schnirelmann. For a lucid exposition of Linnik's work see A. Y. Khinchin, *Three Pearls of Number Theory* (Rochester, New York: Graylock Press, 1952), 64 pp.

The "Ideal" Waring Theorem is that $g(k) = 2^k + [(3/2)^k] - 2$.

13. See Hardy and Wright, *An Introduction to the Theory of Numbers*, p. 322.

Bibliography

Bolker, Ethan D., *Elementary Number Theory, An Algebraic Approach*. New York: W. A. Benjamin, Inc., 1970.

Borevich, Z. I., and I. R. Shafarevich, *Number Theory*, translated by Newcomb Greenleaf. New York and London: Academic Press, Inc., 1966.

Carmichael, Robert D., *The Theory of Numbers*. New York: John Wiley & Sons, Inc., 1914.

———, *Diophantine Analysis*. New York: John Wiley & Sons, Inc., 1915.

Davenport, H., *The Higher Arithmetic*. London: Hutchinson's University Library, 1952. (Reprinted in 1960 as Harper Torchbook 526. New York: Harper & Brothers.)

Dickson, L. E., *History of the Theory of Numbers*, Vol. I, 1919; Vol. II, 1920; Vol. III, 1923. Washington, D. C.: Carnegie Institution of Washington. Reprinted, New York: Chelsea Publishing Company, 1950.

———, *Introduction to the Theory of Numbers*. Chicago: The University of Chicago Press, 1929.

Bibliography

———, *Modern Elementary Theory of Numbers*. Chicago: The University of Chicago Press, 1939.

Dudley, Underwood, *Elementary Number Theory*. San Francisco: W. H. Freeman and Company, 1969.

Gelfond, A. O., *The Solution of Equations in Integers*, translated from the Russian by J. B. Roberts. San Francisco: W. H. Freeman and Company, 1961.

Griffin, Harriet, *Elementary Theory of Numbers*. New York: McGraw-Hill Book Company, Inc., 1954.

Grosswald, E., *Topics from the Theory of Numbers*. New York: The Macmillan Company, 1966. (London: Collier-Macmillan, Ltd.)

Hardy, G. H., and E. M. Wright, *An Introduction to the Theory of Numbers*, Fourth Edition. Oxford: Clarendon Press, 1960.

Hunter, J., *Number Theory*. Edinburgh and London: Oliver and Boyd, Ltd., 1964. (New York: Interscience Publishers Inc.)

Jones, B. W., *The Theory of Numbers*. New York: Holt, Rinehart & Winston, 1955.

Landau, E., *Vorlesungen über Zahlentheorie*, Vols. I, II, III. Leipzig: S. Hirzel, 1927.

LeVeque, W. J., *Topics in Number Theory*, Vols. I, II. Reading, Mass.: Addison-Wesley Publishing Company, Inc., 1956.

———, *Elementary Theory of Numbers*. Reading, Mass.: Addison-Wesley Publishing Company, Inc., 1962.

Long, C. T., *Elementary Introduction to Number Theory*. Boston: D. C. Heath and Company, 1965.

Mathews, G. B., *Theory of Numbers*, Part I. Cambridge, Eng.: Deighton Bell, 1892. Reprinted, New York: G. E. Stechert & Co., 1927. Second Edition (reprint of first edition with corrections), New York: Chelsea Publishing Company.

McCoy, Neal H., *The Theory of Numbers*. New York: The Macmillan Company, 1965.

Mordell, L. J., *Diophantine Equations*. London and New York: Academic Press Inc., 1969.

Nagell, T., *Introduction to Number Theory*. New York: John Wiley & Sons, Inc., 1951; second ed., New York: Chelsea Publishing Co., 1964.

Niven, I., and H. S. Zuckerman, *An Introduction to the Theory of Numbers*, Second Edition. New York: John Wiley & Sons, Inc., 1966.

Ore, Oystein, *Number Theory and Its History*. New York: McGraw-Hill Book Company, Inc., 1948.

Rademacher, Hans, *Lectures on Elementary Number Theory*. New York: Blaisdell Publishing Company, 1964.

Shanks, D., *Solved and Unsolved Problems in Number Theory*, Vol. I. Washington, D. C.: Spartan Books, 1962.

Shockley, James E., *Introduction to Number Theory*. New York: Holt, Rinehart and Winston, Inc., 1967.

Sierpiński, Waclaw, *Elementary Theory of Numbers*, translated from the Polish by A. Hulanicki. New York: Hafner Publishing Company, 1964.

Stewart, B. M., *Theory of Numbers*, Second Edition. New York: The Macmillan Company, 1964.

Uspensky, J. V., and M. A. Heaslet, *Elementary Number Theory*. New York: McGraw-Hill Book Company, Inc., 1939.

Vinogradov, I. M., *Elements of Number Theory*, translated from the Fifth (Russian) Revised Edition by Saul Kravetz. New York: Dover Publications, Inc., 1954.

———, *An Introduction to the Theory of Numbers*, translated from the Sixth Russian Edition by Helen Popova. London and New York: Pergamon Press, 1955.

Wright, H. N., *First Course in Theory of Numbers*. New York: John Wiley & Sons, Inc., 1939.

Appendix

Table of Primes*

(r denotes the least positive primitive root of the prime p)

p	r	p	r	p	r	p	r
2	1	31	3	73	5	127	3
3	2	37	2	79	3	131	2
5	2	41	6	83	2	137	3
7	3	43	3	89	3	139	2
11	2	47	5	97	5	149	2
13	2	53	2	101	2	151	6
17	3	59	2	103	5	157	5
19	2	61	2	107	2	163	2
23	5	67	2	109	6	167	5
29	2	71	7	113	3	173	2

*See tables of least primitive roots given by G. Wertheim, Acta Mathematica, XVII (1893), 315-319 and XX (1897), 153-157; also more extensive table by A. Cunningham, H. J. Woodall, and T. G. Creak, Proceedings of the London Mathematical Society, 2nd Ser. XXI (1923), 350-358.

p	r	p	r	p	r	p	r
179	2	419	2	661	2	947	2
181	2	421	2	673	5	953	3
191	19	431	7	677	2	967	5
193	5	433	5	683	5	971	6
197	2	439	15	691	3	977	3
199	3	443	2	701	2	983	5
211	2	449	3	709	2	991	6
223	3	457	13	719	11	997	7
227	2	461	2	727	5	1009	11
229	6	463	3	733	6	1013	3
233	3	467	2	739	3	1019	2
239	7	479	13	743	5	1021	10
241	7	487	3	751	3	1031	14
251	6	491	2	757	2	1033	5
257	3	499	7	761	6	1039	3
263	5	503	5	769	11	1049	3
269	2	509	2	773	2	1051	7
271	6	521	3	787	2	1061	2
277	5	523	2	797	2	1063	3
281	3	541	2	809	3	1069	6
283	3	547	2	811	3	1087	3
293	2	557	2	821	2	1091	2
307	5	563	2	823	3	1093	5
311	17	569	3	827	2	1097	3
313	10	571	3	829	2	1103	5
317	2	577	5	839	11	1109	2
331	3	587	2	853	2	1117	2
337	10	593	3	857	3	1123	2
347	2	599	7	859	2	1129	11
349	2	601	7	863	5	1151	17
353	3	607	3	877	2	1153	5
359	7	613	2	881	3	1163	5
367	6	617	3	883	2	1171	2
373	2	619	2	887	5	1181	7
379	2	631	3	907	2	1187	2
383	5	641	3	911	17	1193	3
389	2	643	11	919	7	1201	11
397	5	647	5	929	3	1213	2
401	3	653	2	937	5	1217	3
409	21	659	2	941	2	1223	5

Table of Primes

p	r	p	r	p	r	p	r
1229	2	1523	2	1823	5	2131	2
1231	3	1531	2	1831	3	2137	10
1237	2	1543	5	1847	5	2141	2
1249	7	1549	2	1861	2	2143	3
1259	2	1553	3	1867	2	2153	3
1277	2	1559	19	1871	14	2161	23
1279	3	1567	3	1873	10	2179	7
1283	2	1571	2	1877	2	2203	5
1289	6	1579	3	1879	6	2207	5
1291	2	1583	5	1889	3	2213	2
1297	10	1597	11	1901	2	2221	2
1301	2	1601	3	1907	2	2237	2
1303	6	1607	5	1913	3	2239	3
1307	2	1609	7	1931	2	2243	2
1319	13	1613	3	1933	5	2251	7
1321	13	1619	2	1949	2	2267	2
1327	3	1621	2	1951	3	2269	2
1361	3	1627	3	1973	2	2273	3
1367	5	1637	2	1979	2	2281	7
1373	2	1657	11	1987	2	2287	19
1381	2	1663	3	1993	5	2293	2
1399	13	1667	2	1997	2	2297	5
1409	3	1669	2	1999	3	2309	2
1423	3	1693	2	2003	5	2311	3
1427	2	1697	3	2011	3	2333	2
1429	6	1699	3	2017	5	2339	2
1433	3	1709	3	2027	2	2341	7
1439	7	1721	3	2029	2	2347	3
1447	3	1723	3	2039	7	2351	13
1451	2	1733	2	2053	2	2357	2
1453	2	1741	2	2063	5	2371	2
1459	3	1747	2	2069	2	2377	5
1471	6	1753	7	2081	3	2381	3
1481	3	1759	6	2083	2	2383	5
1483	2	1777	5	2087	5	2389	2
1487	5	1783	10	2089	7	2393	3
1489	14	1787	2	2099	2	2399	11
1493	2	1789	6	2111	7	2411	6
1499	2	1801	11	2113	5	2417	3
1511	11	1811	6	2129	3	2423	5

p	r	p	r	p	r	p	r
2437	2	2749	6	3083	2	3433	5
2441	6	2753	3	3089	3	3449	3
2447	5	2767	3	3109	6	3457	7
2459	2	2777	3	3119	7	3461	2
2467	2	2789	2	3121	7	3463	3
2473	5	2791	6	3137	3	3467	2
2477	2	2797	2	3163	3	3469	2
2503	3	2801	3	3167	5	3491	2
2521	17	2803	2	3169	7	3499	2
2531	2	2819	2	3181	7	3511	7
2539	2	2833	5	3187	2	3517	2
2543	5	2837	2	3191	11	3527	5
2549	2	2843	2	3203	2	3529	17
2551	6	2851	2	3209	3	3533	2
2557	2	2857	11	3217	5	3539	2
2579	2	2861	2	3221	10	3541	7
2591	7	2879	7	3229	6	3547	2
2593	7	2887	5	3251	6	3557	2
2609	3	2897	3	3253	2	3559	3
2617	5	2903	5	3257	3	3571	2
2621	2	2909	2	3259	3	3581	2
2633	3	2917	5	3271	3	3583	3
2647	3	2927	5	3299	2	3593	3
2657	3	2939	2	3301	6	3607	5
2659	2	2953	13	3307	2	3613	2
2663	5	2957	2	3313	10	3617	3
2671	7	2963	2	3319	6	3623	5
2677	2	2969	3	3323	2	3631	15
2683	2	2971	10	3329	3	3637	2
2687	5	2999	17	3331	3	3643	2
2689	19	3001	14	3343	5	3659	2
2693	2	3011	2	3347	2	3671	13
2699	2	3019	2	3359	11	3673	5
2707	2	3023	5	3361	22	3677	2
2711	7	3037	2	3371	2	3691	2
2713	5	3041	3	3373	5	3697	5
2719	3	3049	11	3389	3	3701	2
2729	3	3061	6	3391	3	3709	2
2731	3	3067	2	3407	5	3719	7
2741	2	3079	6	3413	2	3727	3

Table of Primes

p	r	p	r	p	r	p	r
3733	2	4073	3	4421	3	4759	3
3739	7	4079	11	4423	3	4783	6
3761	3	4091	2	4441	21	4787	2
3767	5	4093	2	4447	3	4789	2
3769	7	4099	2	4451	2	4793	3
3779	2	4111	12	4457	3	4799	7
3793	5	4127	5	4463	5	4801	7
3797	2	4129	13	4481	3	4813	2
3803	2	4133	2	4483	2	4817	3
3821	3	4139	2	4493	2	4831	3
3823	3	4153	5	4507	2	4861	11
3833	3	4157	2	4513	7	4871	11
3847	5	4159	3	4517	2	4877	2
3851	2	4177	5	4519	3	4889	3
3853	2	4201	11	4523	5	4903	3
3863	5	4211	6	4547	2	4909	6
3877	2	4217	3	4549	6	4919	13
3881	13	4219	2	4561	11	4931	6
3889	11	4229	2	4567	3	4933	2
3907	2	4231	3	4583	5	4937	3
3911	13	4241	3	4591	11	4943	7
3917	2	4243	2	4597	5	4951	6
3919	3	4253	2	4603	2	4957	2
3923	2	4259	2	4621	2	4967	5
3929	3	4261	2	4637	2	4969	11
3931	2	4271	7	4639	3	4973	2
3943	3	4273	5	4643	5	4987	2
3947	2	4283	2	4649	3	4993	5
3967	6	4289	3	4651	3	4999	3
3989	2	4297	5	4657	15	5003	2
4001	3	4327	3	4663	3		
4003	2	4337	3	4673	3		
4007	5	4339	10	4679	11		
4013	2	4349	2	4691	2		
4019	2	4357	2	4703	5		
4021	2	4363	2	4721	6		
4027	3	4373	2	4723	2		
4049	3	4391	14	4729	17		
4051	10	4397	2	4733	5		
4057	5	4409	3	4751	19		

Index

Absolutely pseudoprime number, 42
Abundant number, 85
Algebraic numbers, 5
Algorithm, Euclid's, 18
Aliquot divisor, 85
Amicable numbers, 85
Approximation by rationals, 189, 194, 195, 220
Archibald, Ralph G., 256, 283
Archimedes' principle, 9, 259
Area of integral right triangle, 229, 233
Arithmetic, fundamental theorem of, 27
Arithmetical function, 79, 83, 87, 88, 89, 90 (see also Function)

completely multiplicative, 79, 91
multiplicative, 79
totally multiplicative, 79
Arithmetical progression, primes in, 169
Arndt, F., 269
Asymptotic, 155
Asymptotically equal, 155
Ayoub, R., 282, 283

Bachet, C. G., 250, 256
Bachet's theorem, 250, 256
Bachmann, Paul, 154, 266, 271, 272, 281
Base of notation, 14
Basic theorem of Euclid, 10

Beeger, N. G. W. H., 260
Belonging to an exponent, 103, 104, 105, 109
  sum of numbers belonging to an exponent, 115
Bertrand, J., 281
Bertrand's postulate, 157, 162, 171, 281, 282, 283.
Binary notation, 14
Binomial coefficient, 95, 157
Binomial congruences, solved by indices, 124
Binomial quadratic congruence, 61, 130
Biquadratic residue, 272
Birkhoff, G., 257
Borel, È., 284
Bounds for $\pi(x)$, 163
Braun, J., 25
Buck, R. C., 282

Cahen, E., 281
Calendar problem, 264
Cancellation law for congruences, 32
Canonical decomposition, 27
Canonical form, 27
Carmichael, Robert D., 285
Casting out nines, 33
Chalaux, M., 285
Chinese Remainder Theorem, 48, 49, 64, 66, 68, 142, 260, 262, 266
  generalization of, 49
Clarke, Arthur A., 255
Class, residue, 29
Common divisor, 8, 16
Common multiple, least (L. C. M.), 23
Companion tables, 124
Complete mathematical induction, principle of, 9
Complete quotient, 45, 178, 182
Complete set of least residues, 30
Complete set of residues, 30
Completely multiplicative, 79, 91
Composite number, 8
  highly composite number, 170
Conditional congruence, 52

Congruence symbol, 30
  properties of, 30, 31, 32
Congruences:
  binomial, 61, 124
  conditional, 52
  degree of, 52
  equivalent, 53
  identical, 52
  linear, 43, 44
  multiplicity of solution, 58
  number of solutions of, 57-62
  prime modulus, 44, 63
  properties of, 30
  quadratic, 129
  simple solution, 58
  simultaneous, 48, 49, 64
  solution of, 43, 63
Conjugate, 197
Continued fraction, 44, 45, 175, 176
  convergent, 45, 46, 176
  finite, 176, 177
  finite simple, 176
  infinite simple, 179
  inverse, 182
  kth complete quotient, 45, 178, 182
  kth partial quotient, 45, 178
  periodic, 188, 201, 202
  purely periodic, 202
  symmetric, 182
Criterion, Euler's, 132, 134
Cubic residue, 272
Cunningham, A., 269

Decimal fractions, 269
  infinite, 270
  periodic, 270
  purely periodic, 271
  terminating, 270
Dedekind, R., 272, 283
Deficient number, 85
Degree of a congruence, 52
Denary notation, 14, 33
Descartes, R., 256
Dickson, L. E., 2, 85, 255, 256, 257, 260, 267, 269, 286
Diophantus of Alexandria, 19, 250, 256

Index

Diophantine analysis, 223
Diophantine equation, 19, 223
  cubic, 238
  linear, 19, 20
  Pythagorean, 224
  quadratic, 224, 225, 227, 228, 232, 235, 238, 244, 249, 251, 252, 253
  quartic, 228, 232, 233, 234, 236, 237, 238
  quintic, 238
  simultaneous equations, 232, 238
  $X^4 - 2Y^4 = \pm Z^2$, 238
Dirichlet, P. G. L., 153, 170, 272, 283
Dirichlet's theorem on primes, 170
Disquisitiones Arithmeticae, 30, 255, 269
Distribution of primes, 153, 168
Divisor, 8
  common 8, 16
  greatest common (G. C. D.), 16, 22
  number of, 81
  sum of, 81
Division of polynomials, modulo m, 54
Dudley, Underwood, 282

Eisenstein, F. G., 2, 135, 138
Equal, asymptotically, 155
Equivalence relation, 30
Equivalent congruences, 53
Eratosthenes, 26
  sieve of, 26
Erdös, Paul, 169, 282, 283
Euclid of Alexandria, 10, 18, 24, 25, 85, 169, 224
Euclid's algorithm, 18, 177
Euclid's Elements, 224, 260
Euclid's theorem, 10, 11
Euler, L., 134, 169, 209, 250, 256
Euler's criterion, 132, 134
Euler's generalization of Fermat's theorem, 37, 38
Euler's ϕ-function:
  definition, 35
  generalization of, 78, 79, 84, 94
  properties of, 35, 36, 38, 79, 90, 110, 154

Euler's identity, 250, 251, 252
Even integer, 8
Exponent:
  belonging to, 103, 104, 105, 109
  of power of prime in $n!$, 73

Factor, 8 (*see also* Divisor)
Factor (polynomial), modulo m, 54
Factorial, exponent of highest power of prime in, 73
Factorization into primes, unique, 26, 27
Farey sequence, 218, 220, 221
Feferman, Solomon, 257
Fermat, P., 4, 37, 224, 228, 237, 240, 250, 285
Fermat's equation, 209
Fermat's last theorem, 5, 237, 257
Fermat's method of infinite descent, 224, 228, 237
Fermat's numbers, 28
Fermat's theorem, 37, 41, 52, 59
  converse of, 38
  Euler's generalization of, 37, 38
Fibonacci, 268
Fibonacci numbers, 99, 100, 183, 184, 268
Fibonacci's sequence, 99, 100, 101
Finite continued fraction, 176
Finite simple continued fraction, 45, 176

Footnotes

Section No.	Page	Footnote No.	Footnote on page
1-1	1	1	255
1-1	2	2	255
1-1	2	3	255
1-1	2	4	255
1-1	3	5	255
1-1	3	6	256
1-1	3	7	256
1-2	4	8	256
1-2	4	9	256
1-2	4	10	256
1-2	5	11	256

Section No.	Page	Footnote No.	Footnote on page	Section No.	Page	Footnote No.	Footnote on page
1-2	5	12	257	5-5	115	7	269
1-2	5	13	257	5-6	119	8	269
2-1	7	1	257	5-7	120	9	269
2-2	8	2	257	6-2	130	1	271
2-4	9	3	257	6-6	135	2	272
2-5	10	4	259	6-8	144	3	272
2-8	15	5	259	6-10	150	4	272
2-15	24	6	260	7-2	154	1	281
2-15	24	7	260	7-4	157	2	281
2-15	25	8	260	7-6	163	3	282
2-15	25	9	260	7-7	168	4	282
2-15	25	10	260	7-7	169	5	283
3-7	41	1	260	7-7	169	6	283
3-7	41	2	260	7-8	170	7	283
3-7	42	3	260	7-8	170	8	283
3-11	48	4	260	7-9	170	9	283
3-12	49	5	260	7-9	171	10	283
3-15	53	6	263	7-9	172	11	283
3-15	53	7	263	7-10	172	12	283
3-22	64	8	263	7-10	173	13	283
3-23	69	9	264	7-10	173	14	283
3-23	69	10	264	8-1	176	1	284
4-1	73	1	264	8-3	179	2	284
4-1	76	2	264	8-6	191	3	284
4-2	77	3	264	8-6	195	4	284
4-3	79	4	266	8-9	206	5	284
4-4	83	5	266	8-11	209	6	284
4-6	85	6	266	8-11	210	7	284
4-6	86	7	267	8-11	214	8	284
4-7	89	8	267	8-13	219	9	285
4-8	91	9	267	8-13	220	10	285
4-8	92	10	267	9-1	223	1	285
4-10	95	11	267	9-2	225	2	285
4-10	96	12	267	9-5	236	3	285
4-10	97	13	268	9-7	238	4	285
4-10	98	14	268	9-7	240	5	285
4-11	99	15	268	9-8	246	6	285
4-11	99	16	268	9-8	247	7	285
5-3	110	1	269	9-9	247	8	286
5-3	111	2	269	9-11	250	9	286
5-3	112	3	269	9-11	250	10	286
5-3	112	4	269	9-11	251	11	287
5-3	112	5	269	9-12	253	12	287
5-4	113	6	269	9-12	253	13	287

Four-square theorem, 251
Fraction (*see* Continued fraction)
Fractional part, 71
Frobenius, G., 144, 272
Function (*see also* Arithmetical function):
   $[\alpha]$, 71, 72, 73, 74
   $\lambda(n)$, 91, 92
   $\mu(n)$, 86, 87, 88, 91
   $\pi(x)$, 154
   $\sigma(n)$, 81, 82, 92
   $\sigma_k(n)$, 84
   $\tau(n)$, 81, 82, 91, 92, 153
   $\phi(n)$, 35, 90, 91, 154
   $\psi(n)$, 79, 80, 81, 84, 94
Fundamental principles, 9
Fundamental solution of Pell's equation, 211, 212, 215
Fundamental theorem of arithmetic, 27

Gauss, C. F., 2, 30, 113, 134, 136, 168, 255, 269, 272
Gauss's first proof of the Quadratic Reciprocity Law, 272, 281
Gauss's *Disquisitiones arithmeticae*, 30, 255
Gauss's Lemma, 135, 136, 144
Gauss's *Werke*, 255
Generalized Quadratic Reciprocity Law, 150, 151
Gillies, Donald B., 256
Girard's theorem, 245
Glaisher, J. W. L., 1, 255
Goldbach, Christian, 4
Goldbach's theorem, 5, 256
Greatest common divisor (G. C. D.), 16, 17, 22
Greatest integer function, 71
Gregorian calendar, 264
Griffin, H., 263

Hadamard, J., 153, 169, 283
Hardy, G. H., 3, 172, 253, 256, 267, 283, 284, 287
Haussner, R., 272

Heaslet, M. A., 264, 284
Henkin, L., 257
Highest common factor (H. C. F.), 16
   (*see also* Greatest common divisor)
Highly composite ideals, 173
Highly composite number, 170
   relatively, 172
Hilbert, David, 253, 287
Hurwitz, A., 284

Ideals, 5
   highly composite, 173
Identical congruence, 52
Identity, 244, 249, 250, 286
Incongruent, 30
Indicator, 35
Indices:
   theory of, 122-124
   used to solve congruences, 124
Induction, mathematical, 9, 11
Inequalities, 8, 189, 191, 193, 194, 195, 220, 284
Infinite continued fraction, 179
Infinite descent, 224, 228, 237
Infinite simple continued fraction, 179
Infinitude of primes, 24
Integer, 7
   of specific form, 11, 245, 247, 252
Integers, 7
   amicable, 85
   belonging to an exponent, 103, 115
   congruent, 30
   even, 8
   odd, 8
   perfect, 5, 85
   unique factorization of, 27
Integral polynomial, 54
Inverse of a simple continued fraction, 182
Inversion formula, 87, 89
   application of, 90, 91
Irrational number, expansion of, 184, 186

Jacobi, C. G. J., 147

Jacobi symbol, 147, 148
James, R. D., 256
Julian calendar, 264

Kanold, Hans-Joachim, 266
Kempner, Aubrey, 256
Khinchin, A. Y., 287
Kummer, E., 25

Lagrange, J. L., 57, 209, 250, 252, 256, 260, 285, 286
Lagrange's theorem, 57
Landau, E., 154, 257, 259, 281, 285, 287
Lattice point, 138
Law of Quadratic Reciprocity (*see* Quadratic Reciprocity Law)
Leap year, 264
Least common multiple (L. C. M.), 23
Legendre, A. M., 134, 168, 264
Legendre's symbol, 134, 147, 148
Lehmer, D. H., 260
Leibniz, G. W., 260
Leonardo Pisano, 268 (*see* Fibonacci)
Linear congruence, 43, 44
Linear Diophantine equation, 19, 20
Linnik, Y. V., 287
Liouville, J., 91, 92, 267
Liouville's function $\lambda(n)$, 91, 92
Lucas, E., 38, 79, 266, 285
Lucas numbers, 99
Lucas' sequence, 99, 101

McCarthy, Paul J., 266
Mac Lane, S., 257
Mathematical induction, 11
  principle of, 9
Mediant, 220
Mersenne numbers, 5, 144
Mersenne prime, 3, 51, 256, 257
Method of infinite descent, 224, 228, 237
Mills, W. H., 282
Möbius, A. F., 86
Möbius inversion formula, 87, 90
Möbius μ-function, 86, 91

Modulus, 30
Mordell, L. J., 224, 264
Multiple, 8
  common, 23
  least common, 23
Multiple (polynomial), modulo m, 54
Multiplicative function, 79, 88, 89
  completely, 79, 91
Multiplicity of solutions, 58
Multiply perfect number, 85

Nagell, T., 283, 285
Nature, 255
Nonresidue, quadratic, 130, 131
Norm, 286
Norton, Karl K., 266
Notation (*see* Symbols)
Number,
  composite, 8
  Fermat, 28
  Fibonacci, 99, 100, 183, 184
  highly composite, 170
  Lucas, 99
  Mersenne, 5, 144
  multiply perfect, 85
  natural, 7
  perfect, 5, 85
  prime, 8
  quadratic irrational, 197, 198, 199, 201
  reduced quadratic irrational, 205
  relatively highly composite, 172
Number of:
  divisors, 81, 82
  primes infinite, 24
  primes $\leq x$, $\pi(x)$, 154
  solutions of a congruence, 57–62
Numerical value, 8

Odd integer, 8
Odd perfect number, 85, 122
O-notation, 154, 155, 156
  of lower order than, 155
  of the order of, 155
  same order of magnitude, 155
  o-notation, 154, 155, 156

Index **303**

Parity, 8
Partial quotient, 45
Pearson, Erna H., 269
Pell, John, 209
Pell's equation, 175, 209
   fundamental solution, 211, 212, 215
Perfect number, 5, 85, 93
   odd, 85, 94, 122
Periodic simple continued fraction, 188, 201, 202
   length of period, 202
   period, 202
   purely periodic, 202
   value of, 203
Perron, O., 284
Pisano, Leonardo, 268 (*see* Fibonacci)
Polynomials:
   division of, modulo m, 54
   quotient, modulo m, 54
   remainder, modulo m, 54
Polynomials representing primes, 68, 69
Pope Gregory XIII, 264
Power residue of m, 271
Prachar, Karl, 283
Prime number, 8
Prime number theorem, 153, 168, 169
   generalization, 170
   remarks on, 168
Prime-representing function, 282
Primes, 26
   contained in $n!$, 73
   in arithmetical progression, 169
   infinitude of, 24
   infinitude of form $4M - 1$ and of form $6M - 1$, 28
   infinitude of form $8M + 1$, 110
   infinitude of form $6M + 1$, 143
   infinitude of form $2pM + 1$ (p odd prime), 110
Mersenne, 51, 256, 257
   table of, 291–5
   test for, 38, 52, 144
Primitive root, 110, 112, 113, 119, 291–5
   method of obtaining, 113

Primitive solution, 209, 210, 211, 224
Principles, fundamental, 9
Proper divisor, 85
Proper prime-representing function, 282
Properties of congruences, 30, 31, 32
Prouhet, E., 35
Pseudoprime, 41, 42, 43
   absolutely pseudoprime number, 42
Purely periodic continued fraction, 202
Pythagoras, 85
Pythagorean equation, 224
Pythagorean theorem, 224
Pythagoreans, 224

Quadratic congruence, 129
Quadratic irrational number, 197, 198, 199, 201
   conjugate, 197
   reduced quadratic irrational number, 205
Quadratic nonresidue, 130, 131, 134, 142, 147
Quadratic Reciprocity Law, 134, 137, 139, 147, 272
   another proof, 144
   generalized, 150, 151, 272
Quadratic residue, 130, 131, 134, 141, 148
   criterion for, 132
Quadratic residues and primitive roots, 130
Quaternions, 286
Quotient (polynomial), mod m, 54

Radix, 14
Ramanujan, S., 170, 172, 173, 283
Rational numbers, 269
   approximation by, 189
   finite decimal expansion, 269
   infinite decimal expansion, 269, 270
   periodic, 269, 270, 271
   purely periodic, 269, 271
Recurrence formula, 95

Recurring sequence, 95
Reduced quadratic irrational number, 205
Reduced set of residues, 35
Reflexive property, 30
Relation, equivalence, 30
Relatively highly composite number, 172
Relatively prime, 17, 23
Relatively prime in pairs, 23
Remainder (polynomial), modulo m, 54
Remainder theorem, 57
Representation of irrational numbers, 184
Residue classes, 29
    modulo m, 30
Residues,
    complete set, 30
    complete set of least residues, 30
    quadratic, 130
    reduced set, 35
Rings, 5
Root, primitive, 110, 112, 113, 119, 291-5
Rosser, Barkley, 282
Rotkiewicz, André, 260

Scale of notation, 14
Schnirelmann, L. G., 287
Selberg, Atle, 169, 283
Shapiro, Harold N., 260
Shockley, James E., 284
Siebeck, H., 99, 267
Sierpinski, Waclaw, 266, 282
Sieve of Eratosthenes, 26, 28
Simple continued fraction, 44, 176
    finite, 45, 176
Simple solution of a congruence, 58
Simultaneous congruences, 48, 49, 260-3
Simultaneous Diophantine equations, 232, 238
Skolem, Th., 223, 285
Smith, H. J. S., 2, 255
Smith, W. N., 257

Solomon, Ronald M., 268
Solution of a congruence, 43, 63
    multiple, 58
    number of, 57-62
    of a binomial congruence, 124
    simple, 58
Solution of a Diophantine equation, 209
    positive solution, 209
    primitive solution, 209, 210
Stern, M. A., 269
Stieltjes, T. J., 24, 281
Stone, Marshall, 3, 256
Sum of divisors, $\sigma(n)$, 81
Sum of numbers belonging to an exponent, 115
Sum of four squares, 250
Sum of three squares, 247, 248
Sum of two squares, 244, 245
Sylvester, J. J., 35, 281
Symbols:
    $\equiv$, 30
    $\not\equiv$, 30
    $\equiv$, 52
    $a|b$, 8
    $a \nmid b$, 8
    $\sim$, 155
    $[\alpha]$, 71
    θ, 71
    $\lambda(n)$, 91
    $\mu(n)$, 86
    ξ_j, 186, 200
    ξ', 197
    $\pi(x)$, 154
    $\pi(x; k, l)$, 170
    $\sigma(n)$, 81
    $\tau(n)$, 81
    $\phi(n)$, 35
    $\psi(n)$, 79
    $\binom{m}{n}$, 157
    $(a_1, a_2, \ldots, a_n)$, 22
    $[a_1, a_2, \ldots, a_n]$, 176
    $\left(\dfrac{m}{p}\right)$, 134

Index

$\left(\frac{m}{P}\right)$, 147
C_k, 45, 46
F_n, 28
$g(k)$, 253
$G(k)$, 253
$H_p(y)$, 74
Li(x), 169
M_p, 144
$\prod_{p \leq n} p$, 159
$\sum_{d \mid n} f(d)$, 82
Symmetric property, 30
Symmetric simple continued fraction, 182

Tables of indices, companion, 124
Taylor's expansion, 64, 69
Tchebychef, P. L., 168, 281, 282
Thue, Axel, 223, 244
Thue's theorem, 244, 245
Totally multiplicative, 79
Totient, 35
Touchard, Jacques, 266
Transitive property, 30
Two-square theorem, 245

Unique factorization, 27
Uspensky, J. V., 264, 284

Vallée Poussin, C. J. de la, 153, 169, 283
Varineau, V. J., 257

Walsh, M. J., 257
Waring, E., 252, 256, 260
Waring's problem, 4, 252, 253, 287
Well-ordering principle, 9
  equivalent to principle of mathematical induction, 257, 258
Whole number, 7
Wieferich, Arthur, 256
Wilson, John, 260
Wilson primes, 269
Wilson's theorem, 48, 260
  converse, 52
Woodall, H. J., 269
Wright, E. M., 253, 256, 267, 282, 283, 284, 287

Zeller, Chr., 144, 266, 272